LIMERICK

FLUID MECHANICS FOR
CIVIL ENGINEERS

FLUID MECHANICS
FOR CIVIL ENGINEERS
S.I. Edition

N. B. WEBBER B.Sc.(Eng.), M.I.C.E.,
M.Am.Soc.C.E., M.I.W.E., M.I.Struct.E.
Senior Lecturer in Civil Engineering, University of Southampton

LONDON

CHAPMAN AND HALL

The S.I. edition first published 1971
by Chapman and Hall Ltd
11 New Fetter Lane, London EC4P 4EE
Reprinted three times
Reprinted 1979

Printed by offset in Great Britain by
J. W. Arrowsmith Ltd, Bristol

ISBN 0 412 10600 0

© 1971 N. B. Webber

Distributed in the U.S.A.
by Halsted Press, a Division
of John Wiley & Sons, Inc.
New York

Contents

		page
Preface to S.I. Edition		viii
Preface to Imperial Edition		ix
List of Principal Symbols with Derived Dimensions		xii
Properties of Fluids		xv
Conversion Factors and Useful Constants		xvi

1 Historical Note — 1

2 Properties of Fluids — 6
- 1. General Description — 6
- 2. Density — 6
- 3. Specific Weight — 6
- 4. Specific Gravity — 7
- 5. Bulk Modulus — 7
- 6. Ideal Fluid — 7
- 7. Viscosity — 7
- 8. Surface Tension and Capillarity — 9
- 9. Vapour Pressure — 10
- 10. Atmospheric Pressure — 10

3 Hydrostatic Pressure and Buoyancy — 12
- 1. Hydrostatics — 12
- 2. Pressure Intensity — 12
- 3. Pressure Measurement — 13
- 4. Pressure Force on a Submerged Surface — 16
- 5. Buoyancy — 21
- 6. Stability of Floating Bodies — 21

4 Basic Concepts of Fluid Motion — 25
- 1. Introduction — 25
- 2. Types of Flow — 26
- 3. Stream Lines — 29
- 4. Continuity Equation — 30
- 5. Bernoulli Equation — 31
- 6. Applications of Bernoulli Equation — 33

CONTENTS

7.	Momentum Equation	40
8.	Applications of Momentum Equation	41
9.	Flow in a Curved Path	44
10.	Flow Nets	50
11.	Behaviour of a Real Fluid	53
12.	Dimensional Considerations	68

5 Analysis of Pipe Flow — 76

1.	Introduction	76
2.	Laminar Flow – Poiseuille Equation	76
3.	Turbulent Flow – Darcy-Weisbach Formula	78
4.	The Contribution of Osborne Reynolds	79
5.	Experimental Investigations on Friction Losses in Turbulent Flow	83
6.	Semi-empirical Theory of Pipe Resistance	86
7.	Colebrook and White Transition Formula	95
8.	Empirical Formulae	99
9.	Deterioration of Pipes	103
10.	Non-circular Pipes and Conduits	104
11.	Minor Losses in Pipes	104
12.	Summary of Formulae	108

6 Pipelines and Pipe Systems — 110

1.	Hydraulic and Energy Gradients	110
2.	Power Transmission	111
3.	Discharge under Varying Head	113
4.	Simple Pipe Systems	114
5.	Distribution Mains	119
6.	Transient Behaviour	129

7 Uniform Flow in Channels — 143

1.	Introduction	143
2.	Laminar Flow	144
3.	Fundamental Relationships	146
4.	Empirical Formulae	151
5.	Best Hydraulic Section	158
6.	Enclosed Conduits	161
7.	Scouring and Silting	164

8 Non-uniform Flow in Channels — 168

1.	Introduction	168
2.	Specific Energy and Critical Depth	168
3.	Transition through Critical Depth	177
4.	General Equation of Gradually Varied Flow	185
5.	Classification of Surface Profiles	187
6.	Control Points	191
7.	Outlining of Surface Profiles	193
8.	Profile Evaluation	195
9.	Bridge Piers	202
10.	Translatory Waves in Channels	204

CONTENTS

9 *Hydraulic Structures*   211
   1. Introduction   211
   2. Sluices and Gates   211
   3. Sharp-crested Weirs   214
   4. Solid Weirs   221
   5. Special Types of Weir   224
   6. Throated Flumes   226
   7. Spillways   231
   8. Inflow-outflow Relationship at a Reservoir   246

10 *Pumps and Turbines*   252
   1. Introduction   252
   2. Head   253
   3. Synchronous Speed   256
   4. Types of Pump   257
   5. Types of Turbine   263
   6. Elementary Theory   272
   7. Performance   279
   8. Specific Speed   287
   9. Cavitation Considerations   292
   10. Design of a Pumping Main   295

11 *Hydraulic Models*   297
   1. Introduction   297
   2. Hydraulic Similarity   299
   3. Conformance with Similarity Laws   302
   4. Types of Model Investigation   307
   5. Measuring Instruments and Techniques   322

Appendix   324
   General Publications Relevant to Hydraulic Engineering   324

Name Index   327

Subject Index   330

Preface to S.I. Edition

The decision by Britain and many other countries to adopt the S.I. system of units has made it desirable to publish an edition using these units.

The S.I. system (Système International d'Unités) is a rationalized selection of metric units, whose principal merit lies in the fact that it comprises no more than six basic units. This is in contrast to the non-coherent English system with its multiplicity of units and a convention that is not always logical or consistent. Civil engineering hydraulics makes use of a wide range of quantities and units, and in the process of conversion the author has been impressed by the inherent simplicity of the new system and the economy of working which results.

The opportunity has been taken to up-date and revise the text where this has been considered necessary or desirable.

Again, the author is considerably indebted to Dr T. L. Shaw of Bristol University for his encouragement in the task and for his subsequent critical scrutiny of the manuscript. The advice and co-operation of the publishers is also gratefully acknowledged.

N. B. WEBBER

The University,
Southampton,
July, 1971

NOTE ON REPRINT

In reprinting, the opportunity has been taken to introduce certain desirable amendments and to bring up-to-date the references for further reading.

Preface to Imperial Edition

A glance at the list of textbooks on the subject of fluid mechanics (see appendix) would appear to indicate no lack of suitable material. But the treatment is almost invariably very broad and, with the degree of specialisation unavoidable in our engineering curricula today, there does seem a need for a book which would be concerned primarily with the civil engineering aspect. This was the author's aim in writing the present book, and the experience of seven years of undergraduate teaching preceded by more than ten years of engineering practice has been drawn upon in its compilation.

The subject matter embraces the field normally covered in a complete hydraulics course at undergraduate or comparable level. It provides the groundwork that is necessary before proceeding to the more specialised studies of hydraulic engineering. Without detracting from a consideration of the fundamentals, an attempt has been made to bridge the gap, still too wide, between theory and practice. In this way it is hoped that the book will also appeal to practising engineers, for there is now an increasing awareness of the value of the more soundly based and rational approach afforded by modern fluid mechanics.

Fluid mechanics, as the name implies, is a branch of applied mechanics. Its foundations are the laws of motion enunciated by Newton. The concepts are rational, being based on sound physical reasoning, mathematics, and the results of experiments. By analytical means it seeks to derive equations that are of a general nature, representing these where possible in dimensionless form. This is in contrast with empirical hydraulics where, because of the limited experimental data on which formulae have been established, certain reservations must be placed on their range of application.

The text commences with a short historical note which serves as an introductory background. This is followed by four chapters concerned primarily with the presentation of fundamental principles, embracing both fluid statics and dynamics. Topics include the phenomena associated with turbulent flow and a rational theory for pipe resistance. The remaining chapters deal with the practical application of these principles. By devoting the last chapter to a consideration of model testing recognition is given to its increasing importance as an aid in hydraulic design.

Mathematics of the more sophisticated type is avoided because, whilst it is capable of providing derivations of greater elegance and possibly more general significance, it is nevertheless felt that it would be detrimental to the emphasis on basic physical concepts. Numerous illustrative examples are included. Their purpose is to clarify the text and demonstrate the application to specific problems. Care has been taken to avoid stereotyped solutions, and it is hoped thereby to encourage the development of that analytical ability which is so desirable in the elucidation of the diverse problems encountered in practice. Where appropriate, the notation (see p. xi) is in accordance with the recommendations of B.S. 1991: Part 3: 1961.

The actual observation of a phenomenon is helpful to its understanding and we are fortunate in civil engineering hydraulics that the major interest is with free surface flow. Our rivers, canals, lakes, reservoirs, and coasts afford excellent opportunities for observing a variety of fluid behaviour, ranging from the simple hydraulic jump downstream of a sluice to the impressive cascading of water over the face of a large dam. Indeed, if 'the control of the forces of nature for the benefit of man' is the maxim of the civil engineer, then there can be few more satisfying spheres in which to practise the profession than hydraulic engineering. And for the research worker who is willing to proceed with perseverance and due modesty there is offered a fascinating and potentially rewarding field for study.

The Author is greatly indebted to Dr T. L. Shaw, Lecturer in Civil Engineering, University of Bristol, for his painstaking review of the manuscript. His many constructive comments have resulted in a valuable improvement in the quality of the text. There is grateful appreciation of the willingness with which various firms and organisations have supplied photographic illustrations. Thanks are also due to the publishers and to the many others whose advice and encouragement have helped to bring the book to fruition.

N. B. WEBBER

The University,
Southampton,
February, 1965.

List of Principal Symbols with Derived Dimensions

Symbol	Quantity	S.I.* units	Derived Dimensions M-L-T	F-L-T
A	Area	m²	L^2	L^2
a	Acceleration	m/s²	L/T^2	L/T^2
B	Width of channel at free surface	m	L	L
C	Critical bed gradient			
C	Chézy coefficient	$m^{1/2}/s$	$L^{1/2}/T$	$L^{1/2}/T$
	Overall discharge coefficient			
C_c	Coefficient of contraction			
C_D	Coefficient of drag			
C_d	Coefficient of discharge			
C_v	Coefficient of velocity			
c	Wave celerity	m/s	L/T	L/T
D	Diameter	m	L	L
d	Depth of flow (channel)	m	L	L
d_n	Normal depth of flow (channel)	m	L	L
d_c	Critical depth of flow (channel)	m	L	L
E	Euler number			
E_s	Specific energy	m	L	L
F	Froude number			
F	Force	N	ML/T^2	F
f	Darcy-Weisbach friction factor ($h_t = 4fLV^2/2gD$)			
g	Gravitational acceleration	m/s²	L/T^2	L/T^2
H	Horizontal bed gradient			
H	Total head	m	L	L
h	Piezometric head, head on weir or gate	m	L	L
h_a	Acceleration head	m	L	L
	Afflux	m	L	L
h_f	Friction head loss	m	L	L
h_L	Minor head loss	m	L	L
h_v	Velocity head	m	L	L
I	Second moment of area	m⁴	L^4	L^4
K	Bulk modulus	N/m²	M/LT^2	F/L^2
	Coefficient (minor loss)			
k	Effective excrescence height	m	L	L
L	Length, length of weir crest	m	L	L
l	Mixing length	m	L	L
M	Mach number			
	Mild bed gradient			
M	Mass	kg	M	FT^2/L
	Momentum	kg m/s	ML/T	FT
n	Revolutions per minute	rev/min	$1/T$	$1/T$
	Roughness coefficient (Manning)	$s/m^{1/3}$	$T/L^{1/3}$	$T/L^{1/3}$
	Spacing of stream lines	m	L	L

* In those cases where the numerical values are very large or very small it is customary to introduce prefixes such as mega – (M), kilo – (k) and milli – (m). It is an unfortunate anomaly that kg is a basic unit.

List of Principal Symbols with Derived Dimensions

Symbol	Quantity	S.I. units	Derived Dimensions	
			M-L-T	F-L-T
n_s	Specific speed (pumps)	(see text)	$L^{3/4}/T^{3/2}$	$L^{3/4}/T^{3/2}$
	Specific speed (turbines)	(see text)	$M^{1/2}/$	$F^{1/2}/$
			$L^{1/4}T^{5/2}$	$L^{3/4}T^{3/2}$
P	Total pressure	N	ML/T^2	F
	Power	W	ML^2/T^3	FL/T
	Weir crest height	m	L	L
	Wetted perimeter	m	L	L
p	Pressure	N/m²	M/LT^2	F/L^2
p_a	Absolute pressure	N/m²	M/LT^2	F/L^2
p_v	Vapour pressure	N/m²	M/LT^2	F/L^2
Q	Volume discharge rate	m³/s	L^3/T	L^3/T
q	Volume discharge rate/unit width	(cumecs) m²/s	L^2/T	L^2/T
R	Reynolds number			
R	Reaction force	N	ML/T^2	F
	Hydraulic mean depth	m	L	L
r	Radius	m	L	L
S	Steep bed gradient			
S	Longitudinal gradient ($=\sin\theta$)			
S_f	'Friction' gradient			
S_o	Bed gradient			
s	Side slope (s horizontal: 1 vertical)			
	Specific gravity			
T	Period (cyclic phenomena)	s	T	T
t	Time	s	T	T
u	Peripheral velocity	m/s	L/T	L/T
V	Mean or characteristic velocity	m/s	L/T	L/T
$\dot{V}$	Volume	m³	L^3	L^3
v	Filament velocity	m/s	L/T	L/T
v_*	Shear velocity $[=(\tau_0/\rho)^{1/2}]$	m/s	L/T	L/T
W	Weber number			
w	Specific weight	N/m³	M/L^2T^2	F/L^3
x, y, z	Co-ordinate lengths	m	L	L
z	Elevation head	m	L	L
α	Angle			
	Velocity head coefficient			
β	Angle			
	Momentum coefficient			
δ	Thickness of boundary layer	m	L	L
δ_L	Thickness of laminar sub-layer	m	L	L
η	Efficiency			
	Dynamic eddy viscosity	N s/m²	M/LT	FT/L^2
θ	Angle			
κ	Coefficient			
λ	Darcy-Weisbach friction factor ($h_f = \lambda L V^2/2gD$)			
μ	Dynamic viscosity	N s/m²	M/LT	FT/L^2
ν	Kinematic viscosity	m²/s	L^2/T	L^2/T
ρ	Density	kg/m³	M/L^3	FT^2/L^4

List of Principal Symbols with Derived Dimensions

Symbol	Quantity	S.I. units	Derived Dimensions	
			M-L-T	F-L-T
Σ	Summation sign			
σ	Surface tension	N/m	M/T^2	F/L
	Cavitation factor			
τ	Shear stress	N/m²	M/LT^2	F/L^2
τ_0	Boundary shear stress	N/m²	M/LT^2	F/L^2
ϕ	Angle			
	Function of			
ψ	Function of			
ω	Angular velocity	rad/s	$1/T$	$1/T$

Properties of Fluids (at Atmospheric Pressure)

(a) WATER

Temperature °C	Density ρ (kg/m³)	Specific weight w (kN/m³)	Specific gravity s	Bulk modulus K (N/mm²)	Kinematic viscosity ν (mm²/s)	Surface tension (air contact) σ (mN/m)	Vapour pressure p_v (kN/m²)
0	1000	9·81	1·00	2000	1·79	76	0·62
15	1000	9·81	1·00	2150	1·14	73	1·72
50	990	9·71	0·99	2290	0·56	69	11·7
100	960	9·42	0·96	2070	0·30	58	101·2

(b) OTHER FLUIDS (AT 15°C)

Fluid	Density ρ (kg/m³)	Specific weight w (kN/m³)	Specific gravity s	Kinematic viscosity ν (mm²/s)	Surface tension (air contact) σ (mN/m)	Vapour pressure p_v (kN/m²)
Air	1·23	0·01205	0·00123	14·5	—	—
Sea water*	1025	10·05	1·025	1·17	74	1·72
Oil (crude)*	860	8·44	0·86	18·6	28	39·3
Petrol*	730	7·16	0·73	0·77	25	68·9
Mercury	13 570	133·0	13·57	0·12	481	0·00010

Conversion Factors and Useful Constants

1 metre (m) = 3·281 ft

1 hectare (ha) = 10 000 m^2 = 2·471 acres

1 cubic metre (m^3) = 1·308 cubic yards (yd^3) = 35·31 ft^3
1 litre (l) = 10^{-3} m^3 = 0·220 imperial gallons (gal)
1 US gal = 0·833 imperial gal = 3·785 litres

1 kilometre/hour (km/h) = 0·278 m/s = 0·621 mile/h
1 knot = 1 Brit. nautical mile/h = 0·515 m/s

Standard acceleration due to gravity = 9·81 m/s^2 = 32·2 ft/s^2

1 kilogramme (kg) = 2·205 lb = 0·0684 slug

1 cumec (m^3/s) = 35·31 cusecs (ft^3/s) = 19·01 × 10^6 gal/day
1 litre/second (l/s) = 10^{-3} cumecs = 0·0353 cusecs = 13·2 gal/min

1 newton (N) = 0·102 kgf = 0·225 lbf = 7·23 poundals
1 kgf or 1 kilopond (kp) = 9·81 N = 9·81 × 10^5 dyn = 2·20 lbf
1 tonne = 1000 kgf = 9·81 kN = 0·984 Brit. tons

1 metre head of water = 9810 N/m^2 = 0·00981 N/mm^2 = 9·810 kN/m^2
1 bar = 10^{5}N/m^2 = 14·5 lbf/in^2
Standard atmospheric pressure = 1·013 bar = 101·3 kN/m^2
$\qquad\qquad\qquad\qquad\qquad$ = 0·76 m of mercury = 10·32 m of
$\qquad\qquad\qquad\qquad\qquad\qquad\qquad\qquad$ water

1 centipoise (cP) = 0·001 Ns/m^2 = 2·09 × 10^{-5} lbf s/ft^2 = 2·09 × 10^{-5} slug/ft s
1 centistoke (cSt) = 10^{-6} m^2/s = 1·076 × 10^{-5} ft^2/s

1 joule (J) = 1 Nm = 0·738 ft lbf

1 kilowatt (kW) = 1000 J/s = 1·341 horsepower

To convert degrees Celsius to degrees Fahrenheit multiply by 9/5 and
$\quad$ add 32.

CHAPTER ONE

Historical Note

There is much that can be learned from a study of the past, for it enables us to view the present in its true perspective and is a basis for speculation as to the future. This is as true of science as of mankind generally; and the science of hydraulics, especially, has a great history and one that is well worth recalling.

The present note consists of a short descriptive summary of the stages of advancement. Naturally, there have been many contributors, and space permits only a few of the more distinguished to be mentioned. Amplification is provided in the subsequent text where it is felt that a knowledge of the detailed historical background would be helpful to the understanding of a particular analysis or empirical formula.

Irrigation is the earliest form of hydraulics and has been practised by mankind from the dawn of history. Archaeological discoveries have shown that canals, dams, and reservoirs existed in Egypt and Mesopotamia as early as 4000 B.C. Wells also date from the same period, but primitive mechanical devices (e.g. paddle wheels) for raising water seem to have appeared later (1000 B.C.). The design of these works was almost entirely intuitive, being based on experience and simple trial and error.

The Greeks contributed much that was of outstanding quality and interest in the fields of philosophy, science, and art. In the 3rd century B.C. Archimedes established the elementary principles of buoyancy and flotation. The works of Euclid and Aristotle embraced a wide field, some of which was of relevance to hydraulics.

Whereas the Greeks were imaginative innovators, the Romans were skilful adaptors. Numerous reclamation and drainage works were undertaken. Large water wheels were constructed for grinding corn. Water supply and sanitation received particular attention; for instance, Rome was supplied by nine aqueducts and there was a sewerage system. A useful treatise on the methods of water distribution was prepared by Frontinus. However, in spite of greatly advancing the practice of the art, the Romans did little or nothing to promote the fundamental

1

knowledge of fluid behaviour – in fact their concepts, generally con-
jectural and often fallacious, were those of their Greek mentors.

Although the Romans may perhaps be criticised for their somewhat
unimaginative and entirely practical approach, it must be remembered
that at this stage there was insufficient academic background to establish
hydraulics as a science – for the basic principles of mechanics were as yet
unknown, as well as the mathematical techniques by which they might
be applied.

After the lapse of many unfruitful centuries the dawn of enlighten-
ment came with the Renaissance in the latter half of the 15th century. It
was the Italians who were the scientific pioneers. Leonardo da Vinci, a
man of incredible genius, sketched and commented upon many hydraulic
phenomena as well as supervising constructional works. The astronomer
Galileo conducted experiments on the fall of bodies, but in the field of
hydraulics is probably best known for his alleged remark that 'more is
known about the movements of heavenly bodies than of the fluids
encountered in terrestrial life'. At a time when hydraulic science was
still in its infancy, this was certainly fair commentary, and it is even
germane to our own space age. The mathematician Torricelli is noteworthy for his correct interpretation of the gravitational behaviour of
an issuing jet. But undoubtedly the greatest significance of this first
century or so following the Renaissance lies in the initiation of laboratory experiments and purposeful field observations – and above all in
the questing interest in physical science which was stimulated.

The urge for progress also manifested itself in practical form, for in
the 16th and 17th centuries many important hydraulic projects, such as
harbours, canals, and reclamation schemes were undertaken. Indeed it
was during the mid-17th century that a large part of the fenland of
East Anglia was drained, the major credit being due to the Dutch
engineer, Vermuyden.

English genius could not be said to be lacking, for it was represented
by the scientist Newton, who, as a mathematician and physicist, was
outstanding amongst his contemporaries. In the *Principia* (1687) he set
down concisely the basic laws of motion now named after him. Viscous
resistance was also one of his main topics of study.

During this period of rudimentary development, mathematics also
progressed (e.g. calculus devised), so that by the end of the 17th century
the necessary foundations for a broad advancement in knowledge had
been firmly laid.

In the 18th century the scientific lead quite definitely passed to the
French, although other nations contributed. Bernoulli and Euler are

noteworthy for their interpretation of the pressure-velocity relationship in fluid motion. Lagrange and Gerstner presented mathematical analyses of simple wave motion. Venturi conducted experiments on converging and diverging flow pieces.

Practising engineers were also active in experimenting and in their attack on design problems. Pitot devised a simple instrument for determining velocity. To Chézy is due the credit for being the first to put forward a practical formula for channel flow. Du Buat, besides conducting numerous experiments, published an informative treatise on hydraulics generally.

The progress of hydraulics in the 19th century was along two largely independent paths. On the one hand there was theoretical or classical hydraulics which was often impractical because it involved the rigorous analysis of an ideal fluid behaving without turbulence. On the other hand there was empirical hydraulics which had considerable experimental but little rational basis. In fact, one observer has been led to comment somewhat cynically that 'in the 19th century fluid dynamicists were divided into mathematicians who explained things that could not be observed and hydraulic engineers who observed things that could not be explained'. But this is not entirely true, for there were some who by their broad treatment could claim membership of both schools.

The contributions of the hydrodynamicists are not to be belittled for there are many instances where the simplifying assumptions inherent in the concept of an ideal fluid do lead to results that are very near to the truth. Notable researchers in this field were Navier, Airy, Saint-Venant, Stokes, Kelvin, Rayleigh, and Lamb, not all of whom were mathematicians. Also worthy of mention are Hagen and Poiseuille for their experimental investigations of laminar flow.

The exponents of empirical hydraulics were mostly practising engineers. With the great constructional works stemming from the Industrial Revolution there was an increasing awareness of the need for the formulation of rules and equations which would assist in hydraulic design, particularly with respect to channel and pipe flow. Darcy, Bazin, Ganguillet, Kutter, and Manning were but a few of the engineers who developed workable formulae to fit the field and experimental data that by now were beginning to accumulate.

Many of these formulae have not survived the test of time, and it is fascinating today to turn the pages of some 19th-century treatise on harbour or river engineering and contemplate the simplicity of formulae purporting to describe the most complex sorts of phenomena. One cannot avoid the conclusion that the great engineers of this period wisely

3

placed much more reliance on their own experienced judgement than on the naive application of some scientifically unproven formula.

In the laboratory sphere, Reynolds is noteworthy for his many experimental achievements and in particular for his investigations on the transition between laminar and turbulent flow, the results of which were published in 1883. Weisbach, besides undertaking useful experimental work on various aspects of hydraulics, presented a valuable engineering treatise on the dynamics of fluids which was for many years unsurpassed.

Thus at the close of the 19th century there was still an appreciable lacuna between theory and observed behaviour. Investigators were probing diligently from the two extremities, the purely theoretical and the strictly empirical, but there was little common ground. It was only later, consequent upon the fundamental physical researches initiated by Prandtl at the beginning of the 20th century, that the present unified approach known as fluid mechanics was developed. This seeks to establish the true physical basis of fluid phenomena by a combined process of analytical reasoning and experimentation.

Prandtl's convincing theory, published in 1904, for the behaviour of turbulent flow past a solid boundary earned him a chair at Göttingen University. The new concept opened up a rich field for study which he, his associates, and pupils – noteworthy among whom were Blasius, Kármán, and Nikuradse – proceeded zealously to explore. Turbulence, velocity distribution, form drag, and skin friction drag were all subjected to searching analytical and experimental investigation. Very soon research work in this field was proceeding in many scientific institutions throughout the world, but quite understandably the major effort was directed towards the by now pressing problems of aeronautics, where, because of the streamlined profiles and the absence of a free surface, results of positive practical value were to be quickly forthcoming. Needless to say, the application of the new concept to hydraulics was by no means neglected, and in the first three decades of the 20th century the fundamentals of pipe flow were successfully investigated.

In the wider sense too there was much valuable progress, particularly in dimensional analysis and the related field of model testing. Noteworthy contributors of the early 20th century were Bakhmeteff, Boussinesq, Engels, Forcheimer, Freeman, and Gibson.

Our fleeting journey through history deliberately stops short at this point, because a discussion of the present rightly belongs to the text which follows. The evolution of hydraulic science has been long and laborious. For our present knowledge we are indebted to a multitude of investigators, many unknown, and following diverse callings, but all imbued with a common desire to know more about the mysteries of fluid behaviour.

What does the future hold? It is dangerous and presumptuous to predict the forthcoming fruits of research. Suffice it to say that there are many gaps in our knowledge – problems concerning turbulent separation, air-entraining vortices, sediment transport, density currents and interfacial mixing, to name but a few. Although the great fundamental truths may all have been discovered, their application to complex problems affords much scope for the exercise of human intellect. In recent years the number of research workers throughout the world engaged in the study of hydraulic phenomena has greatly increased; also, improved instrumentation techniques (electronic gadgetry) and mathematical tools (computers) have become available, so it would be a pessimist indeed who did not look forward with confidence to a continuing and possibly exciting advancement in the years that lie ahead.

But whilst recognising the scientific importance of these researches we must not lose sight of the fact that it is engineers who are required to turn the fruits to practical advantage. The facility with which they are able to do so rests on their understanding of present knowledge and their willingness to participate in remedying its inadequacies.

Further Reading

ROUSE, H. and INCE, S. (1957) *History of Hydraulics*. Iowa Inst. Hyd. Res., State Univ. of Iowa. (Dover reprint.)

SINGER, C. J. and OTHERS (Eds.) (1954–58) *A History of Technology* (5 vols.). Clarendon Press.

CHAPTER TWO

Properties of Fluids

2.1 General Description

A *fluid*, as the name implies, is characterised by its ability to flow. It differs from a solid in that it suffers deformation due to shear stress, however small the shear stress may be. The only criterion is that sufficient time should elapse for the deformation to take place. In this sense a fluid is shapeless.

Fluids may be divided into *liquids* and *gases*. A liquid is only slightly compressible, and there is a free surface when it is placed in an open vessel. On the other hand a gas always expands to fill its container. A vapour is a gas which is near the liquid state.

The liquid with which the Civil Engineer is mainly concerned is water. It may contain up to 3 per cent of air in solution which at sub-atmospheric pressures tends to be released. Provision must be made for this when designing pipelines.

The principal physical properties of fluids are described in the sections which follow. Tabulated values are given on page XV.

2.2 Density

The *density*, ρ, of a fluid is its mass per unit volume. In the S.I. system it is expressed in kg/m³ (ML^{-3}), and in the British system in slugs per cubic foot.

For water, ρ is 1000 kg/m³ at 4°C. In British units it is 1·94 slug/ft³ at the same temperature. There is a slight decrease in density with increasing temperature, but for normal practical purposes the value is constant.

2.3 Specific Weight

The *specific weight*, w, of a fluid is its weight per unit volume. In the S.I. system it is expressed in kN/m³ ($ML^{-2}T^{-2}$), and in the British system in pounds per cubic foot. The relationship between w, ρ, and the acceleration due to gravity g is

$$w = \rho g \qquad (2.1)$$

At normal temperatures w is 9·81 kN/m³ or 62·4 lbf/ft³.

2.4 Specific Gravity

The *specific gravity*, s, of a fluid is the ratio of its density to that of pure water at the same temperature (normally 15°C).

2.5 Bulk Modulus

For most practical purposes liquids may be regarded as incompressible. However, there are certain cases, such as unsteady flow in pipes, where the compressibility should be taken into account. The *bulk modulus of elasticity*, K, is given by

$$K = \frac{\Delta p}{\Delta \dot{V}/\dot{V}} \qquad (2.2)$$

where Δp is the increase in pressure, which when applied to a volume $\dot{V}$, results in a decrease in volume $\Delta \dot{V}$. Since a decrease in volume must be associated with a proportionate increase in density, Eq. (2.2) may be expressed as

$$K = \frac{\Delta p}{\Delta \rho/\rho} \qquad (2.3)$$

For water, K is approximately 2150 N/mm² at normal temperatures and pressures. It follows that water is about 100 times more compressible than steel.

2.6 Ideal Fluid

An *ideal* or *perfect fluid* is one in which there are no tangential or shear stresses between the fluid particles. The forces always act normally at a section and are limited to pressure and accelerative forces. No *real fluid* fully complies with this concept, and for all fluids in motion there are tangential stresses present which have a dampening effect on the motion. However, some liquids, including water, are near to an ideal fluid, and this simplifying assumption enables mathematical or graphical methods to be adopted in the solution of certain flow problems.

2.7 Viscosity

The *viscosity* of a fluid is a measure of its resistance to tangential or shear stress. It arises from the interaction and cohesion of fluid

7

molecules. All real fluids possess viscosity, though to varying degrees. The shear stress in a solid is proportional to strain whereas the shear stress in a fluid is proportional to the *rate* of shearing strain. It follows that there can be no shear stress in a fluid which is at rest.

Consider a fluid confined between two plates which are situated a very short distance y apart (Fig. 2.1). The lower plate is stationary whilst the upper plate is moving at velocity v. The fluid motion is assumed to take place in a series of infinitely thin layers or laminae,

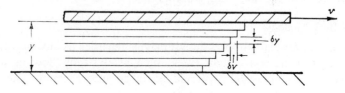

Figure 2.1   Viscous deformation

free to slide one over the other. There is no cross-flow or turbulence. The layer adjacent to the stationary plate is at rest whilst the layer adjacent to the moving plate has a velocity v. The rate of shearing strain or velocity gradient is dv/dy. The *dynamic viscosity* or more simply the *viscosity*, μ, is given by

$$\mu = \frac{\text{shearing stress}}{\text{rate of shearing strain}} = \frac{\tau}{dv/dy}$$

so that

$$\tau = \mu \frac{dv}{dy} \qquad (2.4)$$

This expression for the viscous stress was first postulated by Newton and is known as *Newton's equation of viscosity*. Almost all fluids have a constant coefficient of proportionality and are referred to as *Newtonian*

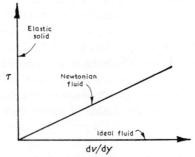

Figure 2.2   Relationship between shearing
stress and rate of shearing strain

fluids. Fig. 2.2 is a graphical representation of Eq. (2.4) and demonstrates the different behaviour of solids and liquids under shearing stress.

Viscosity is expressed in mN s/m² or centipoises $(ML^{-1}T^{-1})$ in the S.I. system and slug/ft-s in the British system.

In many problems concerning fluid motion the viscosity appears with the density in the form μ/ρ and it is convenient to employ a single term ν, known as the *kinematic viscosity*, and so called because the units mm²/s (L^2T^{-1}) or centistokes in the S.I. system, are independent of force. The corresponding units are ft²/s in the British system.

The value of ν for a heavy oil may be as high as 900 mm²/s, whereas for water, which has a relatively low viscosity, it is only 1·14 mm²/s at 15°C. The kinematic viscosity of a liquid diminishes with increasing temperature. At room temperature the kinematic viscosity of air is about 13 times that of water.

2.8 Surface Tension and Capillarity

Surface tension is the physical property which enables a drop of water to be held in suspension at a tap, a vessel to be filled with liquid slightly above the brim and yet not spill, or a needle to float on the surface of a liquid. All these phenomena are due to the cohesion between molecules at the surface of a liquid which adjoins another immiscible liquid or gas. It is as though the surface consists of an elastic membrane, uniformly stressed, which tends always to contract the superficial area. Thus we find that bubbles of gas in a liquid and droplets of moisture in the atmosphere are approximately spherical in shape.

The surface tension force across any imaginary line at a free surface is proportional to the length of the line and acts in a direction perpendicular to it. The *surface tension per unit length*, σ, is expressed in mN/m or lbf according to the system of units. Its magnitude is quite small, being approximately 73 mN/m for water in contact with air at room temperature. There is a slight decrease in surface tension with increasing temperature.

In most spheres of hydraulics surface tension is of little significance since the associated forces are generally negligible in comparison with the hydrostatic and dynamic forces. Surface tension is only of importance where there is a free surface and the boundary dimensions are small. Thus in the case of hydraulic models, surface tension effects, which are of no consequence in the prototype, may influence the flow behaviour in the model, and this source of error in simulation must be taken into consideration when interpreting the results.

Surface tension effects are very pronounced in the case of tubes of

9

small bore open to the atmosphere. These may take the form of manometer tubes in the laboratory or open pores in the soil. For instance,
when a small glass tube is dipped into water, it will be found that the
water rises inside the tube, as shown in Fig. 2.3(a). The water surface in
the tube, or meniscus as it is called, is concave upwards. The phenomenon is known as *capillarity*, and the tangential contact between the
water and the glass indicates that the internal cohesion of the water is
less than the adhesion between the water and the glass. The pressure of
the water within the tube adjacent to the free surface is less than
atmospheric.

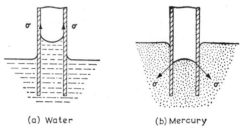

(a) Water   (b) Mercury

Figure 2.3   Capillarity

Mercury behaves rather differently, as indicated in Fig. 2.3(b).
Since the forces of cohesion are greater than the forces of adhesion, the
angle of contact is larger, and the meniscus has a convex face to the
atmosphere and is depressed. The pressure adjacent to the free surface
is greater than atmospheric.

Capillarity effects in manometers and gauge glasses may be avoided
by employing tubes which are not less than 10 mm diameter.

2.9 Vapour Pressure

Liquid molecules which possess sufficient kinetic energy are projected
out of the main body of a liquid at its free surface and pass into the
vapour. The pressure exerted by this vapour is known as the *vapour
pressure*, p_v. An increase in temperature is associated with a greater
molecular agitation and thus an increase in vapour pressure. When the
vapour pressure is equal to the pressure of the gas above it the liquid
boils. The vapour pressure of water at 15°C is 1·72 kN/m².

2.10 Atmospheric Pressure

The pressure of the atmosphere at the earth's surface is measured by a
barometer. At sea level the atmospheric pressure averages 101 kN/m²
and is standardised at this value. There is a decrease in atmospheric

10

pressure with altitude; for instance, at 1500 m it is reduced to 88 kN/m². The water column equivalent has a height of 10·3 m and is often referred to as the water barometer. The height is hypothetical, since the vapour pressure of water would preclude a complete vacuum being attained. Mercury is a much superior barometric liquid, since it has a negligible vapour pressure. Also, its high density results in a column of reasonable height – about 0·76 m.

As most pressures encountered in hydraulics are above atmospheric pressure and are measured by instruments which record relatively, it is convenient to regard atmospheric pressure as the datum. Pressures are then referred to as *gauge pressures* when above atmospheric and *vacuum pressures* when below it. If true zero pressure is taken as datum, pressures are said to be *absolute*.

Further Reading

DRYSDALE, C. V. and OTHERS (1936) *The Mechanical Properties of Fluids.* Blackie (2nd Edition).

CHAPTER THREE

Hydrostatic Pressure and Buoyancy

3.1 Hydrostatics

Hydrostatics is the branch of fluid mechanics which is concerned with fluids at rest. As stated in the previous chapter, no tangential or shear stress exists between stationary fluid particles. Thus in hydrostatics all forces act normally to a boundary surface and are independent of viscosity. As a result, the controlling laws are relatively simple, and analysis is based on a straightforward application of the mechanical principles of force and moment. Solutions are exact and there is no need to have recourse to experiment.

3.2 Pressure Intensity

The *pressure intensity* or more simply the *pressure* on a surface is the pressure force per unit area. In Fig. 3.1 the vertical downward pressure

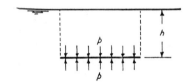

Figure 3.1   Pressure forces on a submerged
horizontal lamina

force acting on the horizontal lamina is equal to the weight of the prism of fluid which is vertically above it plus the pressure intensity at the interface with another fluid. For static equilibrium there must be a corresponding upward vertical pressure below the lamina. In the case of an incompressible liquid in contact with the atmosphere, the gauge pressure p is given by

$$p = wh \tag{3.1}$$

where w is the specific weight of the liquid and h is the depth below the free surface. The latter is referred to as the *pressure head* and is generally stated in feet of liquid. The form of the equation shows that the pressure increases linearly with depth.

As gravity is the physical property which is concerned, the free surface of a still liquid is always horizontal and the pressure intensity is the same on any horizontal plane within the body of the liquid. Moreover, it may be shown that the pressure intensity on any elemental particle is the same in all directions. This follows from a consideration of the

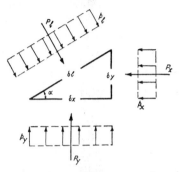

Figure 3.2   Pressure forces on a submerged triangular-shaped prism

pressure forces acting on an elemental triangular-shaped prism (Fig. 3.2) of unit horizontal length, with sectional dimensions δl, δx, δy, and weight δW.

For equilibrium in the horizontal direction, $P_x = P_l \sin \alpha$, or $p_x \, \delta y = p_l \, \delta l \sin \alpha$, so that $p_x = p_l$. Similarly in the vertical direction, $P_y = P_l \cos \alpha + \delta W$, or $p_y \, \delta x = p \, \delta l \cos \alpha + w \, \delta x \, \delta y / 2$; neglecting second-order terms of very small quantities, $p_y = p_l$. Thus

$$p_x = p_y = p_l \qquad (3.2)$$

The pressure intensity is therefore independent of the angle of inclination of the elemental surface and is the same in all directions.

3.3 Pressure Measurement

3.3.1 *Types of Device*

In the case of liquids with a free surface the pressure at any point is represented by the depth below the surface. When the liquid is totally enclosed, such as in pipes and pressure conduits, the pressure cannot readily be ascertained and a suitable measurement device is required. There are three principal types: (a) piezometer, (b) manometer, and

13

(c) Bourdon gauge. These are shown fitted to a pipeline (Fig. 3.3), and a brief description follows.

3.3.2 *Piezometer*

If a tapping is made in the boundary surface and a sufficiently long tube connected, the liquid will rise in the tube until balanced by atmospheric pressure. The pressure in the main body of the liquid is represented by the vertical height of the liquid column. Clearly, the device is only suitable for moderate pressures, otherwise the liquid will rise too high in the piezometer tube for convenient measurement.

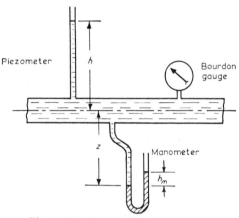

Figure 3.3   Pressure measurement devices

When the liquid is flowing, the piezometer tapping should not exceed ⅛ in. diameter and should be flush with the boundary surface. For greater accuracy a piezometer ring may be fitted. This consists of an annular chamber surrounding the pipe and connected to it by a number of equally spaced tappings.

3.3.3 *Manometer*

The principle is the same as that described above, but the difficulties associated with an excessively long tube are overcome by fitting a U-tube containing an immiscible liquid. Mercury (specific gravity 13·6) is the manometer liquid usually employed for measuring water pressure.

The gauge pressure p in the pipeline is given by

$$p = w_m h_m - wz \qquad (3.3)$$

where h_m is the difference in level of the manometer liquid in the two

14

limbs, z is the height of the pipe centre line above the meniscus in the limb on the pipe side and w, w_m are the specific weights of the pipe and manometer liquids respectively.

Owing to the fluctuating positions of the menisci, direct calibration is not possible. However, this may be achieved if the limb on the pipe side is greatly enlarged so that the level of the meniscus remains virtually constant. Pressures can then be read off a graduated scale attached to the other limb.

A quantitative evaluation of pipe flow is often based on a measurement of pressure difference between nearby tappings. A differential

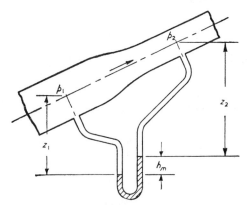

Figure 3.4   Differential manometer

manometer (Fig. 3.4) is employed and again the manometer liquid is usually mercury. When the pressure differences are small, a lighter immiscible liquid will yield more accurate results.

The pressure difference $p_1 - p_2$ is given by

$$p_1 - p_2 = w_m h_m + w(z_2 - z_1) \qquad (3.4)$$

where the symbols have the same meaning as previously.

If the pipe is horizontal, $z_1 = z_2 + h_m$, and

$$p_1 - p_2 = h_m(w_m - w) \qquad (3.5)$$

Differential manometers of a more intricate nature have been devised to meet the specific needs of commercial and laboratory practice.

3.3.4 *Bourdon Gauge*

This is a commercial instrument which is fitted either directly to the pipe itself or to the end of a piezometer line. It consists of a bent tube,

15

suspended freely in the curved portion but held rigidly at the stem. An increase in internal pressure tends to straighten the tube and, as the deflection is directly proportional to the applied pressure, a simple mechanism enables the latter to be directly recorded. Since the pressure on the outside of the tube is atmospheric, a gauge pressure is registered and this is normally applicable to the centre point of the instrument.

The Bourdon gauge is useful as a general indicator of pressure but is not suitable where considerable accuracy is demanded, as is generally the case when differential pressures are to be measured.

3.4 Pressure Force on a Submerged Surface

3.4.1 *Plane Surface*

The determination of the magnitude and location of the resultant pressure force on a plane surface may be explained by reference to Fig. 3.5. The area of the submerged surface is A and it is inclined at angle α

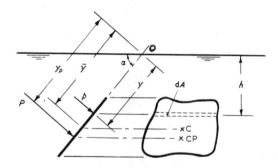

Figure 3.5   Pressure forces on a submerged plane surface

to the horizontal. Clearly, due to the linear increase in pressure intensity with depth, the point of application of this resultant, known as the *centre of pressure*, must be situated below the centroid of area. It is only coincident with it when the surface is horizontal.

The pressure force dP on an elemental horizontal strip of the surface, area dA, where the pressure intensity is p, is given by $dP = p\,dA = wh\,dA$ or $dP = w\sin\alpha\,y\,dA$. Thus the total pressure force is $P = w\sin\alpha \int y\,dA$. Now $\int y\,dA$ is equal to $A\bar{y}$ where $\bar{y}$ is the distance of the centroid of area from the intersection point O. Thus

$$P = w\sin\alpha\,A\bar{y} \tag{3.6}$$

Taking moments about O in order to determine the centre of pressure,

16

we have $Py_P = w \sin \alpha \int y^2 \, dA$. Now $\int y^2 \, dA$ is the second moment of area I_o of the lamina about O. Also we know that $I_o = I_c + A\bar{y}^2 = Ak_c^2 + A\bar{y}^2$, where I_c is the second moment of area about the centroid and k_c is the radius of gyration. Thus

$$y_P = \frac{w \sin \alpha (Ak_c^2 + A\bar{y}^2)}{w \sin \alpha \, A\bar{y}}$$

or

$$y_P = \frac{k_c^2}{\bar{y}} + \bar{y} \qquad (3.7)$$

The centre of pressure is therefore located at a depth $(k_c^2/\bar{y}) \sin \alpha$ below the centroid of area.

The centre of pressure of irregular areas is determined by application of the usual principles of applied mechanics.

Example 3.1

The outfall of a watercourse to an estuary is controlled by a 914 mm (36 in.) diameter flap valve hinged at the top. In the closed position it is inclined at 10 degrees to the vertical. The weight of the flap, which may be assumed uniformly distributed, is 3 kN and the specific gravity is 7·5.

If the water is at the level of the hinges on the seaward side, determine the differential head when the flap is on the point of opening.

$\alpha = 80°$; $k_c^2 = 0.914^2/16 = 0.0522 \ \text{m}^2$; $A = (\pi/4) \times 0.914^2 = 0.656 \ \text{m}^2$

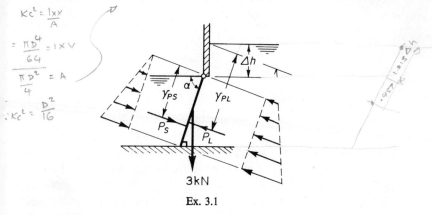

Ex. 3.1

Landward side:

$\bar{y}_L = 0.457 + \Delta h/\sin 80 = 0.457 + 1.015\Delta h$ m;

$P_L = 9.81 \times \sin 80 \times 0.656 \, (0.457 + 1.015\Delta h) = 6.33 \, (0.457 + 1.015\Delta h)$ kN, located at $y_{PL} = [0.0522/(0.457 + 1.015\Delta h)] + 0.457 + 1.015\Delta h$, or $[0.0522/(0.457 + 1.015\Delta h)] + 0.457$ m from the hinge.

Seaward side:

$$\bar{y}_S = 0\cdot457 \text{ m}; P_S = 1\cdot025 \times 9\cdot81 \times \sin 80 \times 0\cdot656 \times 0\cdot457 = 2\cdot97 \text{ kN},$$
located at $y_{PS} = (0\cdot0522/0\cdot457) + 0\cdot457 = 0\cdot57$ m from the hinge.

When the flap is on the point of opening, the net moment about the hinge is zero, or

$$6\cdot33(0\cdot457 + 1\cdot015\varDelta h)[0\cdot0522/(0\cdot457 + 1\cdot015\varDelta h) + 0\cdot457]$$
$$= (2\cdot97 \times 0\cdot57) + 3\cdot0[(7\cdot5 - 1\cdot01)/7\cdot5] \times 0\cdot457 \times \sin 10*$$

from which

$$\varDelta h = \mathbf{0\cdot0824} \text{ m}$$

3.4.2 *Curved Surface*

In the case of a submerged curved surface, the normal pressure on the surface varies in direction and a straightforward integration procedure can no longer be applied. The simplest approach is to assess the pressure forces acting on projected vertical and horizontal planes. These components may be subsequently combined into a resultant pressure force, although in many problems this is unnecessary. Thus in Fig. 3.6 the

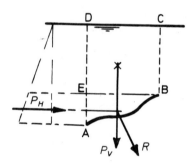

Figure 3.6   Pressure forces on a submerged
curved surface

vertical component P_V of the pressure force per unit length is represented by the weight per unit length of the section ABCDEA. Similarly, the horizontal component P_H is represented by the pressure distribution on EA which is in the form of a trapezium. The resultant R is equal to $\sqrt{P_V^2 + P_H^2}$, acting at an angle of $\tan^{-1} P_V/P_H$ with the horizontal.

Movable control gates are often designed with a curved upstream face. As the water pressure is always normal to the surface, a convex

---

* On the assumption that the specific gravity of sea water is 1·02, the average specific gravity of the two fluids is 1·01.

curvature generally serves to reduce the moment which is exerted on the supporting framework. By means of counterweights and float chambers these gates may be adapted to automatic control. There are three principal types of gate (Fig. 3.7) namely—

(*a*) *Radial or Tainter Gate.* This gate is widely used and is suitable for controlling both large and small flows. Iᴛ consists of a smooth

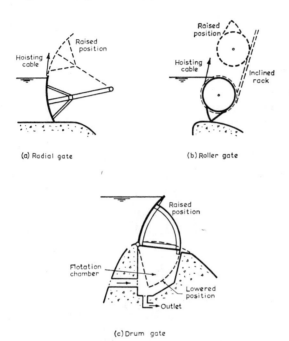

(a) Radial gate   (b) Roller gate

(c) Drum gate

Figure 3.7   Types of movable gate

plated face, shaped to a circular arc and supported by a framework pivoted at or near to the centre of curvature. Cables or chains are employed to raise and lower the gate, the water passing on the underside.

(*b*) *Roller Gate.* In the conventional form this consists of a hollow steel cylinder spanning between piers. A plated segmental framework on the underside increases the effective height of the gate. Cables raise the gate along inclined racks on the piers. It is primarily employed for the control of flow over relatively long spillway crests.

(c) *Drum Gate.* This gate type is suitable for long spillway crests. It approximates in form to a segmental portion of a hollow cylinder, skin-plated around the entire periphery and pivoted at the lower upstream edge. Under full discharge conditions the entire gate framework is housed in a compartment, known as a flotation chamber, and the curvature of the upstream face must therefore be made to conform with the profile of the spillway crest. Water has free access to the flotation chamber and the gate raising and lowering is regulated by valve adjustment in the outlet pipe from the chamber. Automatic drum gates are installed at Pitlochry dam, Scotland, and water passing over the gates and spillway makes an impressive sight.

Example 3.2

The drum gate illustrated in Fig. 3.7(c) is 12·2 m long and has the dimensions given in the inset diagram. The vertical through the centre of gravity of the

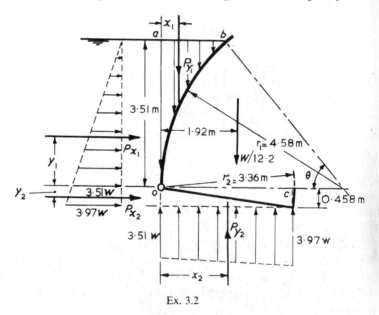

Ex. 3.2

gate is distant 1·92 m horizontally from the hinge. Neglecting friction at the hinge and seal, determine the weight of the gate for static equilibrium.

$$\sin \theta = 3\cdot51/4\cdot58 = 0\cdot767, \quad \text{or} \quad \theta = 50° \, 3';$$
$$ab = 4\cdot58(1 - \cos \theta) = 1\cdot64 \text{ m}; \quad oc = 3\cdot33 \text{ m}.$$

The pressure distribution varies linearly from zero at the surface to

$3.51w$ at the hinge and $3.97w$ at the lowest point of the gate. The component pressure forces acting on the gate per metre run are:

$$P_{z_1} = \tfrac{1}{2} \times 3.51 \times 3.51w = 6.16w$$

$$P_{y_1} = \left(\frac{1.64 + 4.58}{2}\right) \times 3.51w - \tfrac{1}{2} \times 4.58^2 \times 0.874w = 1.76w$$

$$P_{z_2} = \left(\frac{3.51 + 3.97}{2}\right) \times 0.458w = 1.71w$$

$$P_{y_2} = \left(\frac{3.51 + 3.97}{2}\right) \times 3.33w = 12.4w$$

By analytical or geometrical means the locations of the lines of action about the hinge are evaluated as:

$$y_1 = 1.17 \text{ m}, \quad y_2 = 0.235 \text{ m}, \quad x_1 = 0.47 \text{ m}, \quad x_2 = 1.70 \text{ m}.$$

Taking moments about the hinge,

$$\frac{W}{12.2} \times 1.92 = P_{z_2} \times y_2 + P_{y_2} \times x_2 - P_{z_1} \times y_1 - P_{y_1} \times x_1$$

or
$$0.157W = 9.81(1.71 \times 0.235 + 12.4$$
$$\times 1.70 - 6.16 \times 1.17 - 1.76 \times 0.47)$$

from which
$$W = \mathbf{838} \text{ kN}$$

3.5 Buoyancy

In accordance with Archimedes' principle the upthrust or buoyant force on an immersed body is equal to the weight of liquid which it displaces. A submerged body will rise to the surface provided that the weight of the body is less than the weight of the displaced liquid. Under reversed conditions it will sink to the bottom, although in exceptional cases the relative compressibility of body and liquid may result in continued suspension at a greater depth.

The centre of gravity of the displaced liquid is known as the *centre of buoyancy*, and a submerged body orientates itself so that its centre of gravity is located vertically above its centre of buoyancy. If the centre of gravity and centre of buoyancy coincide, the body will remain in any position and is said to be in neutral equilibrium.

3.6 Stability of Floating Bodies

The depth of immersion of a floating body is dependent on its weight (or volumetric displacement) and the shape of its hull. The stability is determined by the forces acting when it has been disturbed from the position of static equilibrium.

Fig. 3.8 shows the cross-section of a vessel in static equilibrium. The centre of gravity G is situated vertically above the centre of buoyancy B – only in the case of a vessel very low in the water-line is B above G.

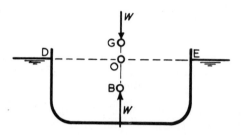

Figure 3.8   Vessel in static equilibrium

Fig. 3.9 shows the vessel heeled over at a small angle θ. With a fixed cargo the relative position of the centre of gravity remains unchanged but, due to the redistribution of pressure on the hull, the centre of buoyancy shifts from B to B'. The displacement is of course unchanged

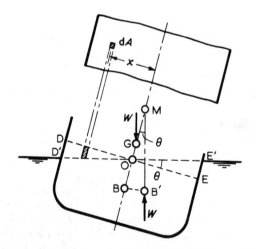

Figure 3.9   Vessel with a small angle of heel

and in effect what has happened is that a wedge-shaped volume of water represented in section by DOD' has shifted across the central axis to EOE'.

The intersection point M of the new line of upthrust through B' and the original axis of symmetry is known as the *metacentre*, and its position relative to G governs the equilibrium of the vessel. If M is above G, the equal and opposite forces at G and B' constitute a couple which tends to restore equilibrium. Conversely, if M is below G, the vessel will list still further and capsize. The distance GM is called the *metacentric height*, which for a small angle of heel may be determined analytically in the following manner.

The volume of each of the 'wedges' is obtained by integration over the entire water-line surface of the elemental horizontal areas dA (Fig. 3.9) multiplied by the height $x \tan \theta$. Therefore the moment about the central axis is $w \tan \theta \int x^2 \, dA$. But $\int x^2 \, dA$ is the second moment of area I of the water-line surface about the longitudinal axis, and thus taking moments about the central axis for the buoyancy forces we obtain

$$w\dot{V} \times \text{BM} \sin \theta = w \tan \theta \, I \tag{3.8}$$

where $\dot{V}$ is the volume of water displaced by the vessel. As θ is very small, $\sin \theta \simeq \tan \theta \simeq \theta$, so that

$$\text{BM} = \frac{I}{\dot{V}} \tag{3.9}$$

and the metacentric height is

$$\text{GM} = \frac{I}{\dot{V}} - \text{BG} \tag{3.10}$$

When G is below B, BG is added, in which case the metacentric height must be positive and the equilibrium stable.

The degree of stability increases with the metacentric height, but on the other hand the period of roll also depends primarily on GM and too large a value tends to result in undesirably rapid rolling. As might be expected, the metacentric height is of the greatest importance in naval architecture.

When mobile cargo or ballast is carried, the centre of gravity of the vessel is displaced in the same direction as the centre of buoyancy, thus decreasing the stability. Tanks or bulkhead compartments will mitigate the adverse effect. In civil engineering marine work, pontoons are often employed for lifting purposes and the depth of immersion is controlled by a carefully designed system of water ballast tanks.

A vessel fitted with a central longitudinal bulkhead and carrying the same liquid as that in which it is floating is shown in cross-section in Fig. 3.10. The reduction in restoring moment due to the change in position of the wedge prism in each compartment is $w \tan \theta_1 \, I_B$, where

23

I_B is the second moment of area of the free surface in the compartment about its longitudinal axis of oscillation. Eq. (3.8) is amended to

$$w\dot{V}_1 \times \mathrm{B}_1\mathrm{M}_1 \sin \theta_1 = w \tan \theta_1(I - 2I_\mathrm{B})$$

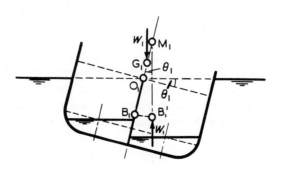

Figure 3.10   Vessel with ballast

where $\dot{V}_1$ is the displacement volume of the vessel plus ballast. Approximating as before,

$$\mathrm{B}_1\mathrm{M}_1 = \frac{I - 2I_\mathrm{B}}{\dot{V}_1} \qquad (3.11)$$

and

$$\mathrm{G}_1\mathrm{M}_1 = \frac{I - 2I_\mathrm{B}}{\dot{V}_1} - \mathrm{B}_1\mathrm{G}_1 \qquad (3.12)$$

Further Reading

MAYER, P. R. and BOWMAN, J. R. (1969) 'Spillway Crest Gates', Sect. 21 of *Handbook of Applied Hydraulics* (Eds. Davis, C. V. and Sorensen, K. E.). McGraw-Hill (3rd Edition).

CHAPTER FOUR

Basic Concepts of Fluid Motion

4.1 Introduction

In the previous chapter it was shown that exact mathematical solutions for the forces exerted by fluids at rest could be readily obtained. This is because in hydrostatics only simple pressure forces are involved. When a fluid in motion is considered, the problem of analysis at once becomes much more difficult. Not only have the particle velocity's magnitude and direction to be taken into account, but also there is the complex influence of viscosity causing a shear or frictional stress to exist both between the moving fluid particles and at the containing boundaries. The relative motion which is possible between different elements of the fluid body means that the pressure and shear stress may vary considerably from one point to another according to the flow conditions.

Owing to the complexities associated with the flow phenomenon, a precise mathematical analysis is only possible in a few, and from the civil engineering point of view, somewhat impractical, cases. It is therefore necessary to solve flow problems either by experimentation or by making certain simplifying assumptions sufficient to facilitate a theoretical solution. The two approaches are not mutually exclusive, since the fundamental laws of mechanics are always valid and enable partially theoretical methods to be adopted in several important cases. Also, it is important to ascertain experimentally the extent of the deviation from the true conditions consequent upon a simplified analysis.

The most common simplifying assumption is that the fluid is ideal or perfect, thus eliminating the complicating viscous effects. This is the basis of classical hydrodynamics, a branch of applied mathematics that has received attention from such eminent scholars as Stokes, Rayleigh, Rankine, Kelvin, and Lamb. There are serious inherent limitations in the classical theory, but due to the fact that water has a relatively low density, it is found to behave in many situations like an ideal fluid. For this reason, classical hydrodynamics may be regarded as a most valuable background to the study of the characteristics of fluid motion.

The present chapter is concerned with the fundamental dynamics of

25

fluid motion and serves as a basic introduction to succeeding chapters dealing with the more specific problems encountered in civil engineering hydraulics. The three important basic equations of fluid motion – namely, the continuity, Bernoulli, and momentum equations – are derived and their significance explained. Later, the limitations of the classical theory are pointed out and the behaviour of a real fluid described. Throughout, an incompressible fluid is assumed.

4.2 Types of Flow

The various types of fluid motion may be classified as follows:

    (a) Turbulent and Laminar   (b) Rotational and Irrotational
    (c) Steady and Unsteady   (d) Uniform and Non-uniform *

(*a*) *Turbulent and Laminar Flow.* These terms describe the physical nature of the flow.

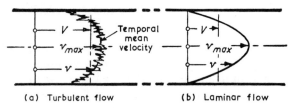

    (a) Turbulent flow   (b) Laminar flow

Figure 4.1   Velocity distribution for turbulent and laminar
flow in a pipe

In *turbulent flow*, the progression of the fluid particles is irregular and there is a seemingly haphazard interchange of position. Individual particles are subject to fluctuating transverse velocities so that the motion is eddying and sinuous rather than rectilinear. If dye is injected at a certain point, it will rapidly diffuse throughout the flow stream. In the case of turbulent flow in say a pipe, an instantaneous recording of the velocity at a section would reveal a distribution somewhat as indicated in Fig. 4.1(a). The steady velocity, as would be recorded by normal measuring instruments, is indicated in dotted outline, and it is apparent that turbulent flow is characterised by an unsteady fluctuating velocity superimposed on a temporal steady mean.

In *laminar flow* all the fluid particles proceed along parallel paths and there is no transverse component of velocity. The orderly progression is such that each particle follows exactly the path of the particle preceding it without any deviation. Thus a thin filament of dye will remain as such without diffusion. There is a much greater transverse velocity gradient

---

  * Sometimes referred to as 'varied'

in laminar flow (Fig. 4.1(b)) than in turbulent flow. In fact, for a pipe the ratio of the mean velocity V and the maximum velocity v_{max} is 0·5 and about 0·85 in the respective cases.

Laminar flow is associated with low velocities and viscous sluggish fluids. In pipeline and open-channel hydraulics, the velocities are nearly always sufficiently high to ensure turbulent flow, although a thin laminar layer persists in proximity to a solid boundary. The laws of laminar flow are fully understood and for simple boundary conditions the velocity distribution can be analysed mathematically. Due to its irregular pulsating nature, turbulent flow has defied rigorous mathematical treatment, and for the solution of practical problems it is necessary to rely largely on empirical or semi-empirical relationships.

(b) *Rotational and Irrotational Flow.* The flow is said to be *rotational* if each fluid particle has an angular velocity about its own mass centre.

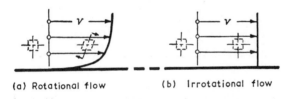

(a) Rotational flow   (b) Irrotational flow

Figure 4.2   Flow adjacent to a straight boundary

Fig. 4.2(a) shows a typical velocity distribution associated with turbulent flow past a straight boundary. Due to the non-uniform velocity distribution, a particle with its two axes originally perpendicular suffers deformation with a small degree of rotation. In Fig. 4.3(a) flow in a circular path is depicted, with the velocity directly proportional to the radius. The two particle axes rotate in the same direction so that again the flow is rotational.

For the flow to be *irrotational*, the velocity distribution adjacent to the straight boundary must be uniform (Fig. 4.2(b)). In the case of flow in a circular path, it may be shown that irrotational flow will only pertain provided that the velocity is inversely proportional to the radius. From a first glance at Fig. 4.3(b) this appears erroneous, but a closer examination reveals that the two axes rotate in opposite directions so that there is a compensating effect producing an *average* orientation of the axes which is unchanged from the initial state.

Because all fluids possess viscosity, the flow of a real fluid is never truly irrotational, and laminar flow is of course highly rotational. Thus irrotational flow is a hypothetical condition which would be of academic interest only were it not for the fact that in many instances of turbulent

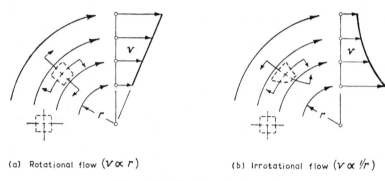

(a) Rotational flow $(V \propto r)$ (b) Irrotational flow $(V \propto 1/r)$

Figure 4.3   Flow in a circular path

flow the rotational characteristics are so insignificant that they may be neglected. This is convenient because it is possible to analyse irrotational flow by means of the mathematical concepts of classical hydrodynamics referred to earlier.

(c) *Steady and Unsteady Flow.* The flow is said to be *steady* when the conditions at any point are constant with respect to time. A strict interpretation of this definition would lead to the conclusion that turbulent flow was never truly steady. However, for the present purpose it is convenient to regard the general fluid motion as the criterion and the erratic fluctuations associated with the turbulence as only a secondary influence. An obvious example of steady flow is a constant discharge in a conduit or open channel.

As a corollary it follows that the flow is *unsteady* when conditions vary with respect to time. An example of unsteady flow is a varying discharge in a conduit or open channel; this is usually a transient phenomenon being successive to or followed by a steady discharge. Other familiar examples of a more periodic nature are wave motion and the cyclic movement of large bodies of water in tidal flow.

Most of the practical problems in hydraulic engineering are concerned with steady flow. This is fortunate, since the time variable in unsteady flow considerably complicates the analysis. Accordingly, in the present treatise, consideration of unsteady flow will be restricted to a few relatively simple cases. It is important to bear in mind, however, that several common instances of unsteady flow may be reduced to the steady state by virtue of the principle of relative motion. Thus, a problem involving a vessel moving through still water may be re-phrased so that the vessel is stationary and the water is in motion; the only criterion for similarity of fluid behaviour is that the relative velocity shall be the same. Again, wave motion in deep water may be reduced to the steady state by assuming that an observer travels with the waves at the same velocity.

(*d*) *Uniform and Non-uniform Flow.* The flow is said to be *uniform* when there is no variation in the magnitude and direction of the velocity vector from one point to another along the path of flow. For compliance with this definition both the area of flow and the velocity must be the same at every cross-section. *Non-uniform* flow occurs when the velocity vector varies with location, a typical example being flow between converging or diverging boundaries.

Both of these alternative conditions of flow are common in open-channel hydraulics, although, strictly speaking, since uniform flow is always approached asymptotically, it is an ideal state which is only approximated to and never actually attained. It should be noted that the conditions relate to space rather than time and therefore in cases of enclosed flow (e.g. pipes under pressure) they are quite independent of the steady or unsteady nature of the flow.

4.3 Stream Lines

A useful visual picture of the pattern of steady (or unsteady uniform) fluid motion is obtained by means of *stream lines*. These are imaginary lines traced out by successive fluid particles throughout the flow stream, neglecting of course the secondary fluctuations superimposed by the turbulence. The tangent to a stream line indicates the direction of the velocity vector at the particular point. When drawing stream lines it is customary to include only a sufficient number to make the flow pattern quite clear. It follows from the definition that there can be no intersection or joining up of stream lines. Also, since there is no flow across a stream line, one stream line always coincides with a free surface and a solid boundary. Fig. 4.4 illustrates the stream lines for a conical connecting piece, and it is evident that a convergence of stream lines is associated with an increase in velocity and of course vice versa. Similarly, it also follows that in uniform straight flow all the stream lines are parallel.

29

In many cases, the flow is two-dimensional, that is to say the velocity component in the third dimension is zero. Typical examples are flow over a weir of uniform shape or under a rectangular sluice gate, each spanning the full channel width. The omission of the third dimension considerably simplifies the diagrammatic representation of the stream

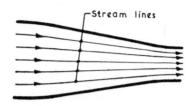

Figure 4.4   Stream lines for a converging
pipe section

lines and it is often convenient to space the lines an equal distance apart in the zone of uniform flow. If the velocity at one point in the diagram is known, the velocity at any other point may be determined from a knowledge of the spacing. Nevertheless stream lines tend to be employed more for qualitative than quantitative purposes.

For an ideal fluid and certain simple boundary conditions, classical hydrodynamicists have been able to determine the pattern of stream lines by mathematical methods. More complicated boundary conditions can be evaluated by means of a flow net, which is a graphical construction with a sound theoretical basis. Flow nets are explained in Sect. 4.10.

4.4 Continuity Equation

The elemental 'stream tube' in Fig. 4.5 comprises a number of enveloping stream lines. Since no flow takes place across the stream lines, the fluid

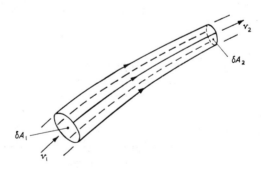

Figure 4.5   Elemental stream tube

must enter and leave the tube only at the end sections. Let us suppose that the end sectional areas are δA_1 and δA_2, and the corresponding velocities, assumed uniform, are v_1 and v_2. It is evident that the elemental discharge δQ is given by

$$\delta Q = v_1 \, \delta A_1 = v_2 \, \delta A_2$$

Integrating for the total discharge Q, we obtain

$$Q = V_1 A_1 = V_2 A_2$$

where V_1 and V_2 are the mean velocities at the end sections with areas A_1 and A_2 respectively. The general *continuity equation* may be expressed in the form

$$Q = VA = \text{constant} \tag{4.1}$$

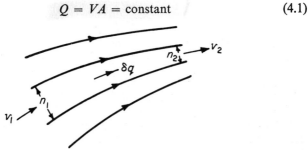

Figure 4.6   Flow between adjacent stream lines

In the case of two-dimensional flow with the stream lines spaced so as to give uniform discharge δq per unit width between adjacent lines (Fig. 4.6), Eq. (4.1) may be modified to

$$\delta q = v_1 n_1 = v_2 n_2$$

where v_1, v_2, and n_1, n_2 are the respective mean velocities and spacings along the same pair of stream lines. The equation may be expressed in the general form

$$vn = \text{constant} \tag{4.2}$$

showing that the velocity is inversely proportional to the spacing, which is a confirmation of the statement made previously that converging stream lines indicate an increase in velocity and vice versa.

4.5 Bernoulli Equation

In Fig. 4.7, a cylindrical element of a stream tube is shown in motion along a stream line. The length, sectional area, and unit weight are δs, δA, and w respectively so that the weight is $w \, \delta s \, \delta A$. The pressure force acting on the rear face is $p\,\delta A$ and on the leading face is $[p + (\mathrm{d}p/\mathrm{d}s)\,\delta s]\,\delta A$. The normal forces acting on the side faces are in equilibrium and, as the fluid is assumed non-viscous, there is no shear stress. The velocity varies along the stream line and therefore there is an

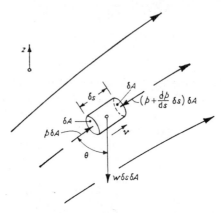

Figure 4.7   Forces acting on a cylindrical element

accelerative force present which it is necessary to take into account when considering the longitudinal balance of forces. An expression for this accelerative force may be obtained by utilising Newton's second law (force = rate of change of momentum):

$$-w\,\delta s\,\delta A\cos\theta + p\,\delta A - \left(p + \frac{\mathrm{d}p}{\mathrm{d}s}\,\delta s\right)\delta A = \frac{w}{g}\,\delta s\,\delta A\,\frac{\mathrm{d}v}{\mathrm{d}t}$$

Simplifying,

$$-w\cos\theta - \frac{\mathrm{d}p}{\mathrm{d}s} = \frac{w}{g}\,\frac{\mathrm{d}v}{\mathrm{d}t}$$

Now $\cos\theta$ may be expressed in the form $\mathrm{d}z/\mathrm{d}s$ where z is the vertical coordinate; also, as the flow is steady (independent of t), $\mathrm{d}v/\mathrm{d}t = v(\mathrm{d}v/\mathrm{d}s)$. Thus

$$w\frac{\mathrm{d}z}{\mathrm{d}s} + \frac{\mathrm{d}p}{\mathrm{d}s} + \frac{w}{g}v\frac{\mathrm{d}v}{\mathrm{d}s} = 0$$

or

$$\frac{\mathrm{d}}{\mathrm{d}s}\left(\mathrm{d}z + \frac{\mathrm{d}p}{w} + \frac{v}{g}\,\mathrm{d}v\right) = 0$$

Integrating along the stream line we obtain

$$z + \frac{p}{w} + \frac{v^2}{2g} = \text{constant} \qquad (4.3)$$

This expression, perhaps the best known in fluid mechanics, is called the *Bernoulli equation*, in recognition of the mathematical physicist Daniel Bernoulli (1700–1782). In the case of irrotational flow between straight parallel boundaries the velocity distribution is uniform ($v = V$) so that $z + p/w$ is constant for all the stream lines.

Some further consideration of the nature of the terms in Eq. (4.3) is merited. Each term has the dimension of energy per unit weight, indicating a scalar quantity. It is thus convenient to refer to each term as a *head*.

The first term z is an elevation head and may be regarded as potential energy. The second term p/w is a pressure head, the energy form of which follows from a consideration of a piston operating in a cylinder. If p is the sustained pressure and $\dot{V}$ is the swept volume, the pressure energy per unit weight or work done per unit weight is evidently $p\dot{V}/w\dot{V}$ or p/w. The last term is a velocity head and represents the velocity or kinetic energy ($mv^2/2$).

The constant term is the *total energy head*, denoted by H, and is the sum of the potential, pressure, and velocity heads. The same general conclusion could have been reached from a consideration of the laws of conservation of energy. There is thus a constant total energy head along a stream line. Furthermore, the concept of an ideal fluid always leads to the conclusion that the total energy head is constant throughout the entire region of flow.

4.6 Applications of Bernoulli Equation

4.6.1

The Bernoulli equation is the basis for the solution of a wide range of hydraulic problems and it is frequently referred to in the text. At this stage our consideration will be limited to the pressure-velocity relationship which the equation reveals, and to a few of the simpler applications.

4.6.2 *Relationship between Pressure and Velocity*

For two points along a stream line the Bernoulli equation may be expressed in the form

$$(z_1 - z_2) + \frac{(p_1 - p_2)}{w} + \frac{(v_1{}^2 - v_2{}^2)}{2g} = 0 \qquad (4.4)$$

For uniform flow ($v_1 = v_2$), the third term is zero, so that the elevation head and pressure head are directly related. Consequently, an increase in elevation head produces a decrease in pressure head, and vice versa. With non-uniform flow an increase in velocity head is associated with a decrease in the sum of the elevation and pressure heads (sometimes referred to as the piezometric head), and vice versa. In many cases of non-uniform flow, where the reference points are only a relatively short distance apart, the difference in elevation head is very small or negligible in comparison with the other head differences. Under these conditions, there is an almost direct relationship between velocity and pressure, and, as we have learnt earlier that a convergence of stream lines indicates an increase in velocity, it now follows that it also denotes a decrease of pressure. Correspondingly, a divergence of stream lines is associated with an increase of pressure.

4.6.3 *Stagnation Pressure*

Fig. 4.8 shows the pattern of stream lines around the upstream face of a long cylindrical-shaped body aligned at right angles to the flow stream.

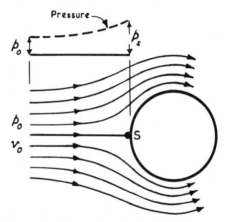

Figure 4.8   Stream lines for flow around the
upstream face of a long cylinder

The divergence of the stream lines indicates a reduction in velocity and an increase in pressure. Due to the symmetrical nature of the flow pattern the central stream line remains undeviated and intersects the cylinder surface at the point S, where the tangent is at right angles to the stream line. The velocity at S is therefore zero, the fluid being brought to rest. Appropriately, it is called a *stagnation point* and the corresponding pressure is the *stagnation pressure*. In this case the stagnation pressure p_s is given by

$$p_s = w\left(\frac{p_0}{w} + \frac{v_0^2}{2g}\right)$$

34

where p_0 and v_0 are the pressure and velocity in the main flow stream. Simplifying,

$$p_s = p_0 + \tfrac{1}{2}\rho v_0{}^2 \tag{4.5}$$

The curve of pressure rise along the central stream line is shown superimposed in Fig. 4.8.

For enclosed flow, it may be stated generally that a stagnation point exists if the boundary stream line abruptly changes direction such that the included angle is less than 180°. Thus in the case of irrotational flow

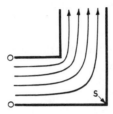

Figure 4.9   Stream lines for flow
around a mitred pipe bend

at a mitre bend, shown in Fig. 4.9, there is a stagnation point at S and the pressure here is greater than that at OO by the velocity head at the latter section.

4.6.4 *Measuring Devices*

(*a*) *Velocity*. The stagnation principle is the basis of a velocity measuring device called a *Pitot tube*, which was first proposed by Henri Pitot in 1732. In its elementary form (Fig. 4.10) it consists of an L-shaped tube with unsealed ends. One limb is inserted in the flow stream and aligned

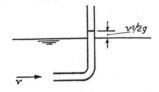

Figure 4.10   Elementary Pitot tube

directly upstream, whilst the other is vertical and open to the atmosphere. A stagnation point exists at the upstream tip of the tube, and, as the static pressure head here is represented by the depth, the liquid level in the vertical limb rises above the free surface by an amount equal

35

to the velocity head $v^2/2g$. Clearly, a measurement of this differential head will yield by simple calculation the velocity v at the Pitot tube tip. Unfortunately, unless the velocity is relatively large, the head rise is very small – for instance at 1 m/s it is only 50 mm – so that sensitive measuring instruments are required.

In order to determine the point velocity in enclosed flow by the Pitot tube method it is necessary to obtain a differential measurement of the dynamic plus static pressure and the static pressure heads. An instrument of this type is known as a *combined Pitot-static tube* (Fig. 4.11) and is particularly suitable for measuring air velocities. The tube is so shaped that there is the minimum disturbance to the flow. The dynamic plus static pressure is admitted at the tip and the static pressure through small holes at the sides. Separate passages within the outer tube

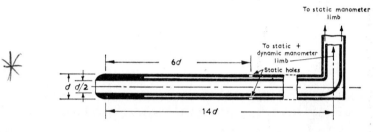

Figure 4.11   N.P.L. Pitot-static tube with hemispherical nose

connect to the limbs of a manometer. The velocity v of the flow stream at the point of insertion is given by

$$v = \sqrt{2g}\left(\frac{p_s - p}{w}\right)^{1/2} \tag{4.6}$$

where $(p_s - p)/w$ is the differential pressure head. The withdrawable form of Pitot tube (without lower limb) is specially designed for traversing a pipe section and has the practical advantage that it is unnecessary to shut off the flow while the tube is being fitted.

For accurate velocity measurement it is necessary to introduce a coefficient C (very nearly equal to unity[1]) in Eq. (4.6) so as to make allowance for the small effects of nose shape and other characteristics.

(b) *Discharge*. The convergence of stream lines and the associated pressure drop are the characteristic features which enable constrictions to be utilised for the purpose of measuring discharge.

---

[1] Actually, $C = 1.0$ for the N.P.L. tube with modified ellipsoidal nose.

36

Fig. 4.12 shows a form of pipeline constriction called a *Venturi tube*. Upstream and downstream of the short parallel throat, conical portions connect with the full pipe section. Assuming a horizontal alignment and applying the Bernoulli equation to centre line points upstream and at the throat we have

$$\frac{p_0}{w} + \frac{V_0^2}{2g} = \frac{p_c}{w} + \frac{V_c^2}{2g}$$

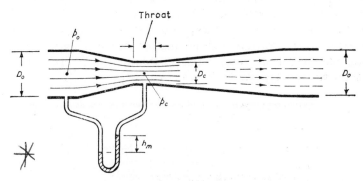

Figure 4.12   Flow through a Venturi tube

Substituting Q/A_0 for V_0 and Q/A_c for V_c, where A_0 and A_c are the respective sectional areas, we obtain for the discharge:

$$Q = \frac{\sqrt{2g}A_c}{[1 - (A_c/A_0)^2]^{1/2}} \left(\frac{p_0 - p_c}{w}\right)^{1/2}$$

or

$$Q = \frac{\sqrt{2g}(\pi/4)D_c^2}{(1 - m^2)^{1/2}} \left(\frac{p_0 - p_c}{w}\right)^{1/2} \tag{4.7}$$

where $m = (D_c/D_0)^2$, D_0 and D_c being the respective diameters.

The value of Q may be determined since both D_0 and D_c are known and the differential head $(p_0 - p_c)/w$ is measured by means of a suitable manometer. For a real fluid, it is necessary to multiply the theoretical discharge by a coefficient C (about 0·97) in order to take account of the very small energy head loss in the converging portion and the non-uniform velocity distribution upstream. The diameter of the constriction should be large enough to ensure that under the maximum flow condition the pressure head at the throat does not fall below 2 m absolute, otherwise there will be the tendency for air bubbles to enter the piezometer line, thus vitiating the readings. For minimum energy loss in the diverging portion, an included angle between 5 and 7 degrees is the optimum, representing a compromise between the eddy losses associated with expanding flow and the skin friction loss along boundaries subjected to relatively high velocity.

Another form of constriction device, and one which is widely employed is the *orifice plate* (Fig. 4.13). It consists of a flat diaphragm or plate, with circular aperture, fitted between two pipe flanges. Whilst having the merit of simplicity, there is the disadvantage of relatively high energy losses due to the abrupt expansion downstream (see Ch. 5, Sect. 5.11.2). The coefficient C has a much lower value (about 0·61) than that for the Venturi tube, since D_c in Eq. (4.7) refers to the diameter of the aperture and not that of the minimum flow stream area, the latter being located further downstream in the vicinity of the low pressure tapping.

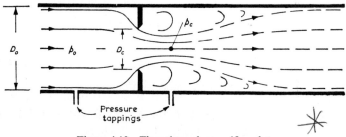

Figure 4.13   Flow through an orifice plate

Design and installation recommendations with calibration data are given in Part 1 (1964) of B.S. 1042, *Methods for the Measurement of Fluid Flow in Pipes*. The standard of accuracy of flow measurement is normally well within ± 2 per cent.

4.6.5 *Orifices*

An *orifice* is a geometric opening in the side of a thin-walled tank or vessel. The usual shape is circular with either a sharp edge or a square upstream edge and downstream bevel. In Fig. 4.14 a fluid is shown issuing from a circular orifice in a tank, the centre of the orifice being situated at depth h below the free surface. Due to the vertical component of the flow near the walls, the stream lines continue to converge for a short distance downstream of the orifice. They become parallel at a place of minimum cross-section called the *vena contracta*. The pressure here is atmospheric and, since the stream lines are very close together, the velocity v_c, sometimes called the *spouting velocity*, is sensibly constant over the section. This velocity may be determined by applying

Bernoulli's equation to the free surface and the vena contracta; thus $h = v_c^2/2g$, or

$$v_c = \sqrt{2gh} \qquad (4.8)$$

In the case of a real fluid there is a small energy loss due to viscous effects and this necessitates the introduction of a *coefficient of velocity*, C_v (about 0·97).

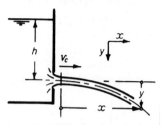

Figure 4.14   Jet issuing from a
circular orifice in a tank

The discharge Q is equal to $a_0 v_c$, where a_0 is the sectional area of the jet at the vena contracta. It is more convenient, however, to express the discharge in terms of the orifice area a_o, which is constant and easily measured, the areas a_o and a_c being related by a *coefficient of contraction*, C_c (about 0·65). The discharge is thus given by

$$Q = C_c C_v a_o \sqrt{2gh} \qquad (4.9)$$

or

$$Q = C_d a_o \sqrt{2gh} \qquad (4.10)$$

where C_d (about 0·63[1]) is known as the *coefficient of discharge*. Various forms of mouthpiece may be fitted which will alter the pattern of the stream lines and the values of the coefficients.

If x and y are the co-ordinates of the jet trajectory and the origin is at the vena contracta, then

$$v_x = v_c, \quad v_y = gt; \quad x = v_x t, \quad y = \tfrac{1}{2}gt^2$$

where t is the time of particle travel from the vena contracta to the reference point. Eliminating t we obtain

$$y = \frac{g}{2v_c^2} x^2 = \frac{1}{4h} x^2 \qquad (4.11)$$

[1] The value of C_d is dependent primarily on the boundary geometry (incl. the orifice edge), but may also be influenced by fluid viscosity and surface tension effects.

showing that the trajectory is parabolic. Knowing the co-ordinates at one point on the trajectory, the spouting velocity and hence the discharge may be determined. It may be noted that trajectory analysis is very similar to that in elementary ballistics.

A practical problem is the determination of the jet trajectory when water is discharged from an orifice near the base of a high dam. Owing to the combined effects of friction and dispersion the actual trajectory is always less than the theoretical.

4.7 Momentum Equation

The impulse-momentum principle of solid mechanics is also applicable to fluids, but there is the important distinction that in the former case

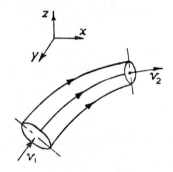

Figure 4.15   Derivation diagram for
the momentum equation

the action is completed in a finite time (e.g. force due to arresting of a vehicle) whereas in the case of the steady flow of a fluid it is continuous.

Let us consider the equilibrium of the stream tube represented in Fig. 4.15. The co-ordinate directions are x, y, and z. If δt is the time taken for an element of fluid to traverse the stream tube, then the mass (δM) of the tube is $\rho \, \delta Q \, \delta t$.

Now from Newton's third law the internal pressures (caused by acceleration) cancel out, so that when assessing the change of momentum of the whole mass it is only necessary to take account of the change in the velocity vector at entry and exit. Thus, in accordance with Newton's second law, the resultant force in the x direction is given by

$$\delta F_x = \rho \, \delta Q \, \delta t \, \frac{\delta v_x}{\delta t}$$

or

$$\delta F_x = \rho \, \delta Q [(v_x)_2 - (v_x)_1]$$

40

Then for the entire flow stream, assuming uniform velocity distribution at the two relevant sections, we have

$$F_x = \rho Q[(V_x)_2 - (V_x)_1] \tag{4.12}$$

where Q is the total discharge and $(V_x)_1$, $(V_x)_2$ are the respective velocity components.

Similarly,

$$F_y = \rho Q[(V_y)_2 - (V_y)_1] \tag{4.13}$$

and

$$F_z = \rho Q[(V_z)_2 - (V_z)_1] \tag{4.14}$$

These are three forms of the *momentum equation*. The equation states that the resultant component force acting on a free body of fluid is equal to the difference between the momentum components at the entry and exit sections.

In cases where the velocity distribution is not uniform (and this is always so with a real fluid) it is convenient to express the momentum equation in terms of the mean velocities. A rigorous analysis demands that each mean velocity be modified by a correction factor[1] in order to take account of the non-uniform distribution, but in practice the error which results is generally negligible.

The momentum equation is of fundamental importance in fluid mechanics. It concerns simple vector quantities of force and velocity, and is independent of internal energy changes. For this reason it is of particular value when there exists between the two reference sections a complex flow condition which would be difficult to analyse.

4.8 Applications of Momentum Equation

4.8.1

The momentum equation finds an application in many hydraulic problems. Generally, it is employed in conjunction with the continuity equation, and often additionally with the Bernoulli equation. Problems which involve a marked change in flow velocity or direction are particularly appropriate. The analysis of the hydraulic jump and the determination of the force exerted by a jet of water impinging on a

---

[1] The correction factor β is obtained by integrating the momentum of the elemental stream tubes over the entire section and dividing by the momentum based on the mean velocity. Thus

$$\beta = \frac{1}{VQ} \int v \, dQ = \frac{1}{V^2 A} \int v^2 \, dA$$

rotating vane are both in this category and are dealt with in later chapters. Our present consideration will be limited to two simple cases. When applying the equation to a real fluid it is important to remember that there are longitudinal frictional stresses present which can only be neglected when the distance between the reference sections is relatively short.

4.8.2 *Change in Velocity*

A nozzle, attached to a pipeline, and discharging to the atmosphere provides a good example of a rapid change in velocity. The fluid exerts a force on the nozzle and in accordance with Newton's third law there is a similar force, of opposite sign, exerted by the nozzle on the fluid. This is the force which the tension bolts must be designed to withstand. The force could be evaluated by an analysis of the pressures acting on

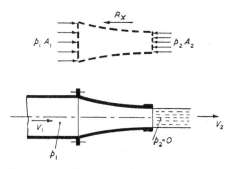

Figure 4.16   Nozzle discharging to atmosphere

the surface of the nozzle but this procedure would, to say the least, be extremely tedious. By contrast, a simple application of the momentum equation between upstream and downstream reference sections will yield a direct solution. In Fig. 4.16 the component forces are the hydrostatic forces $p_1 A_1$ and $p_2 A_2$ and the force R_x exerted by the nozzle on the fluid. The rate of change of momentum is $\rho Q (V_2 - V_1)$ so that the momentum equation may be expressed as

$$p_1 A_1 - p_2 A_2 - R_x = \rho Q (V_2 - V_1)$$

As the discharge is to the atmosphere, $p_2 = 0$, and thus

$$R_x = p_1 A_1 - \rho Q (V_2 - V_1) \qquad (4.15)$$

Further simple cases where the momentum equation may be profitably applied include the determination of the force acting on a pipeline at a contraction and that on a sluice gate (see Ch. 9, Ex. 9.1).

Example 4.1

Calculate the tension force on the flanged connection between a 64 mm ($2\frac{1}{2}$ in.) diameter pipe and a nozzle discharging a jet with velocity of 30 m/s and diameter 19 mm ($\frac{3}{4}$ in.).

$$Q = \frac{\pi}{4} D_2^2 V_2 = \frac{\pi}{4} \times 19^2 \times 30/1000 = 8\cdot5 \text{ l/s.} = 0\cdot307 \text{ cusecs.}$$

For continuity, $D_1^2 V_1 = D_2^2 V_2$, so that $V_1 = 2\cdot65$ m/s.

Applying Bernoulli's equation to entry to and exit from the nozzle, and neglecting losses, we have $p_1/w + V_1^2/2g = V_2^2/2g$, so that

$$p_1 = \frac{w}{2g}(V_2^2 - V_1^2) = \tfrac{1}{2}(30^2 - 2\cdot65^2) = 446\cdot5 \text{ kN/m}^2.$$

Substituting in Eq. (4.15),

$$R_z = 446\cdot5 \times \frac{\pi}{4} \times \frac{64^2}{10^6} - 8\cdot5(30 - 2\cdot65)/1000 = \mathbf{1\cdot21} \text{ kN.}$$

4.8.3 *Change in Velocity and Direction*

A reducing bend with deviation in the vertical plane is shown in Fig. 4.17. Due to the hydrostatic and dynamic pressures a force is exerted by

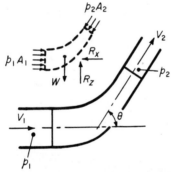

Figure 4.17   Forces on a vertical reducing bend

the fluid on the bend which has to be resisted by a thrust block or other suitable means. This force could be evaluated by plotting the stream lines and thus determining the pressure distribution. However, by a simple application of the momentum equation, and quite independently of any energy losses associated with turbulent eddying (real fluid), we obtain:

For the *x* direction:

$$p_1 A_1 - p_2 A_2 \cos\theta - R_z = \rho Q(V_2 \cos\theta - V_1) \qquad (4.16)$$

43

And for the z direction:

$$R_z - W - p_2 A_2 \sin \theta = \rho Q V_2 \sin \theta \qquad (4.17)$$

where W is the weight of fluid between the reference sections.

From these equations R_x and R_z may be determined and hence the resultant $R = \sqrt{R_x{}^2 + R_z{}^2}$.

It is to be noted that the momentum equation gives no information concerning the location of the resultant, which necessitates an analysis involving forces and moments.

4.9 Flow in a Curved Path

4.9.1

We have not so far given consideration to the forces acting on a fluid element due to its motion along a curved path. In the present section the general expression for the pressure gradient is derived and the two special cases of curvilinear motion – free vortex motion and forced vortex motion – are analysed.

4.9.2 General Equation

Consider the radial forces acting on a small water element moving along a curved stream line as in Fig. 4.18. At a certain instant of time the

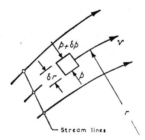

Figure 4.18   Flow in a curvilinear path

velocity is v and the radius of curvature r. The radial acceleration induced by the curvature is v^2/r and the centrifugal force F is equal to the mass multiplied by the acceleration, or

$$F = \frac{w}{g} \delta A \, \delta r \, \frac{v^2}{r} \qquad (4.18)$$

where δr is the width of the element in the plane of the stream lines and δA is the sectional area in a direction normal to this plane. For equilibrium the centrifugal force must be resisted by the difference in

44

pressure force between the inner and outer faces which is $\delta p \, \delta A$. Substituting in Eq. (4.18) we have

$$\delta p \, \delta A = \frac{w}{g} \, \delta A \, \delta r \, \frac{v^2}{r}$$

or in the limit,

$$\frac{dp}{dr} = \frac{w}{g} \frac{v^2}{r} \qquad (4.19)$$

Eq. (4.19) is the general expression for the rate of change of pressure in a direction normal to that of the flow. If the stream lines are straight $r = \infty$ and $dp/dr = 0$. The equation cannot be evaluated unless both v and r are known.

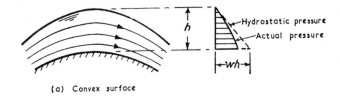

(a) Convex surface

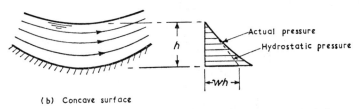

(b) Concave surface

Figure 4.19 Pressure distribution in curvilinear flow in a vertical plane

Curvilinear motion in the vertical plane has an important significance with respect to the pressure-depth relationship below a free surface. Flow over a convex and a concave guiding surface is shown in Fig. 4.19; conditions are not very different to those at the crest and toe of a dam spillway. The effect of the centrifugal force is to modify, in the manner indicated, the normal hydrostatic linear relationship between pressure and depth. In the case of the convex boundary the fluid pressure is reduced below hydrostatic, and may even be sub-atmospheric if the curvature is sharp and the velocity high. In the case of the concave boundary, however, the pressure is raised above the hydrostatic. The water level in a piezometer tube will thus be depressed below the free surface or raised above it according to the boundary curvature. Of course, the pressure

45

at the free surface must always be atmospheric. In practice, with a real fluid, the flow stream may separate from the convex surface (see Sect. 4.11.6).

4.9.3 Free Vortex

It is an essential characteristic of a *free vortex* in an ideal fluid that it does not require the application of external energy for its continuance. However, some initial disturbance is required to set it in motion, although in fact the rotation of the earth may suffice.

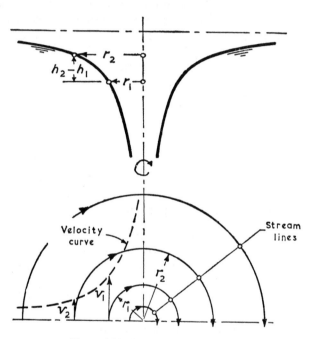

Figure 4.20   Free cylindrical vortex

In a *free cylindrical vortex* (Fig. 4.20) the stream lines are concentric circles and, because there is no energy variation throughout the system, not only is the energy constant along the stream lines but also at all points in the flow stream on a horizontal plane. We therefore have $p/w + v^2/2g = $ constant. Differentiating with respect to the radius r,

$$\frac{\mathrm{d}p}{\mathrm{d}r} = -\frac{w}{g}v\frac{\mathrm{d}v}{\mathrm{d}r} \tag{4.20}$$

Equating (4.19) and (4.20), $v^2/r = -v(\mathrm{d}v/\mathrm{d}r)$,   or $\mathrm{d}v/v + \mathrm{d}r/r = 0$.

Integrating, $\ln v + \ln r = $ constant. Thus $\ln vr = $ constant, or

$$vr = \text{constant} \qquad (4.21)$$

The velocity is therefore inversely proportional to the radius. It follows that the velocity is infinite when the radius is zero, but this is an impossible condition and is a defect in the analysis arising from the assumption of an ideal fluid.

That Eq. (4.21) satisfies the condition of constant energy also follows from a consideration of momentum. As there is no external force there can be no torque at the vortex centre. Now, since torque is equal to the change of moment of momentum with respect to time, we have for the mass of water m at radius r: $(d/dt)(mvr) = 0$. Integrating, we obtain $vr = $ constant as before.

In order to determine the pressure head distribution in a horizontal plane, we proceed to apply Bernoulli's equation to radii r_1 and r_2 so that

$$p_1/w + v_1^2/2g = p_2/w + v_2^2/2g$$

or

$$\frac{p_2 - p_1}{w} = \frac{v_1^2}{2g}\left[1 - \left(\frac{r_1}{r_2}\right)^2\right] \qquad (4.22)$$

The form of this equation indicates a hyperbolic variation of pressure with radius. At the surface the pressure is everywhere atmospheric ($p_1 = p_2 = p_a$) but the elevation head varies, so that the corresponding difference in free surface level is given by

$$h_2 - h_1 = \frac{v_1^2}{2g}\left[1 - \left(\frac{r_1}{r_2}\right)^2\right] \qquad (4.23)$$

The surface elevation curve is asymptotic at the two ends and at the centre -- when $r_2 = \infty$, $h_2 - h_1 = v_1^2/2g$ and when $r_2 = 0$, $h_2 - h_1 = -\infty$. As mentioned in connection with the velocity, this latter condition is impossible and in fact the pressure head cannot fall below the vapour pressure without the water passing into the gaseous state; viscous resistance would prevent this condition being reached. But it is certainly of some significance that in flowing water the pressure tends to be lower near the centre of large swirls and eddies produced by boundary irregularities.

The flow behaviour of an ideal fluid at a long radiused bend of a pressure conduit or open channel is in accordance with the above relationships. If such a conduit of rectangular section, with transverse dimension b and inner and outer radii r_1 and r_2 carries a discharge Q, then integrating radially we have

$$Q = \int_{r_1}^{r_2} bv \, dr = \int_{r_1}^{r_2} b\frac{\kappa}{r} \, dr = b\kappa \ln \frac{r_2}{r_1}$$

47

where κ is the constant term in Eq. (4.21). Thus in Eq. (4.22), which is applicable to a bend in the horizontal plane,

$$v_1 = \kappa/r_1 = Q/[br_1 \ln (r_2/r_1)]$$

so that

$$\frac{p_2 - p_1}{w} = \frac{Q^2}{2g[br_1 \ln (r_2/r_1)]^2} \left[1 - \left(\frac{r_1}{r_2}\right)^2\right] \qquad (4.24)$$

For a bend in a circular pressure conduit, diameter D and centre line radius r_m [i.e. $r_m = (r_1 + r_2)/2$], it may be shown that

$$\frac{p_2 - p_1}{w} = \frac{Q^2 r_m D}{4g\pi^2[r_m{}^2 - (D/2)^2]^2[r_m - \sqrt{r_m{}^2 - (D/2)^2}]^2} \qquad (4.25)$$

Likewise, for an open channel with mean flow depth h, the difference between the depths at the inner and outer radii (r_1 and r_2) is given by

$$h_2 - h_1 = \frac{Q^2}{2g[hr_1 \ln (r_2/r_1)]^2} \left[1 - \left(\frac{r_1}{r_2}\right)^2\right] \qquad (4.26)$$

There is some significant difference in the behaviour of a real fluid at a bend, and this is discussed in Sect. 4.11.8.

The *free spiral vortex* is another type of vortex motion, the direction of flow being part radial and part rotational. For radial flow between parallel plates spaced distance b apart the discharge at any radius r is given by $Q = 2\pi rbv$. For continuity ($Q = $ constant) it is essential that $vr = $ constant. This is the same condition as for a cylindrical vortex (Eq. (4.21)) and therefore the free surface profiles must be similar. It may be shown that the stream lines are in the form of logarithmic spirals and the resultant velocity v at any point may be obtained from

$$v = \sqrt{v_t{}^2 + v_r{}^2} \qquad (4.27)$$

where v_t and v_r are the velocities for the free rotational and radial flows acting independently.

The spiral vortex is well represented by the familiar circulatory behaviour which often occurs when water is discharged through an outlet at the base of a shallow tank. It is met with in nature in the form of waterspouts and tornadoes; the velocities in the low pressure 'eyes' of these tropical phenomena are exceedingly high and can cause great damage.

4.9.4 Forced Vortex

As the description implies, a *forced vortex* is produced and maintained by the application of some external force. For instance, it is the form of motion which results if a fluid is forcibly rotated by a paddle wheel operating in a stationary vessel, there being no relative velocity between the fluid and paddles after the motion has settled down. The result is the same if a cylindrical vessel, partly filled with a real fluid, is rotated about a vertical axis, since the contents will rotate, after the initial viscous resistance has been overcome, at the same angular velocity as the vessel. Strictly speaking then, forced vortex motion is only possible with a real fluid.

The stream lines are concentric circles (Fig. 4.21) and since the angular velocity ω is constant throughout we have

$$v = \omega r \tag{4.28}$$

The velocity therefore increases with the radius, which is the reverse of the condition pertaining to a free vortex. Substituting in the general equation (Eq. (4.19)) for the pressure gradient normal to the curved stream lines, $dp/dr = (w/g)\omega^2 r$. Integrating, $\int dp = (w/g)\omega^2 \int r \, dr$, so that

$$\frac{p}{w} = \frac{\omega^2 r^2}{2g} + \text{constant}$$

If $p = p_0$ when $r = 0$, we may eliminate the constant and obtain

$$\frac{p - p_0}{w} = \frac{\omega^2 r^2}{2g} \tag{4.29}$$

giving the pressure distribution across a horizontal plane, which is seen to be parabolic. For the free surface profile:

$$h - h_0 = \frac{\omega^2 r^2}{2g} \tag{4.30}$$

The free surface of a forced vortex in a cylindrical vessel or tank will thus take up the profile of a paraboloid of revolution as indicated in Fig. 4.21. Now, it is characteristic of a free surface that the slope at any point is perpendicular to the resultant force at that point; as the forces per unit volume are w in a vertical direction and $(w/g)\omega^2 r$ horizontally, the slope of the free surface is given by

$$\tan \theta = \frac{\omega^2 r}{g} \tag{4.31}$$

The same result is obtained by differentiating Eq. (4.30) with respect to r.

Since there is no vertical acceleration the pressure variation with depth follows the hydrostatic law If the cylinder is closed at the top,

49

restricting the upward movement of the free surface, the same equations for pressure distribution are still valid but an upward force is exerted on the roof.

Substituting v for ωr in Eq. (4.29) it is evident that an increase in pressure head from p_1/w to p_2/w is associated with an increase in velocity head from $v_1{}^2/2g$ to $v_2{}^2/2g$. Thus, with the creation of a velocity

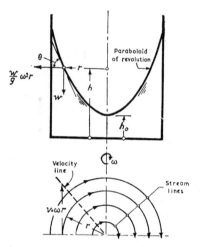

Figure 4.21   Forced vortex

head by external means, the forced vortex so formed results in the generation of an equivalent pressure head. This principle is utilised in the centrifugal pump where water enters with a low pressure and velocity at the centre of the vane wheel or impeller and is discharged at the outer periphery with a much higher pressure and velocity (see Ch. 10).

4.10 Flow Nets

The *flow net* is a useful device for the graphical determination of the pattern of the stream lines in irrotational flow. In two-dimensional flow between straight parallel boundaries the velocity distribution is uniform and it is conventional to space the stream lines an equal distance (n_0) apart as shown in Fig. 4.22. Lines may now be drawn normal to the stream lines with the same spacing (n_0) so that a square-meshed grid is formed. These normal lines are called *equipotential lines*. The grid of perfect squares in a reference zone of uniform flow is the basis for the construction of the flow net in a subsequent zone of non-uniform flow.

It may be demonstrated mathematically[1] that whatever the boundary profile the network will always be composed of squares. Furthermore, for any given boundary conditions and initial spacing, only one network solution is possible. However, because the theory assumes an infinite

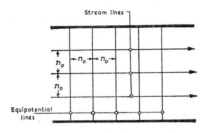

Figure 4.22   Square-meshed flow net for
flow between straight parallel boundaries

number of stream lines and only a finite number can be conveniently reproduced, the actual meshes in the non-uniform zone are only approximately squared. This means, in effect, that the sides may be curved although the intersections must always be at right angles.

If the velocity and pressure heads are known at any given point (not necessarily the same point) in the system, the values elsewhere may be determined from a knowledge of the dimensions of the meshes. This follows, because for the stream lines:

$$v_1 = v_0 \left(\frac{n_0}{n_1}\right) \qquad (4.32)$$

and, utilising the Bernoulli equation, we obtain for the piezometric head:

$$\left(z_1 + \frac{p_1}{w}\right) = \left(z_0 + \frac{p_0}{w}\right) + \frac{v_0^2}{2g}\left[1 - \left(\frac{n_0}{n_1}\right)^2\right] \qquad (4.33)$$

The general mode of procedure may be explained by reference to Fig. 4.23, which shows the flow net for a conduit bend of small radius. In this particular case, abruptness of the deviation precludes the free vortex behaviour from being well developed; nevertheless, the pattern of stream lines in the central portion does exhibit a close resemblance.

[1] The basic equations for two-dimensional flow are

$$\frac{\partial^2 \phi}{\partial x^2} + \frac{\partial^2 \phi}{\partial y^2} = 0 \quad \text{and} \quad \frac{\partial^2 \psi}{\partial x^2} + \frac{\partial^2 \psi}{\partial y^2} = 0$$

where ϕ = potential function, and ψ = stream function. These two equations (Laplace), together with the boundary conditions, define the flow net for a given situation (see Hydrodynamics Literature, p. 75).

Starting well back in the straight portion, which is the reference zone, the meshes are formulated progressively by a graphical process of trial and error. A fair amount of practice, involving much use of the eraser, is required before flow nets can be drawn with facility. In those regions where considerable distortion is evident, a more accurate representation is obtained by the insertion of intermediate meshes. The drawing of diagonals is a useful check, since these should also form a square network. It should be noted that the flow net is dimensionless in the sense that it is independent of the linear scale and of the discharge. In the case of enclosed flow, the latter can be in either direction. Experimental

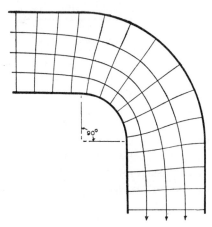

Figure 4.23   Flow net for uni-planar flow at
a bend

techniques (e.g. electrical analogy) and electronic computers are useful aids to analysis.

The construction of the flow net is a convenient means of determining the flow pattern around a body immersed in a fluid of infinite extent. When it is applied to problems which involve a free surface, it is necessary to make use of the Bernoulli equation in order to locate the surface profile. At a free surface the pressure head is zero and thus there is a direct relationship between the elevation and velocity heads at corresponding points. Ex. 4.2 deals with a typical case.

Example 4.2

A sharp-crested rectangular weir is located in a wide channel. The water level in the channel 3 m upstream of the weir is 0·457 m above the crest and the depth at this point is 1·373 m. Draw the flow net for irrotational flow over the weir.

If the discharge is 0·59 cumecs per metre width, determine the total force (per unit width) on the weir plate and the pressure distribution.

The flow net is shown in the inset diagram. Upper and lower surfaces of the overflowing jet are located by trial and error procedure.

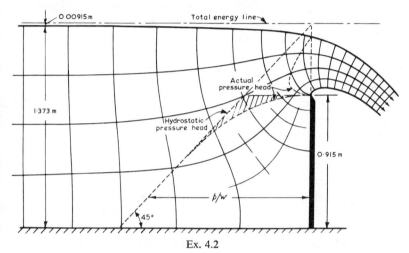

Ex. 4.2

The total pressure force on the weir plate is found to be **8·05** kN as compared with 8·30 kN assuming hydrostatic conditions (or 8·23 kN if a depth of 1·373 m is taken).

4.11 Behaviour of a Real Fluid

4.11.1

There are some important differences in flow behaviour which distinguish a real from an ideal fluid. These differences necessitate certain modifications and limitations being applied to the ideal fluid concept if the latter is to be of value in the solution of practical problems.

4.11.2 *Turbulence Theory*

Turbulent flow is characterised by fluctuating pressures and particle velocities. There is a random formation and decay of multitudes of small eddies throughout the flow stream and the resulting intermixing gives rise to shear stresses. This viscous interaction is of a much larger scale than the molecular interaction associated with laminar flow. In consequence, the energy losses are relatively much greater; also, as Fig. 4.1 illustrates, the velocity distribution is more uniform.

A rigorous theoretical treatment of turbulent flow requires a knowledge of the spatial fluctuations of velocity co-ordinated with respect to time. Mathematicians have approached the problem from the standpoint of statistical mechanics, conceiving of a primary steady velocity on which is superimposed a secondary fluctuation. The scientific importance of the subject and its complexities are such that it is now a recognised field in applied mathematics. In spite of simplifying assumptions, such as isotropic turbulence, workable solutions have so far not been forthcoming and in engineering applications it is necessary to have recourse to semi-empirical methods – semi-empirical because an evaluation is dependent on the experimental determination of certain constants.

Boussinesq proposed (1877) a simple expression for the mean shear stress τ_e arising from the eddying fluctuations in turbulent flow, namely

$$\tau_e = \eta \frac{dv}{dy} \qquad (4.34)$$

This expression is comparable to that for laminar flow [shear stress $= \mu(dv/dy)$], and η, which is known as the *eddy viscosity*, has the same dimensions as μ. Unlike μ, however, η is not a fluid property, but is dependent on the turbulence characteristics.

The total shear stress in turbulent flow is thus

$$\tau = \eta \frac{dv}{dy} + \mu \frac{dv}{dy} \qquad (4.35)$$

In practice the dynamic viscosity is nearly always negligible in comparison with the eddy viscosity so that the last term may be omitted. An exception is of course the layer adjacent to a boundary surface where there can be no transverse fluctuation and the shear is due to molecular viscosity only.

Prandtl (1925) based his theory of turbulence on the principle of momentum exchange. The nature of momentum exchange in turbulent flow has been explained by Bakhmeteff[1] in an analogy whereby two bodies proceed at different speeds along parallel tracks. As they pass each other a stream of bullets is fired from one of the bodies to the other. These bullets are brought to rest by becoming embedded in the second body. Because of the difference in speed between the two bodies there is an exchange of momentum in the forward direction which can be calculated from Newton's second law if the relevant data are available. The

[1] BAKHMETEFF, B. A. (1936) *The Mechanics of Turbulent Flow*. Princeton Univ. Press.

result of the momentum exchange is always to tend to reduce the relative speed.

In order to apply this principle to turbulent flow Prandtl introduced the concept of a *mixing length*. This length represents the average distance travelled by particles before their momentum is suddenly absorbed by their new environment – the greater the mixing length, the higher is the degree of turbulence and the more uniform the velocity distribution. There is an obvious conceptual defect here since particle momentum is gradually, not abruptly, changed throughout its path.

A case of two-dimensional turbulent flow parallel to a straight boundary will now be considered. Fig. 4.24 shows diagrammatically the transfer of a small liquid particle from one layer of flow *aa*, whose temporal

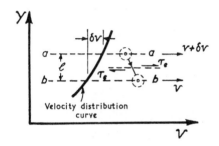

Figure 4.24 Exchange of momentum

mean velocity is $v + \delta v$, to another *bb* whose temporal mean velocity is v. The distance between the two layers is the mixing length l. Both the velocity and the mixing length are influenced by the distance y from the boundary. The differential dv/dy is the velocity gradient and for a small mixing length $\delta v = l(dv/dy)$. Superimposed on the temporal mean velocity are fluctuating velocity components v_x' and v_y' in the axial and transverse directions, respectively.

Now the mass rate of transverse flow per unit area between layers *aa* and *bb* is $\rho v_y'$. The change of mean axial velocity in the mixing length is δv and this is assumed equal to the velocity fluctuation v_x'. Thus in accordance with Newton's second law the shearing stress τ_e per unit area is given by

$$\tau_e = \rho v_y' v_x' \tag{4.36}$$

This is known as the *turbulent shear equation* and was first proposed by Reynolds (1895).

Prandtl made the further assumption that v'_y is of the same order as v'_x. Thus remembering that $\tau \simeq \tau_e$, Eq. (4.36) may be expressed as

$$\tau = \rho l^2 \left(\frac{\mathrm{d}v}{\mathrm{d}y}\right)^2 \tag{4.37}$$

which is the *general equation of turbulence*. Reference to experimental data is necessary in the assessment of l. This aspect is discussed in connection with pipe flow in the next chapter.

Boussinesq's equation is not inconsistent with Eq. (4.37), for it will be observed that $\eta = \rho l^2(\mathrm{d}v/\mathrm{d}y)$ – in other words it is a function of the turbulence. In fact the notion of an eddy viscosity is complementary to the mixing length theory since it constitutes a useful indicator of turbulence intensity.

Turbulence studies are of particular relevance to such problems as heat transfer (e.g. cooling water circulation), interfacial mixing of fluids of different density (e.g. salt and fresh water), sediment suspension, and the dispersion of pollutants. Analysis of these complex phenomena presents a formidable challenge to the research worker.

4.11.3 *Modified Form of Bernoulli Equation*

The constant energy concept is widely employed in practice for deter-mining the variation in pressure and velocity along the path of flow. Some modifications to the basic equation are required in order to make it applicable to the flow of a real fluid. These are necessitated by the rotational nature of the flow and the fact that, in general, it is only the mean velocity (discharge $\div$ area) that is known at any section.

(*a*) *Velocity Head Coefficient.* In the turbulent flow of a real fluid the velocity distribution at any section adjacent to or between solid boundaries is not uniform, and in order to express the velocity head in terms of the mean velocity V it is necessary to introduce a dimensionless coefficient α whereby the velocity head is $\alpha(V^2/2g)$. The value of α is dependent on the velocity distribution and is obtained by dividing the sum of the kinetic energy heads of the elemental stream tubes by the kinetic energy head based on the mean velocity, thus:

$$\alpha = \frac{\int v^2 \, \mathrm{d}Q}{V^2 Q} = \frac{\int v^3 \, \mathrm{d}A}{V^3 A} \tag{4.38}$$

where $\mathrm{d}Q$ and $\mathrm{d}A$ are the discharge and sectional area respectively of the elemental stream tubes. Since the cube of the mean velocity is always less than the weighted mean of the cubes of the filament velocities, the value of α is always greater than unity. The value may be as high as 2·0 (laminar flow in pipes) but for normal turbulent flow in pipes it is usually only slightly in excess of unity, ranging between 1·01 and 1·10. Irregular sections produce higher values. In most flow problems the velocity head is only a small proportion of the total head and the error which results from assuming α to be unity is negligible. However, in those cases where the velocity head is significant some reasonably accurate assessment of α, based either on a velocity traverse or experience, is desirable.

(b) *Energy Head Loss.* There is a loss of energy associated with the flow of a real fluid which it is generally necessary to take into account when equating the total energy head of sections that are situated some little distance apart, or where there is some intermediate discontinuity in flow regime. This loss of head h_f is dependent on a number of complex factors and cannot be determined analytically except for the simplest boundary conditions. The next chapter, which deals with pipe flow, is almost entirely concerned with this important aspect.

(c) *External Energy Conversion.* Any addition or abstraction of energy by an external agency must be accounted for in the relevant total energy equation. Pumps, which convert mechanical energy into hydraulic energy and turbines, which perform the reverse function, are typical examples of this type of agency.

Allowance for these three factors is made in a modified form of the Bernoulli equation as follows:

$$z_1 + \frac{p_1}{w} + \alpha_1 \frac{V_1{}^2}{2g} = z_2 + \frac{p_2}{w} + \alpha_2 \frac{V_2{}^2}{2g} - E + h_f \qquad (4.39)$$

where the suffixes 1 and 2 denote the respective flow sections and E is the energy head that is added. The elevation heads z_1 and z_2 are of course related to a common datum.

In pipe flow, the elevation head and pressure head are generally referred to the pipe centre line. In channel flow, where there is a free surface, it is convenient to refer the elevation head to the bed. The pressure head term may then be replaced by the vertical depth of flow, always provided that the stream lines are reasonably parallel to the bed. Any material departure from this condition is associated with a vertical acceleration head which should be allowed for.

4.11.4 *Cavitation*

An infinite liquid velocity is impossible, since there is an inter-relationship between pressure and velocity, and a liquid will begin to boil (i.e. pockets of vapour form) when the pressure falls to that exerted by the vapour. Actually, well before the vapour pressure is reached, the dissolved gases, which are almost always present, will commence to come out of solution. Since the volume occupied in the gaseous state is greater than the corresponding volume in the liquid state, the pattern of flow suffers some local disturbance and discontinuity.

Vapour bubbles, created in a region of low pressure, are carried downstream with the flow until they reach a region of sufficient pressure for condensation to take place. Their collapse is extremely abrupt and the surrounding liquid rushes in to fill the void with such force that the graphic term 'implosion' is sometimes employed in description. The entire phenomenon is referred to as *cavitation* and is generally to be regarded as of a harmful nature, because it tends to produce:

(a) A change in the flow pattern resulting in a reduction of flow efficiency.
(b) A high-frequency fluctuation in pressure causing instability and consequent noise and vibration.
(c) A violent hammering action on any boundary surface with which the collapsing vapour bubbles come in contact. These very high momentary stresses cause pitting and honeycombing. In the most adverse conditions, structural failure through fatigue may occur very rapidly after the onset of the phenomenon.

The primary cause of cavitation is a low pressure and this may be brought about by either a general lowering in hydrostatic pressure or by a local increase in velocity. In many cases, such as ships' propellers, pump impellers, turbine runners, and some forms of hydraulic structure, the applied pressure cannot conveniently be raised. It is therefore most important to pay very careful attention to the design of flow passages and profiles in order to avoid excessive local velocities or fluctuations in pressure. This subject is discussed more specifically in the chapters on hydraulic structures and turbomachinery (Chs. 9 and 10).

4.11.5 *Boundary Layer*

It is a characteristic of an ideal fluid that it flows past a solid boundary with undiminished velocity, whereas a real fluid, due to the viscous shearing action that is induced, suffers a severe retardation in the vicinity of a boundary and at the boundary itself the velocity is zero. The layer of fluid adjacent to a boundary where the viscous effects are evident is called the *boundary layer* and it is of great significance.

Flow in the boundary layer can be either laminar or turbulent

Fig. 4.25 shows the velocity profile in flow past a thin smooth plate, the dimensions normal to the plate being greatly exaggerated. For a short distance back from the leading edge the flow is laminar with the velocity distribution approximately parabolic and the pressure at any section nearly constant. At a certain point along the plate the laminar flow becomes unstable, some eddying commences, and then after a short transitional zone, turbulence is fully developed. Theoretically, the turbulent boundary layer extends for an infinite distance from the surface and so in accordance with convention the thickness δ is taken as that delineated by $0.99V$, where V is the undisturbed uniform velocity.

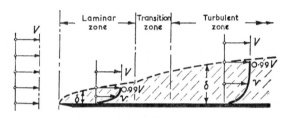

Figure 4.25   Development of the boundary layer on a
thin plate

There is a fairly sharp increase in δ initially, but the rate of increase progressively diminishes with distance along the plate. Due to the turbulent mixing, the velocity distribution is much more uniform than is the case with the laminar boundary layer. Laminar flow nevertheless persists in a very thin film called the *laminar sub-layer* which is in immediate contact with the plate and it is in this sub-layer that the greater part of the velocity change occurs.

The nature of the flow in the turbulent boundary layer is similar to that in a pipe or channel (including a spillway). In these cases the boundary layer occupies the entire sectional area of flow. This is due to the fact that the diameter of the pipe or sectional area of flow is insignificant in comparison with the length; the boundary layer therefore attains its maximum size only a relatively short distance from the inlet (Fig. 4.26). In uniform flow, the shear stresses thereafter remain constant.

The concept of the boundary layer was originally postulated by Prandtl at the beginning of the present century, and it is now the foundation for modern theories of aero- and hydrodynamics. In view of its wide application to many important practical problems in these fields, a very considerable amount of research, both theoretical and experimental, has been carried out with a view to obtaining a greater understanding of its properties and behaviour. In the hydraulic field, the most important practical application of boundary layer theory is to be found in pipe flow and this is discussed at some length in Ch. 5.

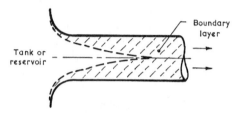

Figure 4.26   Boundary layer development at the entry to a pipeline

4.11.6 *Separation*

We will consider the case (Fig. 4.27) where a convex surface is presented to a flow stream. For irrotational flow (cf. Fig. 4.8) the stream lines would converge towards the section normal to the surface at A, where the velocity attains a maximum, and then diverge again, thus indicating a zone of acceleration and decreasing pressure followed by a zone of

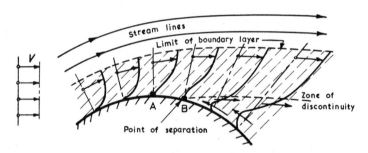

Figure 4.27   Separation at a convex surface

retardation and increasing pressure. With the flow of a real fluid, however, it is necessary to take into account the boundary layer effects. Upstream of A, the accelerating flow tends to offset the retarding influence of viscous stress. Consequently, the boundary layer remains relatively thin instead of markedly expanding in the direction of flow, as in the case of a flat plate. The flow condition is quite stable and the pattern of the stream lines is virtually the same as that for an ideal fluid.

Downstream of A there is an adverse pressure gradient which operates with the viscous stress to reduce the energy and forward momentum of the fluid particles in proximity to the surface, causing the thickness of the boundary layer to increase sharply. Finally, a position is reached where no further retardation can take place without a reversal of flow, and the main flow stream then parts company with the surface. This is the point of *separation* or *break-away*, B, and it occurs where the tangent to the velocity profile is normal to the surface. Pressures tend to be lowest in this region so that separation and cavitation are often associated.

For continuity, the reverse flow downstream of B must be balanced by a forward flow adjacent to the main flow stream. The region of turbulent eddying so formed is called the *wake*, and the approximate dividing line between the wake and the main flow stream is known as the *zone of discontinuity*. There is an appreciable difference in the velocities on either side, although the pressures may be approximately the same. Moreover, the pressure exerted by the wake on the downstream boundary surface is significantly less than where the main flow stream is in contact. As the eddies in the wake move downstream their intensity is damped by viscous resistance, energy being finally dissipated in the form of heat.

For flow over a curved surface, the point of separation and the zone of discontinuity are not readily determined. When the boundary layer is laminar throughout, separation tends to occur further upstream than when, with increasing velocity, turbulence develops. This is explained by the fact that adjacent to the surface the velocity gradient in laminar flow is much less than in turbulent flow, so that the forces of retardation needed to produce a reversal of flow are much less. Also, the transverse mixing of the fluid particles in turbulent flow means that there is a greater relative forward momentum to be overcome. Laminar separation may occur upstream of the point of maximum velocity in irrotational flow, but this is not illogical since the general pattern of stream lines undergoes modification.

From the above remarks it follows that the point of separation is moved further downstream by increasing the general level of turbulence, such as for instance by roughening the surface or by interposing obstacles (wires, rods, etc.). But, of course, the primary factor in determining the location of discontinuity, or indeed whether it will occur at all, is the curvature of the boundary. For straight boundaries it may be said, very generally, that the limiting angle of divergence from the direction of the flow stream is about 4 degrees.

The character of the wake is of some significance. In the case of a long cylinder[1] transversely aligned to the flow stream (e.g. bridge pier deeply immersed), a velocity increasing from zero will first result in the appearance of a pair of large weak vortices symmetrically situated at the rear (Fig. 4.28(a)). At a later stage, the vortices will detach themselves alternately on either side, moving downstream in staggered fashion to form what is known as a *vortex street* (Fig. 4.28(b)). This alternate shedding of the vortices causes the points of separation to fluctuate in position, correspondingly modifying the pressure distribution and thus inducing a variable lateral thrust. If it so happens that the natural frequency of oscillation of the body coincides with that of

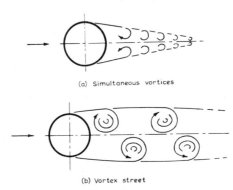

(a) Simultaneous vortices

(b) Vortex street

Figure 4.28   Pattern of the wake behind a circular cylinder

the vortex wake, vibration will be produced. The phenomenon is evidenced by the singing of telephone wires and the aerodynamic instability of structures such as suspension bridges.

As the velocity increases further, the individual vortices merge with the general turbulence. Finally, the stage is reached when the boundary layer becomes turbulent, whereupon the points of separation move downstream to the rear portion of the cylinder, thus narrowing the wake.

Summarising then, we may conclude that the presence of an adverse pressure gradient is conducive to separation and that in this region the pattern of stream lines for rotational and irrotational flow differs

---

[1] The effect of a body interposed in a flow stream may be analysed in terms of an appropriate Reynolds number, which is the parameter of viscosity (see Sect. 4.12 (c), p. 72).

appreciably. In spite of serious limitations, the flow net for an ideal fluid does give a close representation of converging flow and is useful in drawing attention to those localities where discontinuity is likely to occur.

4.11.7 Drag

A viscous fluid in flowing past a partly, or wholly, immersed body exerts a force on the body, the component of which in the direction of the flow is known as the *drag*. An obvious witness to this is the energy that is required to impel a ship or aircraft through a stationary fluid. Similarly, bodies would not remain stationary in a moving stream without the aid of anchors or foundations to resist the drag, emphasising the point that it is the relative velocity which is important. Old London Bridge may be cited as a good example of drag resistance. When this structure, with its numerous piers, was replaced by the wide spans of the present bridge, it was found that the tidal range at and above London was increased by as much as 25 per cent.

We will consider the drag on the bridge pier, with bull-nosed ends, shown in Fig. 4.29. For an ideal fluid, the pattern of stream lines is indicated in diagram (a). There is no viscous shear, and, because of the symmetrical disposition of the stream lines, the net pressure force is zero. As these are the only possible forces acting, the drag must be zero, and in fact this is an erroneous conclusion which always results when the resistance of immersed bodies in an ideal fluid is considered. In the case of a real fluid, we find that the pattern of stream lines will still be symmetrical provided that the flow is everywhere laminar. In this case, however, the flow pattern is disturbed for a greater distance from the pier and energy is absorbed in overcoming shear. The total force derived from this shearing stress makes up the entire drag resistance, since the net pressure force is again zero. This type of wholly viscous resistance is known as *deformation drag*. A practical example is the motion of silt particles in estuaries and rivers.

With the normal case of fully turbulent flow, the stream lines break away from the surface at the rear in the manner indicated in diagram (b). There is a reduced pressure force on the rear of the pier, and the total pressures at the upstream and downstream ends are now out of balance, and, as will be seen from what follows, this is a major contribution to the drag.

The forces acting on an elemental area dA_s of the surface of the pier comprise a pressure force $p \, dA_s$ normal to the surface and a shear force

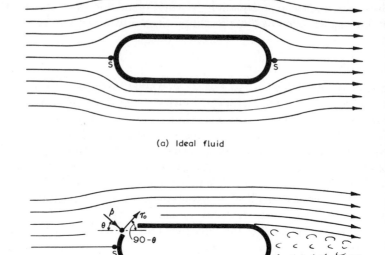

(a) Ideal fluid

(b) Real fluid

Figure 4.29   Flow pattern around a bridge pier

$\tau_0 \, dA_s$ tangential to it. Resolving these forces in a direction parallel to the flow, the resultant force dF is given by

$$dF = p \cos \theta \, dA_s + \tau_0 \sin \theta \, dA_s$$

Integrating around the pier we have

$$F = \int p \cos \theta \, dA_s + \int \tau_0 \sin \theta \, dA_s \qquad (4.40)$$

or

$$F = F_P + F_S \qquad (4.41)$$

The force F is the total drag and is seen to be the sum of a force F_P, known as the *pressure* or *form drag*, and a force F_S, known as the *surface* or *skin friction drag*. The relative magnitude of the two components is largely dependent on the shape and size of the immersed body. Clearly, a short bluff-shaped body will create a considerable wake and will there-

64

fore have a high form drag and a relatively low skin friction drag. As form drag is usually the most important factor, the effect of streamlining is almost always to reduce the total drag. The streamlined aerofoil

Figure 4.30   Aerofoil section

(Fig. 4.30) is found to be the most efficient body shape. Although it might be desirable on hydraulic grounds to design bridge piers in this form, there are, of course, both aesthetic and structural considerations, which would in most situations permit of only a limited degree of symmetrical streamlining.

The *coefficient of drag*, C_D, is given by the ratio of the actual drag force to the dynamic force or

$$C_D = \frac{F}{\frac{1}{2}\rho V^2 A} \qquad (4.42)$$

where V is the relative velocity and A is the projected area of the body in a plane normal to the direction of flow or propulsion. Whilst the value of C_D is dependent primarily on the boundary shape, there are additional influencing factors which include the proximity of other bodies and the creation of waves at a free surface. This means that in the case of partially immersed bodies, such as bridge piers, no definite relationships between drag coefficient and body shape can be established.

A laboratory flume, towing tank, or wind tunnel, facilitates a direct measurement of the drag resistance of small bodies. Wires or threads connected to spring balances or weighing devices are used for this purpose. Large bodies may be dealt with in a similar manner by reduction to convenient model size.

The form drag may be obtained by measuring the pressure variation around the surface of a body immersed in a uniform stream of air or water of known velocity. Integration over the entire surface yields the form drag resistance; Ex. 4.3 illustrates the method of evaluation for a simple case. With bluff-shaped bodies and moderate to high velocities, the skin friction drag is exceedingly small in comparison with the form drag; the latter is then taken as being equal to the total drag and the drag coefficient follows directly. When the pressure distribution is two-dimensional, as is the case with a long cylinder, internal piezometer tappings provide a convenient means of measuring pressure intensities.

Example 4.3

A large pipeline of cylindrical shape is to be laid across the mouth of a narrow tidal inlet and is to be suspended above the bed. As an aid to design, tests are conducted on a 152 mm (6 in.) diameter model in a wind tunnel. When the air velocity is 15·25 m/s the pressures recorded at piezometer tappings around the sectional periphery are as follows:

Angle (degrees clockwise)*	0	30	60	90	120	150	180	210	240	270	300	330
Pressure (N/m²)	441	365	331	303	138	131	124	110	90	69	69	276

* Measured from upstream horizontal centre line.

Neglecting skin friction, determine the coefficient of drag. It may be assumed that the density of air in the wind tunnel is 1·23 kg/m³.

The radial pressures are set off to scale around the circumference and the enveloping line sketched in. Since $P = \int p \cos \theta \, dA_s$, the form drag may be

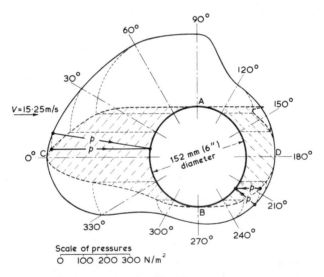

Ex. 4.3

obtained by graphical means as indicated. The hatched area ACB minus the area ADB represents to scale the form drag, and is found to be $F = 28·7$ N per metre run. Neglecting skin friction, the coefficient of drag is given by:

$$C_{\mathrm{D}} = \frac{F}{\frac{1}{2}\rho V^2 A} = \frac{28\cdot7}{0\cdot5 \times 1\cdot23 \times 233 \times 0\cdot152} = \mathbf{1\cdot32}$$

Incidentally, the resultant of the vertical components of pressure represents the lift on the body – in this case it is $-21\cdot9$ N per metre run.

4.11.8 *Flow Behaviour at Bends*

The most abrupt type of bend in a pipeline or channel is the mitred intersection (Fig. 4.9). Separation occurs at the inner upstream boundary and there is a relatively large loss of energy caused by eddying turbulence.

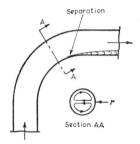

Figure 4.31   Flow around a
pipe bend

The behaviour of an ideal fluid at a conduit bend has been discussed earlier (Sects. 4.9.3 and 4.10), it being found that conditions resemble a free vortex with the velocity inversely proportional to the radius and a complete balance of the centrifugal and centripetal forces. Now, with a real fluid, owing to boundary layer effects, the flow velocity undergoes a marked retardation adjacent to a solid boundary. This means that the transverse equilibrium is upset and a cross-current induced, which when superimposed on the longitudinal velocity, results in a double spiral type of fluid motion (Fig. 4.31), the condition persisting downstream of the bend and being only gradually dissipated by viscous damping. If the bend is sharply radiused, separation occurs at the inner boundary, where the stream lines diverge, resulting in the shedding of eddies and a further departure from the free vortex behaviour. In the case of very large pipe bends, such as the draft tube elbows of sizeable hydraulic turbines, the insertion of fins or guide vanes serves to restrict the transverse currents and thus reduces the energy losses. Advantage is sometimes taken of the pressure head difference which exists between the inner and outer edge of bends to measure the discharge; pressure tappings connected to a manometer are required and careful calibration is essential. Eq. (4.24) or (4.25) may be utilised, but an empirical discharge coefficient must be introduced.

The flow behaviour around channel bends is very similar, but owing to the presence of a free surface, the increase in pressure with radius produces a greater depth of water at the outer boundary. In the case of an ideal fluid the transverse surface profile is hyperbolic (Eq. (4.23)). With a real fluid there is a considerable velocity gradient adjacent to the bed and as a result the hydrostatic force in this region is greater than the centrifugal force. A secondary transverse current is thus induced which is directed inwards at the bed (Fig. 4.32). For continuity, there must exist an outward transverse current near the surface so that the resultant motion is of a spiral or helicoidal nature. This behaviour of the current in an unlined channel or river bend is conducive to the scouring

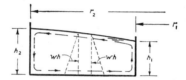

Figure 4.32   Flow around a channel bend

and transportation of material from the bank and bed on the concave side and its deposition in the less disturbed regions both at and downstream of the bend. The transverse current at the surface is weaker and is not capable of returning the scoured material. Nevertheless it does constitute a navigational hazard since vessels tend to be carried towards the concave bank. The whole phenomenon is favourable to the accentuation of a bend and the development of a meander.

When assessing the difference in surface level ($h_2 - h_1$) of the outer and inner radii (r_2 and r_1), it is of sufficient accuracy for practical purposes to assume an ideal fluid and utilise Eq. (4.26).

4.12 Dimensional Considerations

Before closing the present chapter some discussion of the concept called dimensional analysis is appropriate. Briefly, this is a mathematical technique which recognises the dimensional homogeneity of all terms in a physical equation. It constitutes a powerful analytical tool in the study of problems concerning physical phenomena and is particularly relevant to fluid mechanics. Its general utility will be manifested on a number of occasions in subsequent chapters.

Newton's basic laws are universally applicable and they may be expressed dimensionally in terms of force (F), mass (M), length (L), and time (T). In fluid mechanics these dimensions, either singly or suitably combined, are sufficient to define the boundary geometry, flow characteristics, and the influencing physical properties.[1] Moreover, from the second law (force = mass × acceleration), which may be expressed dimensionally as $F = MLT^{-2}$, it follows that three fundamental dimensions are in fact adequate. Normally, it is the practice in dimensional analysis to adopt the M-L-T system, although for most problems in engineering some simplification results from an adoption of the F-L-T system. On page xii the derived dimensions are quoted for the various quantities most commonly encountered, both systems being listed.

A logical application of the principle of dimensional homogeneity is the checking of the qualitative validity of equations. By this means computation errors arising from dimensional inaccuracy may be readily identified. Thus in the expression $\tau = \mu(dv/dy)$ for laminar shearing stress, the insertion of fundamental dimensions leads to $ML^{-1}T^{-2} = [ML^{-1}T^{-1}][LT^{-1}][L^{-1}]$, confirming that the equation is dimensionally correct.

The same procedure is also useful in determining the dimensions of empirical coefficients. An example would be the dimensional evaluation of C in the Chézy formula for channel flow, $V = C\sqrt{RS}$ (Eq. (7.15), p. 152). As R is a characteristic length and S is a slope (i.e. ratio of lengths), the formula is expressed dimensionally as $LT^{-1} = [C]L^{1/2}$ so that the dimensions of C are $L^{1/2}T^{-1}$. A further useful application of the dimensional concept immediately suggests itself – that of the conversion from one system of units to another, say from the English system to the metric system. In this respect, particular care is needed when dealing with dimensional coefficients.

But the most valuable role of dimensional analysis is the guidance which it affords to experimentation. Fluid mechanics is essentially an experimental science, since the variables influencing any particular phenomenon are normally so many or of such a nature that theoretical analysis alone is incapable of providing a thorough evaluation. Dimensional analysis enables a significant relationship to be established between the dependent and independent variables by grouping them into a definite number of independent dimensionless products.

---

[1] In problems concerning heat transfer, an additional dimension – usually the temperature (Θ) – must be introduced.

Very often this is the only mathematical step that can be taken at the beginning of an investigation. The subsequent experimental procedure, instead of involving an exhaustive study of the effects of each independent variable acting singly, may be conducted in a systematic and expeditious fashion in the light of the dimensionless parameters so deduced. An important advantage of the conversion to dimensionless parameters is that the results may be presented in a generalised form. Not only does this serve to widen the scope of experimental procedure – for instance air may be used as a fluid instead of water – but also it is the whole basis of the similarity concept, whereby tests on a model provide sufficient information to enable accurate predictions to be made concerning prototype performance, even though the scalar ratio may be quite considerable. This important subject is discussed in Ch. 11.

The procedure in dimensional analysis is best described by way of example, but prior to this it is desirable to explain the nature of the more significant dimensionless parameters that are commonly formulated. These parameters are derived from a consideration of the forces that may influence a particular phenomenon, such as pressure, gravity, viscosity, surface tension, and elasticity. Since, in accordance with Newton's laws, an accelerative force gives rise to an inertial reaction, a dimensionless term is in each case obtained which is in effect the ratio of a component force to an overall inertial reaction. It is only when the velocity is exceedingly small (i.e. laminar flow) that the latter is relatively insignificant.

Considering each type of force in turn:

(a) *Pressure.* Let us suppose that a pressure difference is the only operative force influencing fluid motion. A good example is steady irrotational flow between solid boundaries, such as through the converging piece depicted in Fig. 4.4. We have seen earlier in this chapter that under these conditions the pattern of stream lines may be ascertained by means of a graphical construction called the flow net; moreover, that the solution is unique in the sense that it is dependent only on the boundary geometry and not on the absolute values of the pressure or discharge, or even the direction of flow. According to Bernoulli's equation, for any two points 1 and 2 in the system, the pressure is related to the velocity by the expression $(p_1 - p_2)/w = (v_2{}^2 - v_1{}^2)/2g$, or in the dimensionless form:

$$\frac{\Delta p}{\rho v_2{}^2/2} = 1 - \left(\frac{v_1}{v_2}\right)^2 \tag{4.43}$$

The term on the left is sometimes called the *pressure coefficient*, C_P, and it is comparable to the coefficient of drag (Eq. (4.42)). It does in fact represent the dimensionless ratio pressure force: inertial reaction, since if we consider the equilibrium of an elemental cube, defined by linear dimension L, the pressure force is $\Delta p\, L^2$ and the inertial reaction is $\rho L^3(L/T^2)$ or $\rho L^2 V^2$, thus giving the ratio $\Delta p/\rho V^2$. By inversion and slight modification we formulate the conventional dimensionless parameter known as the *Euler number*[1], denoted by

$$E = \frac{V}{\sqrt{2\Delta p/\rho}} \qquad (4.44)$$

The Euler number is a constant for confined flow of the simple nature described. Furthermore in the case of flow through a submerged orifice or pipeline constriction it represents a coefficient of discharge. As pressure is generally taken as the dependent variable, the Euler number is implicit but not always readily recognised in the dimensionless groupings pertaining to more intricate phenomena.

(*b*) *Gravity.* In any phenomenon involving the interaction of two fluids of different density (usually air and water) the effect of gravity is nearly always important. Typical instances are channel flow and the flow over weirs and spillways; surface waves (other than capillary waves) are also in this category. Proceeding as before, the gravitational force is represented by wL^3 so that the ratio inertial force : gravitational force is $\rho L^2 V^2/wL^3$, which in the case of an air-water interface ($w = \rho g$) reduces to V^2/gL. In a slightly modified form this dimensionless parameter is known as the *Froude number*[2], denoted by

$$F = \frac{V}{\sqrt{gL}} \qquad (4.45)$$

The length L is taken as a characteristic linear dimension such as the depth of flow in a channel or the head on a weir. The value of F gives a qualitative indication of the gravitational influence and as the numerator is a velocity it might be inferred that high velocities are associated with powerful gravitational forces. This is generally but not always true as a comparison between the flow over a dam spillway and the jet issuing from a nozzle at the end of a pumping main will reveal.

Gravitational forces are compatible with irrotational flow (cf. the Bernoulli equation) and in these circumstances we know that for a given or calculated free surface profile the pattern of hypothetical stream

[1, 2] See footnote on page 73.

71

lines is unique. But of course the surface profile is influenced by gravity, so assuming that no other forces besides pressure and gravity are operative, then there must be a direct relationship between the Euler and Froude numbers.

(c) *Viscosity.* As we have seen earlier, viscous forces are associated with the flow of a real fluid – the velocity gradient adjacent to a solid boundary is evidence of their effect. Viscous force may be represented by τL^2 or $\mu(V/L)L^2$ so that the ratio inertial force : viscous force is given by $\rho L^2 V^2 / \mu V L = \rho L V / \mu$. This dimensionless parameter is known as the *Reynolds number*[1], denoted by

$$R = \frac{\rho L V}{\mu} = \frac{L V}{\nu} \tag{4.46}$$

where ν is the kinematic viscosity. The characteristic length L is often taken as a diameter (pipe flow) or hydraulic mean depth (open channel).

From the presence of L and V in the numerator it may be inferred in the case of water that the viscous force is relatively insignificant in comparison with other forces (i.e. R is large) when the flow section is sizeable and/or the velocity high. In contrast, the viscous forces are prominent (i.e. R is small) when the flow section is restricted and/or the velocity low. This important topic is discussed in detail in relation to pipe flow in the next chapter. Suffice it to say at this stage that with circular pipes and $R < 2000$ the flow is laminar (highly viscous) in character.

(d) *Surface Tension.* By virtue of its nature, surface tension is only operative at a fluid interface. Normally, the forces involved are too weak to exert any influence on fluid behaviour. However, surface tension may be significant where linear dimensions are small, as for instance in flow over a sharp-crested weir under a very low head when it tends to cause the overflowing sheet to cling to the downstream face. Surface tension effects are also evident in the ripples (capillary waves) that are created on an open expanse of water by a faint breeze.

A surface tension force may be represented by σL where σ is the tension force per unit length. The ratio inertial force : surface tension force thus becomes $\rho L^2 V^2 / \sigma L$ or $\rho L V^2 / \sigma$. Introducing conventional modifications we arrive at the *Weber number*,[1] namely

$$W = \frac{V}{\sqrt{\sigma/\rho L}} \tag{4.47}$$

[1] See footnote on page 73.

The constituent arrangement of this dimensionless parameter indicates qualitatively that surface tension effects may be expected to be significant when V and L are small.

(e) *Elasticity*. The compressibility of water is not really a factor meriting consideration in experimental studies. It is only of significance in problems concerning unsteady flow in pipes, and, for these, satisfactory analytical methods of solution are normally available. Of course, compressibility is an important factor in the case of air flow, particularly with regard to the high speed travel of aircraft and rockets. The relevant dimensionless parameter is known as the *Mach number*[1] and is denoted by

$$M = \frac{V}{\sqrt{K/\rho}} \qquad (4.48)$$

where K is the bulk modulus of the fluid. Of some interest is the fact that this parameter also represents the velocity of a fluid (or object immersed in a fluid) to that of the speed of travel of a sound wave in the same medium.

The first step in dimensional analysis is to correctly identify the independent variables. Normally, these comprise the controlling geometry, flow characteristics, and fluid properties. Thus the free discharge from a sluice might be expressed as

$$Q = \phi(a, b, c, V, g, \rho, \mu, \sigma) \qquad (4.49)$$

where a, b, and c are linear dimensions, one of which would be a head. The dimensionless groupings that are formulated usually constitute an expression such as

$$E = \psi\left(\frac{a}{c}, \frac{b}{c}, F, R, W\right) \qquad (4.50)$$

[1] The dimensionless parameters are named after scientists who were noteworthy in the relevant field. Thus Leonhard Euler (1707–1783) deduced relationships concerning pressure and velocity. In this case the appellation is more recent and is due to Hunter Rouse. William Froude (1810–1879) was the pioneer of ship testing by means of models in towing tanks; in the conduct of his experiments he recognised the importance of gravitational forces (wave resistance). Osborne Reynolds (1842–1912), an outstanding scientist and mathematician, investigated the effect of viscosity in pipe flow (see p. 79). Moritz Weber (1871–1951), Professor of Naval Mechanics at Berlin, specifically named the Froude and Reynolds numbers and himself formulated the surface tension number. Ernst Mach (1838–1916) was Professor of Physics at Prague. His valuable experiments on projectiles and shock waves have brought him recognition as a pioneer of supersonics.

although the numbers would not necessarily appear in the conventional form.

The successful employment of dimensional technique requires both skill and experience. For simple problems, involving only three or four variables, the Rayleigh method[1] is quite satisfactory. This method is explained in the example that follows. When dealing with more intricate phenomena a special procedure known as the Buckingham Pi theorem[2] is necessitated.

Summarising, we have found that a consideration of dimensions is helpful to:

(a) The qualitative checking of equations.
(b) A determination of the dimensions of empirical coefficients.
(c) The conversion from one system of units to another.
(d) The setting up and conduct of experiments, and the generalising of results.
(e) The formulation of similarity laws. These are of considerable relevance to model studies.

Example 4.4

Obtain an expression for the resistance F to the steady rate of fall of a small sphere in a viscous fluid.

This phenomenon could be representative of the slow settling of silt particles in quiescent waters. Gravity is not a parameter, since the rate of fall is steady, with the implication that inertial forces are negligible in comparison with viscous shear. The motion may then be reduced to a static equilibrium of forces, the resistance or deformation drag F (equal to the immersed weight of the body) being dependent upon the sphere diameter D_s, the velocity V, and the viscosity μ. Hence we can write

$$F = \phi(D_s, V, \mu)$$

signifying a series where

$$F = \kappa_1 D_s{}^{a_1} V^{b_1} \mu^{c_1} + \kappa_2 D_s{}^{a_2} V^{b_2} \mu^{c_2} + \ldots$$

in which κ_1, κ_2, etc., are unknown dimensionless coefficients and the values of the indices are to be determined. From the principle of dimensional homogeneity it follows that

$$a_1 = a_2 = \ldots = a; \qquad b_1 = b_2 = \ldots = b; \qquad c_1 = c_2 = \ldots = c$$

[1] RAYLEIGH, LORD (1892) 'On the Question of the Stability of the Flow of Fluids', *Phil. Mag.*, **34**, 59.
[2] BUCKINGHAM, E. (1915) 'Model Experiments and the Forms of Empirical Equations', *Trans. Am. Soc. Mech. E.*, **37**, 263.

It is therefore only necessary to select one representative term for dimensional analysis. Inserting the fundamental dimensions,

$$MLT^{-2} = [L]^a[LT^{-1}]^b[ML^{-1}T^{-1}]^c$$

Hence, for M: $1 = c$
      for L: $1 = a + b - c$
      for T: $-2 = -b - c$

from which $a = 1$, $b = 1$, and $c = 1$. Thus

$$F = k\,D_s V \mu$$

It is interesting to note that by a complex mathematical analysis Stokes[1] found that $k = 3\pi$. Also that in accordance with this relationship the coefficient of drag (Eq. (4.42)) is

$$C_D = 24\left(\frac{\mu}{\rho D_s V}\right) = \frac{24}{R}$$

[1] STOKES, G. G. (1856) 'On the Effect of the Internal Friction of Fluids on the Motion of Pendulums', *Trans. Camb. Phil. Soc.*, **9** (Pt. 2), 8.

Further Reading

DUNCAN, W. J. (1953) *Physical Similarity and Dimensional Analysis*. Arnold.

FOCKEN, C. M. (1953) *Dimensional Methods and their Applications*. Arnold.

GOLDSTEIN, S. (1938) *Modern Developments in Fluid Dynamics*, 2 vols. Oxford Univ. Press. (1965) (Dover reprint).

HINZE, J. O. (1975) *Turbulence*. McGraw-Hill (2nd Edition).

IPSEN, D. C. (1960) *Units, Dimensions and Dimensionless Numbers*. McGraw-Hill.

LAMB, SIR HORACE (1932) *Hydrodynamics*. Cambridge Univ. Press (6th Edition).

LANDAU, L. D. and LIFSHITS, E. M. (1959) *Fluid Mechanics*. Pergamon Press.

LANGHAAR, H. L. (1951) *Dimensional Analysis and Theory of Models*. Wiley.

PRANDTL, L. (1952) *Essentials of Fluid Dynamics*. Blackie.

ROBERTSON, J. M. (1966) *Hydrodynamics: in Theory and Application*. Prentice-Hall.

ROUSE, H. (Ed.) (1959) *Advanced Mechanics of Fluids*. Wiley.

SCHLICHTING, H. (1968) *Boundary Layer Theory*, McGraw-Hill (6th Edition).

SEDOV, L. I. (1959) *Similarity and Dimensional Methods in Mechanics*. Academic Press.

SHAMES, I. H. (1962) *Mechanics of Fluids*. McGraw-Hill.

VALLENTINE, H. R. (1967) *Applied Hydrodynamics*. Butterworths (2nd Edition).

WILSON, D. H. (1959) *Hydrodynamics*. Arnold.

CHAPTER FIVE

Analysis of Pipe Flow

5.1 Introduction

Pipes were introduced in the very earliest days of the practice of hydraulics. Their commonplace use today makes it of great importance that the laws governing the flow in them should be fully understood.

As has been demonstrated in the previous chapter, some loss of energy is inevitable in the flow of any real fluid. In the case of flow in a horizontal uniform pipeline this is evidenced by the fall of pressure in the direction of flow. Clearly, an ability to predict the energy loss per unit length is essential to efficient pipeline design.

The present chapter is concerned with an analysis of the mechanics of flow, leading directly to an assessment of losses. Many years of painstaking effort by research workers of diverse disciplines have contributed to our present knowledge. In explaining the fundamental approach a historical sequence will be followed since this appears to afford the most logical and effective means of developing the analysis.

Flow is in every case deemed to be steady – that is, the discharge is constant with respect to time.

5.2 Laminar Flow – Poiseuille Equation

Hagen and Poiseuille, working independently, carried out experiments on the behaviour of fluids in small bore pipes. Empirical relationships based on these experiments were presented in 1839 and 1841 respectively. It was some twenty years later, however, before Hagenbach and Neumann, again independent investigators, were able to present the first theoretical analysis of laminar flow.

The diagram (Fig. 5.1) shows a horizontal pipe, diameter D, in which the flow is laminar. The mass flow may be visualised as being built up of a number of concentric cylinders sliding one upon the other. We will consider the forces acting on any one imaginary cylinder radius r and length ΔL. Due to viscous resistance the pressure (uniformly distributed) falls by Δp in this distance. At the circumference the stress is τ and the velocity is v.

76

The shear force is $\tau \times 2\pi r \, \Delta L$ which is sustained by a pressure force $\Delta p \times \pi r^2$. For dynamic equilibrium these are equal so that

$$\tau = \frac{\Delta p}{\Delta L} \frac{r}{2} \tag{5.1}$$

Now, from Newton's equation of viscosity, the shear stress is related to the velocity by the equation $\tau = -\mu(dv/dr)$, where μ is the dynamic viscosity, the negative sign being adopted because r is measured outwards. Substituting in Eq. (5.1) we obtain

$$dv = -\frac{\Delta p}{\Delta L} \frac{r}{2\mu} \, dr$$

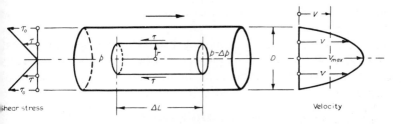

shear stress   Velocity

Figure 5.1   Laminar flow in a pipe

Integrating,

$$v = -\frac{\Delta p}{\Delta L} \frac{r^2}{4\mu} + \text{constant}$$

With viscous flow there can be no slip at the boundary so that $v = 0$ when $r = D/2$. Using this relationship to evaluate the constant in the above equation we obtain

$$v = \frac{\Delta p}{\Delta L} \frac{1}{4\mu} \left(\frac{D^2}{4} - r^2 \right) \tag{5.2}$$

which is the equation of a parabola and the velocity distribution in the pipe is therefore in the form of a paraboloid. The maximum velocity occurs at the centre line where $r = 0$, or

$$v_{\max} = \frac{\Delta p}{\Delta L} \frac{D^2}{16\mu} \tag{5.3}$$

By integration of the individual discharge filaments, or from a knowledge of the geometry of the paraboloid, it is evident that the mean velocity is

$$V = \frac{v_{\max}}{2} = \frac{\Delta p}{\Delta L} \frac{D^2}{32\mu} \tag{5.4}$$

and substituting for $\Delta p/\Delta L$ in Eq. (5.1),

$$\tau = \frac{16\mu Vr}{D^2} \qquad (5.5)$$

which is the equation of a straight line, with $\tau = 0$ at the pipe centre.

From Eq. (5.4) the pressure head loss $\Delta p/w$ is equal to $32\mu V\Delta L/\rho g D^2$, or since the head loss h_{f} is proportional to the length of pipe L, we can write

$$h_{\mathrm{f}} = \frac{32\mu VL}{\rho g D^2} = \frac{32\nu VL}{gD^2} \qquad (5.6)$$

where ν is the kinematic viscosity. This is known as *Poiseuille's equation* and the results of experiments fully confirm its validity. It will be observed that the friction gradient (h_{f}/L) is directly proportional to the mean pipe velocity.

One important point which it is necessary to emphasise is that laminar flow is unaffected by the nature of the boundary surface. Thus we do not find in the analysis any term giving indication of the wall roughness.

5.3 Turbulent Flow – Darcy-Weisbach Formula

In the period around 1850, Darcy, Weisbach, and others, as a result of pipe experiments, deduced a formula for pipe friction loss which may be expressed in the form

$$h_{\mathrm{f}} = \frac{\lambda LV^2}{2gD} \qquad (5.7)$$

where λ is a dimensionless coefficient called the *friction factor*.[1] The expression is known as the *Darcy-Weisbach formula*.

It may be derived by dimensional analysis (see Ch. 4, Sect. 4.12) or from a consideration of the shear stress at the pipe boundary. The procedure in the latter case is as follows:

Frictional resistance = Reduction in pressure force

or

$$\tau_0 \times \pi D \Delta L = \Delta p \times \frac{\pi D^2}{4}$$

where τ_0 = shear stress at the pipe wall. Thus

$$\frac{\Delta p}{w} = \frac{4\tau_0 \Delta L}{\rho g D} \qquad (5.8)$$

---

[1] B.S. 1991: Part 3: 1961 admits of a friction factor symbol f as an alternative to λ, with $4f = \lambda$. But in American terminology f replaces λ; thus to avoid confusion f does not appear in the present text.

The experiments, which were necessarily of somewhat limited scope and accuracy, tended to indicate that the head loss was proportional to V raised to a power which was near to 2. On the assumption then that $\tau_0 = cV^2$, Eq. (5.8) becomes $\Delta p/w = 4cV^2 \Delta L/\rho g D$. Putting $\lambda = 8c/\rho$ and integrating along the pipe we obtain $h_f = \lambda L V^2/2gD$, which is in the desired form.

It also follows from Eqs. (5.7) and (5.8) that

$$\tau_0 = \frac{\lambda \rho V^2}{8} \tag{5.9}$$

or

$$V = \left(\frac{\tau_0}{\rho}\right)^{1/2} \left(\frac{8}{\lambda}\right)^{1/2} \tag{5.10}$$

It was realised at an early stage that the friction factor varied not only with the roughness of the pipe walls but also with the diameter and velocity. This indicated that λ was not a simple coefficient, as once supposed, but an overall coefficient, which had to represent the combined effect of several variables. Furthermore h_f was not dependent on the square of V but on V raised to some slightly lesser power.

5.4 The Contribution of Osborne Reynolds

5.4.1 Dye Experiments

Osborne Reynolds[1] in his classic experiments at Manchester University demonstrated most effectively the characteristics of laminar and turbulent flow. He showed that, under suitable conditions, the two types of flow could be made to occur in the one pipe. His apparatus was extremely simple and consisted essentially of a glass tube through which water could be passed at varying velocities. Provision was made for the insertion of a thin jet of aniline dye into the stream of water at the upstream end.

Commencing with a very low water velocity, it was found that the dye remained intact in the form of a thin slender thread extending the whole length of the tube as in Fig. 5.2(a). This indicated that the particles of liquid were moving in straight parallel paths and that the flow was therefore laminar.

The velocity of the water was then gradually increased and at a certain point the thread broke up, as in Fig. 5.2(b), and the diffused dye

[1] REYNOLDS, O. (1884) 'An Experimental Investigation of the Circumstances which determine whether the Motion of Water shall be Direct or Sinuous, and of the Law of Resistance in Parallel Channels', *Phil. Trans. Roy. Soc.*, **174**, 935.

intermingled with the water in the tube. The flow had evidently passed into the turbulent state and was found to remain in this condition for all higher velocities. By using varying sizes of tube and differing temperatures of water Reynolds showed that the critical velocity at which the change took place depended on both these factors.

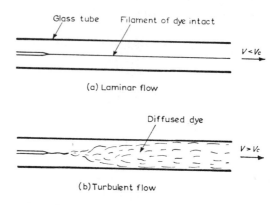

Figure 5.2   Behaviour of dye in a pipe

Both Reynolds and later investigators, who studied the transition region in greater detail, found that even with the same size of pipes and temperature of water, the critical velocity was not rigidly fixed, since with some initial agitation of the water, turbulent flow developed at a lower velocity. Likewise, if care was taken to ensure quiescence of the water before it entered the pipe it was possible to maintain laminar flow at higher velocities. It could thus be concluded that there was an upper and a lower critical velocity – the lower one, which was the more stable, being the true critical velocity.

5.4.2 *Laws of Resistance*

The simple aniline dye apparatus was not suitable for determining the laws of resistance. Accordingly, Reynolds measured the head loss in different-sized lengths of lead pipe with various discharges. The curve of head loss h_f (or in Reynolds case, its equivalent – the pressure gradient) against mean velocity V, when plotted logarithmically for a given pipe diameter, was found to take the form shown in Fig. 5.3.

The lower portion of Fig. 5.3 is a straight line inclined at 45 degrees (equal scales), indicating that h_f is proportional to V, which is characteristic of laminar flow. The upper portion Reynolds interpreted as a straight line with slope just below 2, indicating that h_f is very nearly proportional to V^2, which earlier experimenters such as Darcy had found to be characteristic of turbulent flow. As will be seen later this interpretation by Reynolds of a simple empirical curve of the form $h_f = cV^n$ was not strictly correct.

One very significant point which the resistance curve effectively demonstrated was that the head loss in turbulent flow was much greater than in laminar. This is only to be expected since the eddying motion of the water particles must result in an increased loss of energy.

The plotted points indicated a certain instability in the transition

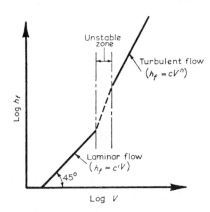

Figure 5.3   Resistance-velocity graph

region and the results of the various tests showed that the magnitude of the critical velocity varied inversely with pipe diameter. There was no immediately apparent explanation for this and so the next logical step in Reynolds' analysis was an evaluation of the factors which governed the critical velocity.

5.4.3 *Reynolds Number*

Reynolds correctly reasoned that there were only three factors which could influence the nature of the flow – the viscosity μ and the density ρ of the fluid, and the diameter D of the pipe. The roughness might be expected to be a factor in turbulent flow but not in laminar. The only possible way in which μ, ρ, and D could be combined to yield the same dimensions as the critical velocity was in the form $\mu/\rho D$.

By dividing the values of the critical velocity obtained in his tests by $\mu/\rho D$ Reynolds found that for circular pipes the dimensionless ratio so obtained was always about 2000. He therefore affirmed that if $\rho DV/\mu$

was below 2000 the flow must be laminar, whereas if it was above the flow must be turbulent. In acknowledgment of Reynolds' most valuable deduction the dimensionless term $\rho DV/\mu$ or DV/ν is referred to universally as *Reynolds number* (R).

We now know that owing to the region of instability which exists at the transition stage a more precise definition would stipulate:

for laminar flow,

$$R = \frac{DV}{\nu} < 2000 \tag{5.11}$$

and for turbulent flow,

$$R = \frac{DV}{\nu} > 4000 \tag{5.12}$$

It is appropriate at this stage to correlate the Poiseuille and Darcy-Weisbach equations. Thus $h_f = 32\mu VL/\rho g D^2 = \lambda LV^2/2gD$, from which λ for laminar flow conditions is given by

$$\lambda = 64\frac{\nu}{DV} = \frac{64}{R} \tag{5.13}$$

5.4.4 *Similarity of Flow*

The Reynolds number has a much wider significance than the mere transition between laminar and turbulent flow, since it may be regarded as a criterion of similarity of motion in pipes of different sizes and conveying different fluids. Also, as discussed in Ch. 4, Sect. 4.12, it represents the ratio of the inertial to the viscous forces and as such has an application extending well beyond that of pipe flow.

Example 5.1

Determine a suitable diameter for a pipeline which is to convey 0·057 cumecs of oil a horizontal distance of 300 m., if the pressure loss is not to exceed 140 kN/m². At the operating temperature the specific gravity of the oil is 0·9 and the dynamic viscosity is 1·43 N s/m². What power output is required from the pump?

$$V = 0\cdot057 \times 4/\pi D^2 = 0\cdot0726/D^2 \text{ m/s}$$

Assuming laminar flow and utilising Eq. (5.6) we obtain

$$V = \frac{pD^2}{32\mu L} \quad \text{or} \quad \frac{0\cdot0726}{D^2} = \frac{140 \times 1000 \times D^2}{32 \times 1\cdot43 \times 300}$$

from which $D = 0.29$ m, or 290 mm. For practical purposes $D = 305$ mm (12 in.) diameter. The corresponding values of V and p are 0·781 m/s and 116 kN/m² respectively. Then

$$R = \frac{0.9 \times 305 \times 0.781}{1.43} = 150$$

which is less than 2000 so that the assumption of laminar flow was correct. The required pump output = $116 \times 0.057 = \mathbf{6.61}$ kW.

5.5 Experimental Investigations on Friction Losses in Turbulent Flow

5.5.1 *Blasius, and Others*

From an examination of the pipe friction data, which were beginning to accumulate at the beginning of the present century, Blasius[1] came to the important conclusion that there were two types of pipe friction in turbulent flow. The first type he associated with smooth pipes where the viscosity effects predominate so that the friction factor is dependent solely on the Reynolds number [$\lambda = \phi(R)$]. The second type was relevant to rough pipes where the viscosity and roughness effects influence the flow and the friction factor is dependent both on the Reynolds number and a parameter of relative roughness.

On the basis of his own researches and of the experimental data supplied by Saph and Schoder[2], Blasius deduced the following expression for the friction in smooth pipes:

$$\lambda = \frac{0.316}{R^{0.25}} \qquad (5.14)$$

Substituting for λ in the Darcy-Weisbach formula, we obtain

$$h_f = 0.316 \left(\frac{\nu}{DV}\right)^{0.25} \left(\frac{LV^2}{2gD}\right)$$

or

$$h_f \propto V^{1.75} \qquad (5.15)$$

Some years later, Stanton and Pannell[3], working at the National Physical Laboratory, investigated in considerable detail the head loss in smooth brass pipes conveying both air and water. On a dimensionless graph of λ (actually τ_0/V^2) against R the plotted points were found

[1] BLASIUS, P. R. H. (1913) 'Das Ähnlichkeitsgesetz bei Reibungsvorgängen in Flussigkeiten', V.D.I.-Forschungsheft, No. 131.
[2] SAPH, A. V. and SCHODER, E. W. (1903) 'An Experimental Study of the Resistances to the Flow of Water in Pipes', Trans. Am. Soc. C.E., 51, 253.
[3] STANTON, T. E. and PANNELL, J. R. (1914) 'Similarity of Motion in Relation to the Surface Friction of Fluids', Phil. Trans. Roy. Soc., 214, 199.

to be compactly situated along a single curve. The curve in the turbulent zone agreed quite well with the Blasius formula for values of R up to 10^5, but after this point there was increasing divergence, indicating that the power by which V should be raised was not constant but increased with R. Confirmation of this characteristic was provided by the experimental results of other investigators.

5.5.2 Nikuradse

The next major experimental contributions were made by Nikuradse at Göttingen in the period around 1930. He obtained data[1] which enabled the λ-R smooth pipe curve to be extended to the very high value of $R = 3 \times 10^6$. Also he carried out velocity traverses across the section of pipes, which as will be seen later, proved of considerable value in developing a convincing semi-empirical theory for turbulent pipe flow. Nikuradse's most notable contribution,[2] however, lay in the field of turbulent flow in rough pipes which had hitherto not been investigated in any detail.

Clearly, it is extremely difficult to make a quantitative assessment of roughness. A full definition needs to take into account not only the height of the surface excrescences, k, but also their spacing and pattern. An additional complication is that in practice the size of the individual excrescences may be expected to vary. Nikuradse was able to eliminate most of these possible variables by roughening pipes artificially. He did this by gluing uniform sand grains in closely packed order to the internal surface.

By so arranging that the excrescence size was proportional to the pipe diameter, it was possible to achieve true geometric similarity between pipes of different size. A judicious combination of k and D resulted in six values of k/D, ranging from $k/D = 1/30$ to $k/D = 1/1014$, being obtained. The test data produced the series of smooth curves shown in Fig. 5.4. These curves were found to be defined not by D but by the ratio k/D.

The diagram illustrates several important points:

(i) The laminar portion is a straight line (inclined at 45 degrees for equal vertical and horizontal scales) and is unaffected by the degree of roughness. This, of course, is in accordance with theory.

[1] NIKURADSE, J. (1932) 'Gesetzmässigkeiten der turbulenten Strömung in glatten Rohren', V.D.I.-Forschungsheft, No. 356.
[2] NIKURADSE, J. (1933) 'Stromungsgesetze in rauhen Rohren', V.D.I.-Forschungsheft, No. 361.

(ii) Above a certain value of R, dependent on the relative roughness, λ is independent of R and is constant for a given value of k/D. This means that the viscosity no longer influences the flow and that h_f is proportional to V^2, which is sometimes referred to as the quadratic relationship.

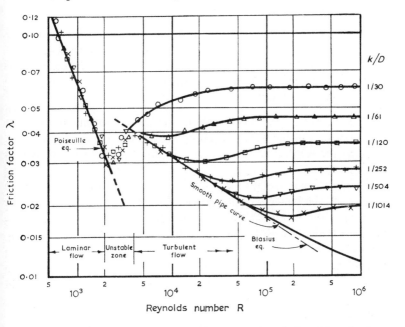

Figure 5.4 Nikuradse's λ-R curves for artificially roughened pipes (plotted points shown diagrammatically only)

(iii) For values of R above 4000 the fully rough state is approached by means of an S-shaped curve branching off the smooth pipe curve. The rougher the pipe the lower the value of R at which the departure occurs.

(iv) The general form of the curves indicates that the relationship between λ, R, and k/D is complex and can only be represented by simple empirical relationships of the type $h_f = cV^n$ over very limited ranges.

As will be seen in the following section, the experimental evidence obtained by Nikuradse provided just that information which Prandtl and von Kármán needed to support and complete their theoretical formulae defining turbulent smooth and turbulent rough flow. The direct practical value of Nikuradse's results was, however, somewhat

85

limited since it was difficult to correlate the uniform artificial roughness with the irregular and wavy type of roughness found in commercial piping.

5.6 Semi-empirical Theory of Pipe Resistance

5.6.1 *Velocity-Deficiency Equation*

In attacking the problem of pipe flow analytically, Prandtl was faced with the difficulty that the general equation of turbulence, $\tau = \rho l^2 (dv/dy)^2$ (Eq. (4.37), p. 56), is of little practical value as it stands since both τ and l are unknown variables. He therefore made two simplifying assumptions as follows:

(a) The shear stress τ is constant, with a value equal to the shear stress τ_0 at the wall.[1] In support of this there is the fact that the major velocity variation occurs in the thin laminar sub-layer at the wall.

(b) The mixing length l has a linear relationship with y, the distance from the boundary (i.e. $l = \kappa y$). This is based on the contention that the turbulent exchange increases the greater the distance from the boundary and that at the boundary it is zero.

Thus

$$\tau_0 = \rho \kappa^2 y^2 \left(\frac{dv}{dy}\right)^2$$

Rearranging,

$$\frac{dv}{dy} = \frac{1}{\kappa}\sqrt{\frac{\tau_0}{\rho}}\frac{1}{y} = \frac{v_*}{\kappa}\frac{1}{y} \tag{5.16}$$

where $v_* \, (= \sqrt{\tau_0/\rho})$ is known as the *shear velocity* (dimensions L/T). Integrating,

$$v = \frac{v_*}{\kappa}\ln y + \text{constant} \tag{5.17}$$

At the pipe axis $v = v_{max}$ and $y = D/2$, so that

$$v_{max} = \frac{v_*}{\kappa}\ln\frac{D}{2} + \text{constant}$$

---

[1] There is an anomaly here in that according to Eq. (5.1) (which is valid for any form of steady flow) $\tau = 0$ at the pipe axis. But of course turbulent flow is of an eddying, pulsating, nature. Further justification is provided by the fact that the combined assumptions serve to produce equations that are shown to have very reasonable agreement with reality.

from which we obtain

$$\frac{v_{max} - v}{v_*} = \frac{1}{\kappa} \ln \frac{D}{2y}$$

Now the velocity traverses of Nikuradse indicated that a suitable value for κ was 0·4. Substituting this value and converting to common logarithms:

$$\frac{v_{max} - v}{v_*} = 5·75 \log \frac{D}{2y} \tag{5.18}$$

This is known as the *velocity-deficiency* ($v_{max} - v$) *equation* and is applicable to both smooth and rough pipes. As depicted in Fig. 5.5, the equation governs the relationship between v_{max} and v. Unless either the maximum or wall velocity is known it does not of itself provide

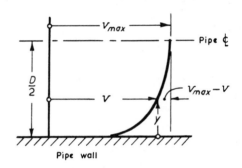

Figure 5.5   Velocity-deficiency distribution

sufficient information to enable the actual velocities across the section to be determined. The logarithmic form of the equation does, however, indicate that the velocity gradient is much steeper adjacent to the walls than in the central portion of the pipe.

Rearranging Eq. (5.18),

$$\frac{v}{v_*} = 5·75 \log \frac{2y}{D} + \frac{v_{max}}{v_*} \tag{5.19}$$

Since v_*, v_{max}, and D are constants for any given pipe and discharge, they may be represented by a single constant y_1, which appears in the logarithm term, so that

$$v = 5·75v_* \log \frac{y}{y_1} \tag{5.20}$$

As v_* has the dimensions of a velocity, y/y_1 is dimensionless and y_1 must therefore be a length. It will be noted that when $y = y_1$, $v = 0$. The significance of this is explained later.

Reverting to Eq. (5.16) and putting $\kappa = 0.4$, we obtain

$$\frac{dv}{dy} = \frac{2.5v_*}{y} \qquad (5.21)^1$$

At the centre of the pipe dv/dy must be zero since $v = v_{max}$, but Eq. (5.21) indicates a finite value. Again when $y = 0$, $dv/dy = \infty$, which is clearly impossible. These anomalies are the result of defects in the theory, which though they might seem important, in fact do little to mar the good agreement that is found in practice.

5.6.2 *Smooth Pipes*

As was stated in the preceding chapter, turbulent flow is separated from a rigid boundary by a thin film or layer in which the flow is laminar. This laminar sub-layer serves the function of transmitting the frictional

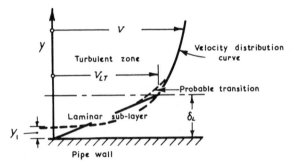

Figure 5.6   Velocity distribution adjacent to pipe wall

drag, induced by the turbulent flow, to the boundary by the normal mechanism of viscous shear.

The laminar sub-layer at the wall of a pipe is illustrated to a greatly enlarged scale in Fig. 5.6. There is no well-defined line of demarcation between laminar and turbulent flow, and the dimension δ_L therefore arbitrarily represents the distance from the boundary at which the flow changes from being predominantly laminar to being predominantly turbulent.

In the turbulent zone the velocity distribution is logarithmic and in accordance with Eq. (5.20). If this curve is extended to the zero velocity

---

[1] Experimental evidence indicates that this equation is applicable to turbulent flow generally.

point it will be found to have an intercept y_1 on the y axis. In the laminar zone the velocity distribution is parabolic in accordance with Eq. (5.2). However, because of the situation on the parabolic curve and the extreme thinness of the laminar film it may be represented with negligible error by a straight line. If the velocity curves are reproduced in this way the intersection at the interface will not be smooth. A much more plausible and likely representation would show a transitional type of curve such as that indicated in dotted outline. In this region both laminar and turbulent flow characteristics are present in complex fashion.

The fact that the velocity at the laminar-turbulent interface is common to both zones was utilised by Prandtl as a basis for deriving an expression for the frictional resistance. The derivation is as follows:

At the common interface $y = \delta_L$, $v = v_{LT}$. Thus from Eq. (5.19) we obtain for the turbulent zone

$$v_{LT} = 5 \cdot 75 v_* \log \frac{2\delta_L}{D} + v_{max} \qquad (5.22)$$

and for the laminar zone, assuming a linear velocity distribution, $\tau_0 = \mu(v_{LT}/\delta_L)$, from which

$$v_{LT} = \frac{\tau_0}{\rho} \frac{\delta_L}{\nu} = v_*{}^2 \frac{\delta_L}{\nu} \qquad (5.23)$$

Equating (5.22) and (5.23),

$$\frac{v_*{}^2 \delta_L}{\nu} = 5 \cdot 75 v_* \log \frac{2\delta_L}{D} + v_{max}$$

or

$$\frac{v_{max}}{v_*} = \frac{v_* \delta_L}{\nu} - 5 \cdot 75 \log \frac{2\delta_L}{D} \qquad (5.24)$$

Eliminating v_{max} by means of Eq. (5.19),

$$\frac{v}{v_*} = 5 \cdot 75 \log \frac{2y}{D} + \frac{v_* \delta_L}{\nu} - 5 \cdot 75 \log \frac{2\delta_L}{D}$$

or

$$\frac{v}{v_*} = 5 \cdot 75 \log \frac{y}{\delta_L} + \frac{v_* \delta_L}{\nu} \qquad (5.25)$$

It is convenient to eliminate δ_L in the logarithm term and as $v_* \delta_L/\nu$ and y/δ_L are dimensionless, Eq. (5.25) may be expressed in the form

$$\frac{v}{v_*} = 5 \cdot 75 \log \frac{v_* y}{\nu} + \text{constant}$$

89

Again utilising the velocity profiles obtained by Nikuradse, the value of the constant is found to be 5·5 so that

$$\frac{v}{v_*} = 5\text{·}75 \log \frac{v_* y}{\nu} + 5\text{·}5 \qquad (5.26)$$

The mean velocity is obtained by integration of the individual velocity filaments, or

$$V = \frac{4}{\pi D^2} \int_0^{D/2} 2\pi \left(\frac{D}{2} - y\right) v \, dy$$

the lower limit of zero being taken because the laminar film is exceedingly thin. Substituting for v we have

$$V = \frac{8v_*}{D^2} \int_0^{D/2} \left(2\text{·}5 \ln \frac{v_* y}{\nu} + 5\text{·}5\right)\left(\frac{D}{2} - y\right) dy$$

the solution to which is

$$\frac{V}{v_*} = 5\text{·}75 \log \frac{v_* D}{\nu} + 0\text{·}17$$

Substituting $v_* = \sqrt{\tau_0/\rho} = \sqrt{\lambda/8}\, V$ and $DV/\nu = \text{R}$, we obtain

$$\frac{1}{\sqrt{\lambda}} = 2\text{·}03 \log \text{R}\sqrt{\lambda} - 0\text{·}86 \qquad (5.27)$$

It is found that better agreement with the experimentally derived λ-R curve is obtained if a minor adjustment is made so that

$$\frac{1}{\sqrt{\lambda}} = 2 \log \text{R}\sqrt{\lambda} - 0\text{·}80$$

or

$$\frac{1}{\sqrt{\lambda}} = 2 \log \frac{\text{R}\sqrt{\lambda}}{2\text{·}51} \qquad (5.28)$$

This is the general equation for the frictional resistance of smooth pipes and is sometimes referred to as the *smooth pipe law*. It provides a much better fit to experimental data than the Blasius formula (Eq. (5.14)). As would be expected the friction factor is dependent on the Reynolds number.

An expression for the thickness δ_L of the laminar sub-layer may be obtained as follows:

Reverting to Eq. (5.26) and inserting $v = v_{LT}$ at $y = \delta_L$, $v_{LT}/v_* = 5\text{·}75 \log (v_* \delta_L/\nu) + 5\text{·}5$. But from Eq. (5.23), $v_{LT}/v_* = v_* \delta_L/\nu$, so that

90

$5 \cdot 75 \log (v_* \delta_L / v) + 5 \cdot 5 - v_* \delta_L / v = 0$, the graphical solution to which is $v_* \delta_L / v = 11 \cdot 6$. Replacing v_* by $\sqrt{\lambda/8} V$ and inserting D on both sides so as to obtain R, we have

$$\frac{\delta_L}{D} = \frac{32 \cdot 8}{R\sqrt{\lambda}} \qquad (5.29)$$

This equation shows that the thickness of the laminar sub-layer varies inversely with the Reynolds number. As the value of R is very dependent on the velocity, δ_L is normally relatively large with low velocities and relatively small with high ones. In order to gain some idea as to the order of δ_L we can examine two typical cases:

(a) $V = 4 \cdot 5$ m/s, $D = 610$ mm (24 in.), $v = 1 \cdot 14$ mm²/s

The value of R is $2 \cdot 41 \times 10^6$, so that from Eq. (5.28), $\lambda = 0 \cdot 01$. Substituting in Eq. (5.29), $\delta_L = 0 \cdot 083$ mm.

(b) $V = 0 \cdot 3$ m/s, $D = 51$ mm (2 in.), $v = 1 \cdot 14$ mm²/s

The values of R and λ are $1 \cdot 34 \times 10^4$ and $0 \cdot 029$, respectively, from which $\delta_L = 0 \cdot 719$ mm.

The relationship between the value y_1 (Eq. (5.20)) of the zero intercept and the thickness of the laminar sub-layer may be obtained by substituting $v_* \delta_L / v = 11 \cdot 6$ in Eq. (5.25), which results in $v/v_* = 5 \cdot 75 \log (y/\delta_L) + 11 \cdot 6$. When $v = 0$, $y = y_1$, whence $5 \cdot 75 \log (\delta_L/y_1) = 11 \cdot 6$, or

$$\frac{\delta_L}{y_1} = 104 \qquad (5.30)$$

The significance of y_1 is that it represents the distance by which the effective or critical boundary is displaced inwards from the actual boundary. At this effective boundary the same conditions are deemed to apply as at the actual boundary. The value of y_1 is exceedingly small, since it is approximately one hundred times smaller than δ_L, which as has been demonstrated earlier is itself very small.

5.6.3 Rough Pipes

From Nikuradse's λ-R diagram in Fig. 5.4 it is apparent that at the higher values of R the friction factor becomes constant and is governed entirely by the relative roughness k/D. This portion of the diagram is referred to as the *turbulent rough zone*. The greater the relative roughness the lower the value of R at which the rough zone commences. In this zone the viscosity has ceased to have any influence on the flow and the frictional resistance has a quadratic relationship with the velocity.

We can make some conjecture as to the physical effect of roughness by referring to the three diagrams of Fig. 5.7:

DIAGRAM (a) ($k < \delta_L$). The excrescences lie well within the laminar sub-layer, which smooths out the flow and prevents any eddies from forming. The pipe behaves as though it were smooth, the actual size, shape and pattern of the excrescences being irrelevant.

DIAGRAM (b) ($k > \delta_L$). The excrescences are slightly larger than the thickness of the sub-layer and therefore protrude out of it into the turbulent region. Some eddying is caused which absorbs additional

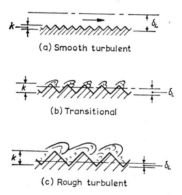

(a) Smooth turbulent

(b) Transitional

(c) Rough turbulent

Figure 5.7   Effect of roughness on the
sub-layer

energy and increases the resistance to flow; the sub-layer remains, however, substantially intact. This condition represents the transition between the smooth and fully rough zones.

DIAGRAM (c) ($k \gg \delta_L$). The excrescences are much larger and have only one fifth or less of their height immersed in the sub-layer. The projections create considerable disturbance, a turbulent wake and a train of vortices being formed. The sub-layer is almost completely disrupted and the form drag due to the roughness is proportional to the square of the velocity.

The derivation of an expression for the friction factor in the rough turbulent zone is the next step in the analysis. This, if correct, should accord with the family of horizontal lines in Nikuradse's λ-R diagram (Fig. 5.4).

We know that λ is independent of R but is governed by the roughness which Nikuradse denoted by the single parameter k for his artificial surface. Since λ is dimensionless we must seek to establish a relationship in terms of λ and some dimensionless representation of the roughness, the logical form being k/D.

Experiment shows that the velocity-deficiency equation (Eq. (5.18)) is equally applicable to rough as well as to smooth pipes, so that

$$\frac{v}{v_*} = 5\cdot75 \log \frac{2y}{D} + \frac{v_{max}}{v_*}$$

Rearranging so as to introduce y/k,

$$\frac{v}{v_*} = 5\cdot75 \log \frac{y}{k} + \left(5\cdot75 \log \frac{2k}{D} + \frac{v_{max}}{v_*}\right)$$

Now Nikuradse's velocity traverses indicated a constant value for the term in the bracket equal to $8\cdot5$, so that

$$\frac{v}{v_*} = 5\cdot75 \log \frac{y}{k} + 8\cdot5 \tag{5.31}$$

Integration across the pipe yields the average velocity which is given by

$$\frac{V}{v_*} = 5\cdot75 \log \frac{D}{2k} + 4\cdot75$$

Substituting $v_* = \sqrt{\tau_0/\rho} = \sqrt{\lambda/8}\, V$,

$$\frac{1}{\sqrt{\lambda}} = 2\cdot03 \log \frac{D}{2k} + 1\cdot68$$

Better agreement is obtained with Nikuradse's λ-R diagram if the equation is slightly modified to

$$\frac{1}{\sqrt{\lambda}} = 2 \log \frac{D}{2k} + 1\cdot74 \tag{5.32}$$

or

$$\frac{1}{\sqrt{\lambda}} = 2 \log \frac{3\cdot7D}{k} \tag{5.33}$$

This is the general equation for the frictional resistance of rough pipes and is sometimes referred to as the *rough pipe law* or *quadratic law*.

The velocity distribution equation (Eq. (5.31)), after being modified to comply with the amendments introduced in Eq. (5.33), becomes

$$v = 5.75 v_* \log \frac{33y}{k} \qquad (5.34)$$

Making comparison with Eq. (5.20), which is equally valid for rough pipes, we find that

$$y_1 = \frac{k}{33} \qquad (5.35)$$

It is extremely difficult to carry out any experimental observations in the very thin laminar sub-layer, but from the hypothetical reasoning previously put forward it would appear that when the excrescence size k exceeds a certain fraction of the thickness δ_L of the laminar sub-layer, the smooth pipe law (Eq. (5.28)) ceases to be valid. Now, we have found that $v_* \delta_L / v = 11.6$, so that if k were to be equal to δ_L we would have

$$\frac{v_* k}{v} = 11.6 \qquad (5.36)$$

The term $v_* k / v$ is a form of Reynolds number and is referred to as the *Reynolds roughness number* denoted by R_*. It is significant in determining the zone to which the particular flow and pipe conditions are appropriate. This is the more readily apparent if $v_* k / v$ is expanded so that

$$R_* = \frac{v_* k}{v} = R \sqrt{\frac{\lambda}{8}} \frac{k}{D} \qquad (5.37)$$

whereupon R_* is seen to be the product of the three most important parameters in pipe flow.

Nikuradse's family of horizontal lines in the rough turbulent zone of the λ-R diagram is reduced to a single line if $2 \log (3.7 D / k) - 1/\sqrt{\lambda}$ is plotted against $\log R_*$ as in Fig. 5.8. The smooth pipe law appears as a line of constant slope – this is because Eq. (5.28) may be expressed as $2 \log R\sqrt{\lambda} - 1/\sqrt{\lambda} = +0.80$, or $2 \log [(VD/v)(v_* \sqrt{8}/V)(k/k)] - 1/\sqrt{\lambda} = +0.80$, whence $2 \log (D/k) + 2 \log (v_* k/v) - 1/\sqrt{\lambda} = $ constant, or

$$2 \log \frac{3.7 D}{k} - \frac{1}{\sqrt{\lambda}} = -2 \log \frac{v_* k}{v} + \text{constant}$$

which is clearly in the form $y = mx + c$ and therefore plots as a straight line with slope $m = -2$.

The departure from the smooth pipe law occurs at $R_* = 4$, indicating that a pipe ceases to be 'smooth' when the excrescence size is about one third the thickness of the laminar sub-layer. The rough pipe law commences approximately where R_* = 60, that is when the excrescence size is about 5 times the thickness of the laminar sub-layer.

5.7 Colebrook and White Transition Formula

5.7.1 Correlation with Commercial Pipes

The work of the Prandtl school represented a very real advancement in the knowledge of pipe flow. The logical next step was to make it possible for the results obtained to be applied to practical design problems.

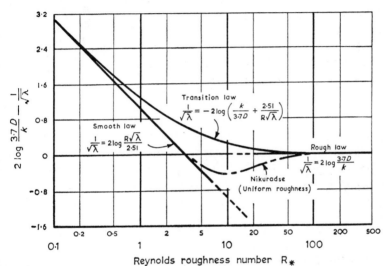

Figure 5.8 Comparison of transition zone curves for commercial and uniform roughness

It was early realised that the roughness of normal commercial pipes could be assessed in terms of an *effective sand roughness, k*, which would correspond to the artificial roughness introduced by Nikuradse. The value of this equivalent roughness is obtained by measuring the loss of head in the turbulent rough zone and solving for k in Eq. (5.33).

It can be inferred that if the roughness characteristics of commercial pipes are similar to those of the Nikuradse pipes, then the flow patterns and λ-R curves will also be similar. Here, however, we encounter a difficulty because commercial pipes have a non-uniform and irregularly distributed type of roughness, which is very different to the uniform roughness created by Nikuradse. This means that, whilst the smooth and rough pipe laws would almost certainly still be valid, the curve in

95

the transitional zone might be expected to differ quite appreciably. The importance of any variation in this zone is accentuated by the fact that nearly all commercial pipes, especially when new, are found to operate within it.

5.7.2 Transition Formula

A much greater understanding of the behaviour of commercial pipes resulted from the studies by Colebrook and White[1] at Imperial College. These investigators carried out tests with air on a pipe 53·5 mm diameter coated with two sizes of sand grain arranged in different patterns. Comparison was made with the test records of various types of commercial pipe and it was found that the closest agreement was obtained when the artificial roughness was composed of isolated grains. The test results when plotted in the graphical form of Fig. 5.8 showed that in spite of the variation in manufacturing technique a single curve could be regarded as reasonably representative. In contrast to the Nikuradse S-shaped curve, there was no inflexion point and the curvature was gradual with asymptotic connections to the smooth law and rough law at each end. A plausible explanation is that with the irregular roughness the laminar sub-layer is disturbed by the larger excrescences at an earlier stage than is the case with the corresponding uniform roughness; however, because these large excrescences are relatively less numerous, the departure from the smooth law is more gradual and remains so throughout.

Colebrook and White[2] found that by judiciously combining the smooth law (Eq. (5.28)) and the rough law (Eq. (5.33)) a single expression could be formulated which would represent the entire turbulent flow range of the λ-R diagram. This expression is:

$$\frac{1}{\sqrt{\lambda}} = -2 \log \left(\frac{k}{3 \cdot 7 D} + \frac{2 \cdot 51}{R \sqrt{\lambda}} \right) \tag{5.38}$$

and is known as the *Colebrook-White transition law*.

The transition curve, indicated in Fig. 5.8, was found to be in reasonable agreement with the experimental data. An essential feature of the expression is that at low values of R the second term in the brackets predominates and the smooth law is approached, whilst at high values of R the first term predominates and the rough law is

[1] COLEBROOK, C. F. and WHITE, C. M. (1937) 'Experiments with Fluid Friction in Roughened Pipes', *Proc. Roy. Soc. Series A*, **161**, 367.
[2] COLEBROOK, C. F. (1938–39) 'Turbulent Flow in Pipes, with Particular Reference to the Transition Region between the Smooth and Rough Pipe Laws', *J. Inst. C.E.*, **11**, 133.

approached. The curve is asymptotic at approximately those points which the experimental results indicate as being appropriate – that is, at about $R_* = 0.3$ to the smooth law and at about $R_* = 60$ to the rough law.

5.7.3 *Importance of the Transition Law*

The Colebrook-White formula was the successful culmination of many years of painstaking effort in the lines of thought first initiated by Prandtl.

The engineer was now presented with a design tool of much greater reliability than the purely empirical formulae which were all that were hitherto available. It should not be lost sight of, however, that the Colebrook-White formula is by no means exact in the same sense as the Poiseuille formula, since certain experimental constants are essential to its derivation. Nevertheless it fits the experimental facts so well, which tests of a wider scope have amply confirmed, that the evidence points overwhelmingly to its validity. The whole history of its development makes a fascinating study and is an excellent example of how theory and experiment can be blended to achieve profoundly successful results.

5.7.4 *Practical Application*

After the publication of the rough pipe law, data concerning the value of the effective roughness k for various classes of pipe started to accumulate. Fairly detailed information is now available and Table 5.1 gives a representative range of values.

Table 5.1
Values of Effective Roughness for Various Pipe Materials

Type of pipe	Effective roughness k (mm)
Copper, lead, brass, alkathene, glass, asbestos cement	smooth
Cast iron, bitumen lined	0·03
Cast iron, concrete lined	0·03
Uncoated steel	0·03
Coated steel	0·06
Galvanised iron	0·16
Coated cast iron	0·16
Uncoated cast iron	0·3
Wet-mix spun precast concrete	0·6
Glazed stoneware	0·6
Precast concrete, mortar not wiped on inside of joint	3·0

Unfortunately, the Colebrook-White formula, as it stands, is in too complex a form to appeal to engineers who wish to obtain a rapid solution to their design problems. These problems usually involve a determination of the pipe size required to handle a certain quantity of liquid at a given friction gradient. In Eq. (5.38), D, R, and λ, although

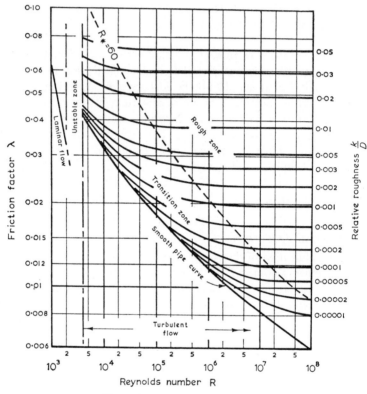

Figure 5.9   Standard λ-R diagram

interrelated, are all initially unknown so that a mathematical solution must proceed on a cumbersome trial and error basis.

Moody[1] in 1944 simplified the mathematical procedure by reproducing the transition law curves on a standard λ-R type of diagram (Fig. 5.9). The relationship between λ, R, and k/D can be read off directly from the diagram. Ingenious nomographic charts have also

[1] MOODY, L. F. (1944) 'Friction Factors for Pipe Flow', *Trans. Am. Soc. Mech. E.*, **66**, 671.

been devised. The 'Universal' design charts,[1] based on dimensionless parameters, permit a direct solution to be effected under a wide range of conditions. In addition, useful tables[2] are available.

5.8 Empirical Formulae

5.8.1 *Exponential Form of Equation*

Prior to the publication of the logarithmic formulae the only design equations available were those of the purely empirical exponential type. Simplicity is their chief merit, since they are particularly amenable to solution by means of nomograms and charts. They have been and still are used extensively. For pipes conveying water they normally take the form

$$V = aD^x S_f^y \qquad (5.39)$$

where S_f is the friction gradient h_f/L, and the coefficient a and the exponents x and y are empirical. The expression is not dimensionless so that care has to be taken in any conversion of units.

It is instructive to investigate the relationship between the friction factor λ and the above terms. To do this we must first write the Darcy-Weisbach formula in the form $S_f = h_f/L = \lambda V^2/2gD$. Substituting in Eq. (5.39),

$$V = aD^x \left(\frac{\lambda V^2}{2gD}\right)^y$$

from which

$$\lambda = 2g \frac{D^{1-x/y}}{a^{1/y}V^{2-1/y}} \qquad (5.40)$$

Since a is normally a constant coefficient which varies with the roughness and viscosity, Eq. (5.40) incorporates all the essential ingredients of roughness and Reynolds number on which λ is known to be dependent.

All exponential formulae when plotted on the standard λ-R diagram (Fig. 5.9) appear as straight lines of varying slope. Since the true form of the friction factor equation, except in the rough turbulent zone, is a logarithmic curve concave upwards, the exponential formula can only be approximately valid over a limited range. At the two ends the straight line will lie below the curve resulting in an under-estimation

[1] ACKERS, P. (1958) 'Charts for the Hydraulic Design of Channels and Pipes', *Hydraulics Research Paper No.* 2. H.M.S.O.
[2] ACKERS, P. (1969) 'Tables for the Hydraulic Design of Storm-drains, Sewers and Pipe-lines', *Hydraulics Research Paper No.* 4 (metric edition). H.M.S.O.

of the frictional loss. It is therefore important to be aware of the range within which each exponential formula may be applied since any attempt to extrapolate may lead to considerable error. Normally, the maximum deviation does not exceed 3 per cent which is within the limits of accuracy of estimating roughness.

Although the application of the transition law has been greatly simplified by the introduction of design charts and tables, the use of exponential formulae normally leads to a more rapid solution. Whatever formula is used the probable error in establishing roughness and other factors is not likely to be much less than 5 per cent.

5.8.2 *Smooth Pipes*

Pipes of aluminium, brass, copper, lead, alkathene, glass, and asbestos cement may generally be classified as smooth. The Blasius formula (Eq. (5.14)) is applicable up to $R = 10^5$ and may be expressed in the exponential form. This follows since

$$S_f = \frac{0 \cdot 316}{R^{0 \cdot 25}} \frac{V^2}{2gD} = 0 \cdot 316 \left(\frac{\nu}{DV}\right)^{0 \cdot 25} \frac{V^2}{2gD}$$

Substituting $\nu = 1 \cdot 14$ mm²/s for water at 15°C, we obtain

$$V = 75 D^{5/7} S_f^{4/7} \tag{5.41}$$

5.8.3 *Pipes in the Transition Zone*

A wide choice of exponential formula is available. The *Hazen-Williams formula*

$$V = 0 \cdot 354 C_H D^{0 \cdot 63} S_f^{0 \cdot 54} \tag{5.42}$$

is probably the most common. The indices are constant for all classes of pipe, and variation in roughness is allowed for by adjustment to the coefficient C_H. For instance with spun iron pipes conveying non-aggressive water the value of C_H is usually taken as 135. When the water is aggressive a value nearer to 100 may be more appropriate.

5.8.4 *Pipes in the Rough Turbulent Zone*

The *Manning formula*, which is extensively employed in connection with channels (see Ch. 7, Sect. 7.4.4), is also applicable to pipes. For pipes the formula becomes[1]:

$$V = \frac{0 \cdot 397}{n} D^{2/3} S_f^{1/2} \tag{5.43}$$

---

[1] The Manning formula is $V = R^{2/3} S_f^{1/2}/n$, where R is the hydraulic radius (sectional area ÷ perimeter). For a circular pipe $R = D/4$.

where n is a roughness coefficient. Because the exponent of S_f is $\frac{1}{2}$, the friction factor λ is independent of V and from Eq. (5.40) is given by

$$\lambda = 124 \frac{n^2}{D^{1/3}} \qquad (5.44)$$

This equation shows that, for a given diameter of pipe, λ is independent of R with the result that the Manning formula is represented on the λ-R diagram by a series of lines parallel to the abscissa – which is exactly the manner in which the logarithmic formula appears in the rough turbulent zone, thus indicating the range of application.[1]

Since n and k are both measures of the roughness, it is important to establish an appropriate relationship. This can be done by utilising the expression for λ in Eq. (5.44) and the rough pipe law (Eq. (5.33)), whence $D^{1/6}/124^{1/2}n = 2 \log(3 \cdot 7 D/k)$, or

$$n = \frac{D^{1/6}}{22 \cdot 3 \log(3 \cdot 7 D/k)} \qquad (5.45)$$

It is convenient to rearrange the equation in the dimensionless form:

$$\frac{k^{1/6}}{n} = \frac{22 \cdot 3 \log(3 \cdot 7 D/k)}{(D/k)^{1/6}} \qquad (5.46)$$

The graph of $k^{1/6}/n$ against k/D over the typical range $k/D = 1/100$ to $1/1000$ is shown in Fig. 5.10. The variation of $k^{1/6}/n$ is remarkably small and if the average value $26 \cdot 0$ is taken the maximum deviation is only about 5 per cent, which is within the limits of accuracy of determining k.

Fig. 5.11 shows the graph of n against k on a logarithmic plot. It will be noted that k is much more sensitive than n to small variations in roughness. Values of n for some representative pipe surfaces are given in Table 7.1 on page 155.

Example 5.2

A cast iron pipe is to convey $0 \cdot 152$ cumecs of water. If $S_f = 1/400$ determine the size of pipe which is required according to the formulae of (a)

---

[1] It is interesting to note that by writing

$$S_f = \frac{h_f}{L} = \frac{\lambda V^2}{2gD} = \frac{\lambda}{2gD} \frac{v^2}{D^2} R^2$$

and utilising Eq. (5.44) in order to eliminate D, we obtain the expression

$$\lambda = 103 \cdot 9 n^{1 \cdot 8} S_f^{0 \cdot 1} v^{-0 \cdot 2} R^{-0 \cdot 2}$$

which will plot as a series of straight inclined lines for various values of n, S_f and v (cf. Fig. 7.3, p. 150).

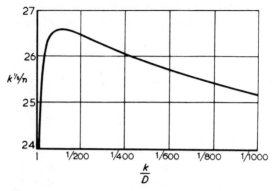

Figure 5.10　Curve of $k^{1/6}/n$ against relative roughness

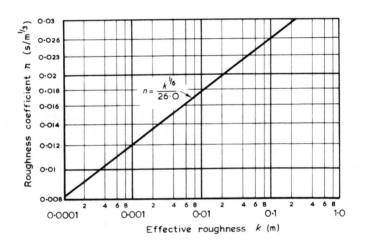

Figure 5.11　Relationship between n and k

Colebrook-White ($k = 0{\cdot}3$ mm), (b) Hazen-Williams ($C_\text{H} = 135$), and (c) Manning ($n = 0{\cdot}01$). Assume $\nu = 1{\cdot}14$ mm²/s.

$$V = 0{\cdot}193/D^2 \quad \text{m/s}$$

(a) Colebrook-White:

Substituting in Eq. (5.38),

$$\frac{1}{\sqrt{\lambda}} = -2\log\left(\frac{81{\cdot}1 \times 10^{-6}}{D} + \frac{0{\cdot}425 \times 10^{-6} \times D}{\sqrt{\lambda}}\right) \tag{i}$$

102

From the Darcy-Weisbach formula, $S_f = \lambda V^2/2gD$ or $1/400 = (\lambda/19\cdot61\,D)(0\cdot193/D^2)^2$, so that $\sqrt{\lambda} = D^{2\cdot5}/0\cdot872$. Substituting in Eq. (i),

$$\frac{0\cdot872}{D^{2\cdot5}} = -2\log\left(\frac{81\cdot1 \times 10^{-6}}{D} + \frac{0\cdot370 \times 10^{-6}}{D^{1\cdot5}}\right)$$

the solution to which by trial and error is $D = 0\cdot425$ m or **425** mm.

(b) Hazen-Williams:

Substituting in Eq. (5.42),

$$\frac{0\cdot193}{D^2} = \frac{0\cdot354 \times 135 \times D^{0\cdot63}}{400^{\,0\cdot54}} \tag{ii}$$

from which $D = 0\cdot420$ m or **420** mm.

(c) Manning:

Substituting in Eq. (5.43),

$$\frac{0\cdot193}{D^2} = \frac{0\cdot397 \times D^{2/3}}{0\cdot01 \times 400^{1/2}} \tag{iii}$$

from which $D = 0\cdot417$ m or **417** mm.

In practice a **457** mm (18 in.) diameter pipe would be appropriate to all the values of D which have been obtained.

For $D = 0\cdot425$ m: $V = 1\cdot07$ m/s and $\lambda = 0\cdot0182$ whence by substituting in Eq. (5.37), $R_* = 13\cdot7$. This value of R_* indicates that the pipe is operating in the transition zone.

5.9 Deterioration of Pipes

Most water mains which have been in service for several years suffer some reduction in carrying capacity due to incrustation or a coating of slime which tends to gather on the internal surface. The rate of deterioration is dependent on the chemical constituents of the water and the pipe material. When designing a water main, therefore, it is prudent to take into consideration its likely condition after a period of years in service.

Colebrook and White[1] showed by a simple application of their transition law that, if the roughness was increased from say $k = 0\cdot254$ mm to $k = 2\cdot54$ mm in a 508 mm (20 in.) pipe, the resulting reduction in carrying capacity would be about 25 per cent. The corresponding reduction in sectional area would, however, only be about 2 per cent. It seems reasonable to conclude, therefore, that the reduction in carrying capacity is

[1] COLEBROOK, C. F. and WHITE, C. M. (1937–38) 'The Reduction of Carrying Capacity of Pipes with Age', *J. Inst. C.E.*, **7**, 99.

almost entirely due to the increase in surface roughness with age; examinations and tests of pipes confirm the soundness of this assessment.

By analysing the data from tests on cast iron pipes Colebrook and White deduced that the roughness increased uniformly with age and could be best expressed by the simple empirical formula

$$k_T = k_0 + \alpha T \tag{5.47}$$

where k_0 is the initial effective roughness, k_T the effective roughness after T years, and α the annual rate of growth of roughness.

Lamont[1] and others have shown that the same form of equation is applicable to other classes of pipe. The value of α may be obtained from tables or from pipe tests carried out at intervals of time.

5.10 Non-circular Pipes and Conduits

In the case of a non-circular section (e.g. box culvert flowing full) it is unfortunately not possible to develop general expressions for the friction factor by the same sort of analytical reasoning as was adopted for circular pipes. This is because the velocity and shear stress gradients are no longer uniformly distributed about the pipe axis. Similar considerations apply to open channel flow and, as explained in the latter connection (Ch. 7, Sect. 7.3.2), the substitution of the hydraulic radius R (sectional area ÷ perimeter) for $D/4$ in the friction factor formulae is an artifice which is found to yield results that are of reasonable accuracy. But of course it is important that the degree of asymmetry should not be excessive.

5.11 Minor Losses in Pipes

5.11.1 *General Expression for Minor Losses*

Apart from the loss of head due to friction, there are other, minor, losses in pipe flow which arise from changes of section, junctions, bends, and valves. The frictional loss in long pipelines is usually far in excess of the minor losses and in these circumstances the latter are often neglected. With short pipelines, however, the minor losses assume a greater relative importance and some appropriate allowance should be made. The fact that the limit of accuracy for assessing frictional loss is about 5 per cent serves as a good guide to the need for their inclusion.

It is usual to regard the minor losses as being additional to the normal type of frictional loss. As they are associated with a turbulent

[1] LAMONT, P. A. (1954) 'The Reduction with Age of the Carrying Capacity of Pipelines', *J. Inst. Wat. E.*, **8**, 53.

dissipation of energy they may be conveniently expressed in terms of the velocity head. Except in the simple cases of a sudden enlargement and contraction, which will be next considered, it is not possible to make a theoretical assessment, and so the general procedure is to introduce a coefficient K, the value of which must be determined by experiment. The expression for energy head loss is then

$$h_L = K \frac{V^2}{2g} \tag{5.48}$$

Appropriate values for K will be found listed in pipe manufacturers' handbooks.[1] Obviously, in order to minimise losses, the transition should be as gradual as possible.

A less accurate but often more convenient method of dealing with minor losses is to express them in terms of the equivalent length of straight pipe L which yields the same loss in head. This involves an assessment of L/D for each pipe fitting; for example, with a 90 degree close radius bend the equivalent length might be 18 pipe diameters. Again, there are appropriate listings available.

5.11.2 Enlargement of Section

An abrupt enlargement in sectional area of flow is shown in Fig. 5.12. As a result, the pressure increases from p_1 to p_2 and the velocity decreases from V_1 to V_2. Separation occurs where the flow stream emerges from

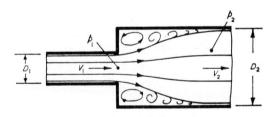

Figure 5.12   Sudden enlargement

the smaller pipe and normal conditions are not restored until some distance downstream. The space around the expanding stream of water is filled with a violent eddying motion, whose maintenance requires a continuous supply of energy from the main body of flow.

[1] Also in Table 34, p. 165 of vol. 2 of (1969) *Manual of British Water Engineering Practice* (Ed. SKEAT, W. O.). Heffer (4th Edition).

Now it is reasonable to assume – and experimental evidence tends to support the contention – that the pressure on the annular area $A_2 - A_1$ at the enlargement is the same as that in the smaller pipe, namely p_1. The frictional force at the pipe wall is relatively small in comparison with the force created by the change of momentum, so we can utilise the momentum equation and obtain $p_2 A_2 - p_1 A_2 = \rho Q(V_1 - V_2)$, from which

$$\frac{p_2 - p_1}{w} = \frac{V_2}{g}(V_1 - V_2) \tag{5.49}$$

and from Bernoulli's equation, $p_1/w + V_1^2/2g = p_2/w + V_2^2/2g + h_L$, whence

$$h_L = \frac{V_1^2 - V_2^2}{2g} - \frac{p_2 - p_1}{w}$$

Substituting for $(p_2 - p_1)/w$ from Eq. (5.49) and simplifying, we obtain

$$h_L = \frac{(V_1 - V_2)^2}{2g} = \left(1 - \frac{A_1}{A_2}\right)^2 \frac{V_1^2}{2g} \tag{5.50}$$

Thus the coefficient K, when applied to the velocity head in the smaller pipe, is given by

$$K = \left(1 - \frac{A_1}{A_2}\right)^2 = \left[1 - \left(\frac{D_1}{D_2}\right)^2\right]^2 \tag{5.51}$$

from which it will be observed that K is dependent on the ratio of the respective diameters. For instance if $D_1 = 102$ mm (4 in.) and $D_2 = 152$ mm (6 in.), then $K = 0.31$. Similarly, for $D_1 = 254$ mm (10 in.) and $D_2 = 305$ mm (12 in.), $K = 0.09$. Experiment confirms the approximate validity of this relationship.

The head loss at a sudden enlargement is relatively high. A tapered connection provides a more efficient transition. The more gradual the taper the less is the loss due to excess turbulence, but beyond a certain stage this is outweighed by higher frictional loss in the upstream portion of the taper. Gibson[1] found the optimum cone angle to be about 6 degrees, and this is roughly the angle adopted in standard waterworks practice. The corresponding value of K is 0.1 or possibly less.

[1] GIBSON, A. H. (1952) *Hydraulics and its Applications*, p. 92. Constable (5th Edition).

A common type of sudden enlargement is the discharge of a pipe into a large tank or reservoir (Fig. 5.13). In Eq. (5.50), $A_2 = \infty$, so that

$$h_L = \frac{V_1^2}{2g} \tag{5.52}$$

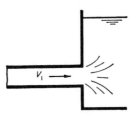

Figure 5.13   Entry to a tank

indicating that the entire velocity energy is dissipated in eddying and turbulence.

5.11.3 *Contraction of Section*

At an abrupt contraction (Fig. 5.14) the stream lines converge upstream of the change of section and the convergence continues at a diminishing rate as far as the vena contracta (see p. 38). Experiment indicates that

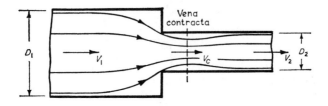

Figure 5.14   Sudden contraction

the sectional area of the latter is generally about $0.6A_2$. Utilising this value and bearing in mind that the entire head loss is accounted for by the subsequent enlargement from the vena contracta, we can substitute in Eq. (5.50) to obtain

$$h_L = (1 - 0.6)^2 \frac{(V_2/0.6)^2}{2g} = 0.44 \frac{V_2^2}{2g} \tag{5.53}$$

Thus the value of K, based on the smaller pipe, is 0.44. In practice it is

107

normally taken as 0·5. Again, tapered connections will effect a considerable reduction in the head loss.

At the entrance to a pipe from a tank or reservoir the loss of head is dependent on the form of entry. For the simple type of inlet (Fig. 5.15a), the value of K is about 0·5; for the re-entrant type (Fig. 5.15b)

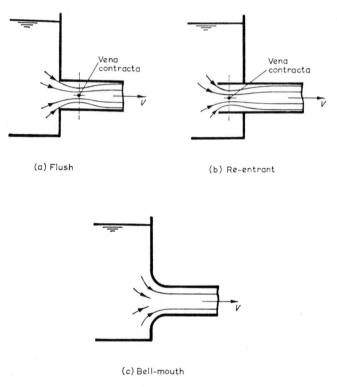

(a) Flush   (b) Re-entrant

(c) Bell-mouth

Figure 5.15   Types of pipe entry

it is approximately 1·0. A bell-mouthed entry (Fig. 5.15c) is most efficient and reduces the value of K to about 0·05; this type is usually employed on the suction branch of pumps.

5.12 Summary of Formulae

An appreciable number of expressions for pipe resistance have been quoted in the text and it may be helpful to conclude by tabulating those that are the more important.

Table 5.2

List of Pipe Resistance Formulae

Title or Description	Formula	Eq. No.	Range of R	Range of R_*
Poiseuille	$h_f = \dfrac{32\nu VL}{gD^2}$	5.6	< 2000	no limit
Darcy-Weisbach	$h_f = \dfrac{\lambda L V^2}{2gD}$	5.7	universal	
Blasius	$\lambda = \dfrac{0\cdot316}{R^{0\cdot25}}$	5.14	4000–10^5	< 0·3
Smooth Law (Kármán-Prandtl)	$\dfrac{1}{\sqrt{\lambda}} = 2 \log \dfrac{R\sqrt{\lambda}}{2\cdot51}$	5.28	> 4000	< 0·3
Rough Law (Kármán-Prandtl)	$\dfrac{1}{\sqrt{\lambda}} = 2 \log \dfrac{3\cdot7D}{k}$	5.33	> 4000	> 60
Transition Law (Colebrook-White)	$\dfrac{1}{\sqrt{\lambda}} = -2 \log\left(\dfrac{k}{3\cdot7D} + \dfrac{2\cdot51}{R\sqrt{\lambda}}\right)$	5.38	> 4000	no limit
Exponential (Hazen-Williams)	$V = 0\cdot354 C_H D^{0\cdot63} S_f^{0\cdot54}$	5.42	> 4000	0·3–60
Exponential (Manning)	$V = \dfrac{0\cdot397}{n} D^{2/3} S_f^{1/2}$	5.43	> 4000	> 40
Minor Losses	$h_L = K \dfrac{V^2}{2g}$	5.48	> 4000	no limit

Further Reading

ACKERS, P. (1958) 'Resistance of Fluids Flowing in Channels and Pipes', *Hydraulics Research Paper No. 1*. H.M.S.O.

BAKHMETEFF, B. A. (1936) *The Mechanics of Turbulent Flow*. Princeton University Press.

PRANDTL, L. (1952) *Essentials of Fluid Dynamics*. Blackie.

CHAPTER SIX

Pipelines and Pipe Systems

6.1 Hydraulic and Energy Gradients

In accordance with the principle of Bernoulli, the total energy head at any point in a pipeline is the sum of the elevation, pressure, and velocity heads. It may be conveniently represented by the *energy grade line*, or more simply the *energy line*, which is superimposed on the pipeline longitudinal section in the manner indicated in Fig. 6.1. Changes of

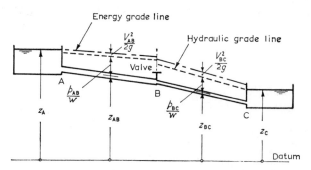

Figure 6.1   Energy and hydraulic grade lines

section and other points where minor losses occur are signified by steps in the energy line. Neglecting minor losses, the slope of the energy line or *energy gradient* is the same as the friction gradient.

The *hydraulic grade line* is determined by the sum of the pressure and elevation heads measured relative to the centroid of the pipe section. It therefore lies below the energy line by an amount equal to the velocity head in the pipe and represents the height to which liquid will rise in an open stand-pipe connected to the pipeline at any point. The slope of the hydraulic grade line is referred to as the *hydraulic gradient*. Unlike the energy gradient which must fall continuously in a downstream direction, the hydraulic gradient may rise at enlargements of section. The velocity head is often exceedingly small in comparison with the pressure head and if it is neglected the energy and hydraulic grade lines coincide.

A knowledge of the hydraulic gradient is of value since it indicates the pressure variation to which the pipeline is subjected. The maximum pressure head governs the stress which the pipeline must be designed to withstand. Apart from the important requirements of service connections, the minimum pressure head may not have any great significance. This arises from the fact that, provided the hydraulic grade line is always above the pipe section, the pressure will not fall below atmospheric. Even if over a certain length the hydraulic grade line does lie below the pipe section, discharge will be maintained by siphonage up to about 7·5 m negative head, at which point vaporisation may be expected to commence. It is good design practice, however, to site the pipeline so that the pressure does not fall below atmospheric. This is to avoid any risk of impurities being sucked into the pipeline.

6.2 Power Transmission

Although hydraulic power transmission has an important role in mechanical engineering (such as the operation of hydraulic servo-mechanisms, lifts, presses, etc.), we will limit our consideration here to

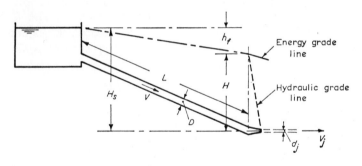

Figure 6.2 Pipeline terminating in a nozzle

the more civil engineering aspect – that of the hydraulic power supply to a water turbine. The pipelines leading to these machines are sometimes of considerable length, and this is more particularly the case with a Pelton wheel turbine, where the terminal feature of the pipeline is a nozzle (or nozzles) discharging a jet (or jets) of water at atmospheric pressure (see Ch. 10, Sect. 10.5.2). As indicated in Fig. 6.2, the energy grade line falls gradually throughout, whereas the hydraulic grade line in the short length of the nozzle falls sharply to the level of the jet.

Assuming that minor losses are either neglected or incorporated in

111

equivalent pipe lengths, the energy head at entry to the nozzle is equal to the static head minus the friction head loss, or

$$H = H_s - h_f \qquad (6.1)$$

Using the Darcy-Weisbach formula and replacing V by $4Q/\pi D^2$ we obtain

$$H = H_s - \frac{16\lambda L Q^2}{2g\pi^2 D^5} \qquad (6.2)$$

Now the power available at the nozzle is $P = wQH$ so that

$$P = wQ \left(H_s - \frac{16\lambda L Q^2}{2g\pi^2 D^5} \right) \qquad (6.3)$$

Normally, the discharge Q is governed by the water resources or the turbine requirements and hence it follows from Eq. (6.3) that for maximum power availability the diameter D of the pipeline must be as large as possible, thereby minimising friction losses. In practice, the optimum size of pipeline must be determined from an economic analysis of power saving set against capital cost. Generally it is found that the most satisfactory operating conditions result when the friction head losses are between 5 and 10 per cent of the static head.

In order to determine the best relationship between Q and H_s for maximum power transmission, Q now being variable, the expression for P in Eq. (6.3) must be differentiated with respect to Q. Assuming that λ remains constant we have

$$\frac{dP}{dQ} = w \left(H_s - 3 \times \frac{16\lambda L Q^2}{2g\pi^2 D^5} \right)$$

and equating to zero for a maximum:

$$H_s = \frac{48\lambda L Q^2}{2g\pi^2 D^5} = 3h_f \qquad (6.4)$$

signifying that the maximum power transmission is obtained when one-third of the gross head is absorbed in friction.

We may proceed a stage further and formulate an expression relating D and the jet diameter d_j. This is obtained from $H = 2h_f$, or $V_j^2/2gC_v^2 = 2\lambda L V^2/2gD$, where C_v is the coefficient of velocity of the nozzle. Substituting $V = V_j(d_j/D)^2$ and simplifying, we obtain

$$D = \sqrt[5]{2C_v^2 \lambda L d_j^4} \qquad (6.5)$$

Although Eqs. (6.4) and (6.5) represent the best conditions for maximum power transmission, the corresponding efficiency of the pipeline is only 67 per cent, and so great a loss in energy head would probably be unacceptable in practice.

6.3 Discharge under Varying Head

A typical problem concerns the rise and fall of water levels in inter-connected service reservoirs or tanks. Fig. 6.3 shows two service reservoirs, surface areas A_{SA} and A_{SB}, connected by a uniform pipeline, length L and diameter D. Let us suppose that it is required to determine the time for the difference in water level to be reduced from H_{t1} to H_{t2}. Strictly speaking, this is a case of unsteady flow, but normally the variation of head is so gradual that negligible error results from assuming that instantaneous conditions are steady – that is to say there are no inertial effects (see Sect. 6.6).

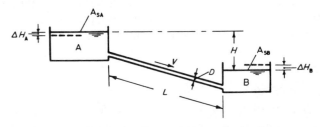

Figure 6.3   Flow under a varying head

We will consider the conditions resulting from a flow Q in a small time interval Δt which reduces the difference in level H by ΔH. During this period the level of A falls by ΔH_A and the level of B rises by ΔH_B. Equating the displacements,

$$A_{SA}\Delta H_A = A_{SB}\Delta H_B = -Q\,\Delta t$$

Now, $\Delta H = \Delta H_A + \Delta H_B$, so that

$$\Delta H = -\frac{A_{SA} + A_{SB}}{A_{SA}A_{SB}} Q\,\Delta t$$

and in the limit $\Delta t \to 0$,

$$dt = -\frac{A_{SA}A_{SB}}{A_{SA} + A_{SB}} \frac{dH}{Q} \qquad (6.6)$$

Treating minor losses in the same way as before, the application of the Darcy-Weisbach formula (λ constant) leads to the relationship

$$H = \frac{16\lambda L Q^2}{2g\pi^2 D^5} \qquad (6.7)$$

113

Substituting for Q in Eq. (6.6),

$$dt = -\frac{4}{\pi}\left(\frac{A_{SA}A_{SB}}{A_{SA}+A_{SB}}\right)\left(\frac{\lambda L}{2gD^5}\right)^{1/2}\frac{dH}{H^{1/2}} = -c\,\frac{dH}{H^{1/2}}$$

where c is a constant coefficient. Integrating between the limits t_2 and t_1,

$$t_2 - t_1 = \int_{t_1}^{t_2} dt = -c\int_{H_{t1}}^{H_{t2}}\frac{dH}{H^{1/2}}$$

or

$$t_2 - t_1 = 2c(H_{t1}^{1/2} - H_{t2}^{1/2}) \qquad (6.8)$$

The pipe velocity and hence the Reynolds number vary in accordance with the head difference. The friction factor λ does not therefore remain constant unless the pipeline is operating throughout in the rough turbulent zone. If it is desired to make provision for variation in λ a step-by-step time interval must be adopted.

6.4 Simple Pipe Systems

6.4.1 *Pipes in Series*

When a single pipeline consists of pipes of different sizes, these pipes are said to be in series. Fig. 6.4 shows a pipeline with three diameters of

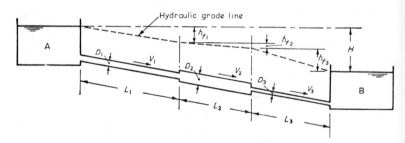

Figure 6.4   Pipes in series

pipe D_1, D_2, and D_3 with friction factors λ_1, λ_2, and λ_3, respectively. Let us suppose that it is required to determine the discharge Q when the difference in level of the two reservoirs is H.

Dealing with minor losses as before,

$$H = \frac{\lambda_1 L_1 V_1^2}{2gD_1} + \frac{\lambda_2 L_2 V_2^2}{2gD_2} + \frac{\lambda_3 L_3 V_3^2}{2gD_3}$$

Since $V = 4Q/\pi D^2$, we have

$$H = \frac{16Q^2}{2g\pi^2}\left(\frac{\lambda_1 L_1}{D_1{}^5} + \frac{\lambda_2 L_2}{D_2{}^5} + \frac{\lambda_3 L_3}{D_3{}^5}\right)$$

from which

$$Q = \frac{\pi\sqrt{2gH}}{4(\lambda_1 L_1/D_1{}^5 + \lambda_2 L_2/D_2{}^5 + \lambda_3 L_3/D_3{}^5)^{1/2}} \qquad (6.9)$$

It is sometimes advantageous to conceive of the pipe system as being replaced by a single equivalent pipe of uniform section. A pipe is said to be equivalent when the loss of head for a given flow is the same as that for the pipe or system of pipes which it replaces. An infinite number of equivalent pipes are possible for any given set of conditions, being represented by various combinations of λ, L, and D. However, it is often convenient to assume a diameter D_e and friction factor λ_e equal to that of the longest pipe in the system. The problem then resolves itself into an assessment of the length L_e of equivalent pipe. Proceeding in this manner,

$$H = \frac{16Q^2}{2g\pi^2}\left(\frac{\lambda_e L_e}{D_e{}^5}\right) \qquad (6.10)$$

and for the particular system above:

$$L_e = \frac{D_e{}^5}{\lambda_e}\left(\frac{\lambda_1 L_1}{D_1{}^5} + \frac{\lambda_2 L_2}{D_2{}^5} + \frac{\lambda_3 L_3}{D_3{}^5}\right) \qquad (6.11)$$

6.4.2 *Pipes in Parallel*

In any pipe system the pressure at the common junction points must be the same for all the various branches which meet there. The flow therefore distributes itself among the branches in accordance with the controlling end pressures. If pressure drop is substituted for voltage and discharge for current, the analogy may be drawn between a pipe system and an electrical circuit. As with the electrical circuit, when pipes are laid in parallel the flow in each branch may be evaluated from equations describing the common pressure drop. The analogy must not be drawn too closely since the resistance law for pipe flow is very different to that for electrical circuits.

Let us consider the case of two reservoirs A and B interconnected by three pipes (Fig. 6.5). The total discharge is Q and the differential head H.

Since there is a common pressure drop,

$$H = \frac{\lambda_1 L_1 V_1{}^2}{2g D_1} = \frac{\lambda_2 L_2 V_2{}^2}{2g D_2} = \frac{\lambda_3 L_3 V_3{}^2}{2g D_3} \qquad (6.12)$$

The sum of the flows in each of the branches is equal to the total flow so that $Q = Q_1 + Q_2 + Q_3$, or

$$Q = \frac{\pi}{4} (D_1^2 V_1 + D_2^2 V_2 + D_3^2 V_3) \tag{6.13}$$

The four separate equations represented by Eqs. (6.12) and (6.13) may be solved provided that the unknowns do not exceed this number.

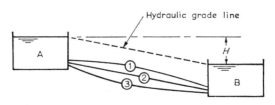

Figure 6.5   Pipes in parallel

The equivalent pipe concept affords a convenient method of analysis. From Eq. (6.10) we have

$$Q = \frac{\pi}{4} \sqrt{2g} \left(\frac{D_e^5}{\lambda_e L_e}\right)^{1/2} H^{1/2}$$

Similarly,

$$Q_1 = \frac{\pi}{4} \sqrt{2g} \left(\frac{D_1^5}{\lambda_1 L_1}\right)^{1/2} H^{1/2}$$

and so on for Q_2 and Q_3.

Substituting in Eq. (6.13), we obtain

$$\left(\frac{D_e^5}{\lambda_e L_e}\right)^{1/2} = \left(\frac{D_1^5}{\lambda_1 L_1}\right)^{1/2} + \left(\frac{D_2^5}{\lambda_2 L_2}\right)^{1/2} + \left(\frac{D_3^5}{\lambda_3 L_3}\right)^{1/2} \tag{6.14}$$

With assumed values of λ_e and L_e, D_e may be evaluated; alternatively, if λ_e and D_e are assumed, L_e can be found.

If the discharge Q and the pipe characteristics are known and it is required to determine H, another method of approach, and one which is sometimes advantageous, is to assume a likely discharge Q_1' for the first pipe and then to evaluate H' from Eq. (6.12). Also using Eq. (6.12), values of Q_2' and Q_3' are obtained. The resulting total flow Q' will normally be found to differ from the actual value Q, but it is reasonable to assume that the actual discharge is distributed among the branches in the proportions of $Q_1' : Q_2' : Q_3'$ so that

$$Q_1 = Q_1' \frac{Q}{Q'} \quad ; \quad Q_2 = Q_2' \frac{Q}{Q'} \quad ; \quad Q_3 = Q_3' \frac{Q}{Q'} \tag{6.15}$$

The values of h_{f1}, h_{f2}, and h_{f3} should now be checked for equality and, if discrepancies are found, the procedure must be repeated using a more appropriate value for Q'_1. However, with experience, a judicious selection may generally be made in the first instance which will obviate the need for any repetition.

Parallel pipes are a device commonly adopted in waterworks practice when an existing main is found to be inadequate to convey the required flow. Duplication over a part or all of the length of the existing main is generally more economical than replacing it with a larger main. The resulting increased capacity can be determined in accordance with the above equations.

Example 6.1

Water is pumped from a reservoir A to a reservoir B through a piping system which consists of one 610 mm (24 in.) diameter pipe 450 m long branching into two pipes of diameter 305 mm (12 in.) and 457 mm (18 in.), each 600 m long. The pumping station is situated adjacent to reservoir A and the surface level of B is 60 m above that of A. Determine the head on the pumps if water is to be transferred at the rate of 0·4 cumecs. Also determine the flow in each of the branched pipes. Take $\lambda = 0·02$ for all the pipes.

The parallel pipes will be replaced by a single equivalent pipe 610 mm diameter. Its equivalent length is obtained from Eq. (6.14):

$$\left(\frac{0·0845}{L_e}\right)^{1/2} = \left(\frac{0·0026}{600}\right)^{1/2} + \left(\frac{0·0199}{600}\right)^{1/2}$$

from which $L_e = 1367$ m.

The total equivalent length of 610 mm pipe is thus 1817 m and the head loss is

$$h_f = \frac{0·02 \times 1817 \times 0·4^2}{2 \times 9·81 \times (\pi/4)^2 \times 0·0845} = 5·69 \text{ m}$$

Hence the pumping head required is **65·69** m.

The head loss in each of the branch pipes is the same so that $Q_2^2/D_2^5 = Q_3^2/D_3^5$, from which $Q_3 = 2·75\, Q_2$. Hence $Q_2 = \mathbf{0·107}$ cumecs and $Q_3 = 0·293$ cumecs.

6.4.3 *Multiple Reservoirs*

A typical problem concerns the analysis of flow in a simple pipe system linking three or more reservoirs. The factors governing flow distribution are the same as previously outlined in connection with branch pipes – that is to say the flows are dictated by the pressures at the common junction points.

Let us consider the system shown in Fig. 6.6. Clearly the flow in pipe 1 is directed towards the junction point, whilst in pipe 3 it is away from it. The direction of flow in pipe 2 is not initially known but depends upon the elevation of the hydraulic grade line at the junction point.

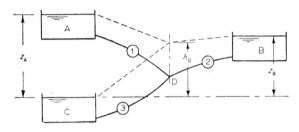

Figure 6.6   Inter-connected reservoirs

The water level in reservoir C is taken as datum, and it is assumed that the grade line elevation at D is below B (i.e. $h_D < z_B$). The equations relating to the flow are:

$$z_A - h_D = \frac{\lambda_1 L_1 V_1{}^2}{2g D_1} \tag{6.16}$$

$$z_B - h_D = \frac{\lambda_2 L_2 V_2{}^2}{2g D_2} \tag{6.17}$$

and

$$h_D = \frac{\lambda_3 L_3 V_3{}^2}{2g D_3} \tag{6.18}$$

Also for continuity, $Q_1 + Q_2 = Q_3$, from which

$$D_1{}^2 V_1 + D_2{}^2 V_2 = D_3{}^2 V_3 \tag{6.19}$$

Up to four unknowns may be determined from these equations. Thus if z_A, z_B and the pipe characteristics are given, h_D, Q_1, Q_2, and Q_3 may be determined. In these circumstances, solution is best carried out on a trial and error basis for various assumed values of h_D. If, when balancing the discharges in Eq. (6.19), it is found that a greater flow towards C is required, then the grade line at D must be raised – conversely, if a lesser flow is required, the grade line must be lowered. Should h_D be found greater than z_B, flow in pipe 2 is towards C and the continuity equation (Eq. (6.19)) must be appropriately amended. Solution is complete when this equation is satisfactorily balanced.

Whilst quite involved pipe systems can be analysed in the above manner, care should be taken to assign an assumed value to only one of the unknowns, otherwise the required convergence in solution will not be obtained. This means that the method becomes cumbersome when applied to complex systems since a very large number of simultaneous equations are involved.

6.5 Distribution Mains

6.5.1 *General Principles of Design*

The network of distribution mains is nearly always the most expensive item of equipment in a water undertaking. Also, the cost of its upkeep generally represents a large proportion of the annual maintenance budget. It is therefore incumbent upon the water engineer to devote some considerable care to the design of the most efficient distribution system and this entails an accurate prediction of the flows and pressures in the various pipe components.

The water demand is dependent upon the population and type of industry served. Present domestic consumption is about 250 litres per head per day, but with rising living standards the trend is upwards. In residential districts the optimum pressure head in the street main is between 28 and 35 metres with lower and upper limiting pressure heads of about 20 and 65 metres. If the pressure head is below 20 metres, there will be difficulty in supplying water above the third storey, whilst with pressure heads in excess of 65 metres, leakage at pipe joints and the wear on domestic valve fittings may prove troublesome.

Distribution systems in urban areas are generally best laid out on a grid-iron or ring main pattern, mains being interconnected at distances not exceeding about 600 m. With this arrangement, water can be fed to any point in the system by more than one route, thus providing greater flexibility in meeting the demand and also enabling repairs to be carried out with the minimum disruption to supply. 'Dead-end' mains are in any event undesirable features since frequent flushing is usually necessary in order to prevent any tendency to stagnation.

In the design of a new supply network the topography of the land is very often a controlling factor, and due regard must be paid to the fact that frictional resistance causes a loss in pressure head. In large urban areas where the ground contours vary by more than 60 m it may be advantageous to divide the supply network into pressure zones arranged so as to avoid excessive pressures at the low points. Pressure relief valves and booster pumps are employed to achieve the pressures required. Contour maps indicating the variation in pressure are informative, a close spacing of contours often signifying inadequate pipe sizes.

Distribution mains are in standard pipe sizes from 76 mm (3 in.) diameter upwards. It is generally the aim to keep frictional losses to between 1 m and 10 m per 1000 m of main, although higher losses are sometimes unavoidable with the smaller pipe sizes. Losses of this order are associated with velocities of between 0·6 and 1·2 m/s – a range which is usually found to be satisfactory on both operational and economic grounds.

6.5.2 *Hydraulic Analysis of Pipe Networks*

The methods of pipe flow analysis employed to solve the simple types of pipe system discussed in Sect. 6.4 are no longer suitable when the system is complex, as is generally the case with distribution networks. It is true that the network may usefully be simplified by the replacement of parallel pipes and pipes in series by single equivalent pipes, but beyond this some other means must be employed to effect a solution.

Several methods are available, ranging from analytical and graphical techniques to the use of an electrical analyser.[1] The last-mentioned involves an electrical circuit model of the system with special types of resistors which comply with the laws of pipe resistance. Another valuable tool is the electronic digital computer which serves to eliminate much laborious calculation.[2]

The Hardy Cross analysis is a practical method that is well suited for application in design offices. The method is essentially one of successive approximation and the approach can be either from the aspect of head balance or quantity balance.

Some abbreviated procedure for expressing friction head loss is essential in any network analysis and a few remarks on this topic form a useful preliminary to a detailed explanation of the Hardy Cross method.

6.5.3 *Expression for Friction Head Loss*

For each pipe of a system or network there is a definite relationship between the head loss and the discharge. This may be expressed in the form

$$h_f = rQ^m \tag{6.20}$$

[1] McILROY, M. S. (July 1952) 'Nonlinear Electrical Analogy for Pipe Networks', *Proc. Am. Soc. C.E.*, **78**.

[2] ADAMS, R. W. (1961) 'Distribution Analysis by Electronic Computer', *J. Inst. Wat. E.*, **15**, 415.

where the exponent m is dependent upon the particular pipe friction formula utilised and the coefficient r is dependent both on the pipe friction formula and on the pipe characteristics. In actual fact the exponent m is not strictly constant for the system unless the pipes are operating in the rough turbulent zone, which is unlikely. However, since the range of velocity is not large and in view of unavoidable limitations in the accuracy of the basic data, it is generally reasonable to assume that m is constant throughout.

For the Darcy-Weisbach formula with λ constant, $m = 2$ and $r = \lambda L/12 \cdot 1 D^5$. For the exponential type formulae quoted in Ch. 5, Sect. 5.8, the appropriate values and expressions for m and r are tabulated below:

Table 6.1
Empirical Formulae for Pipe Friction with m and r Values

Formula	h_f	m	r
Blasius: $V = 75\,D^{5/7}S_f^{4/7}$	$\dfrac{LQ^{1\cdot75}}{1253\,D^{4\cdot75}}$	$1\cdot75$	$\dfrac{L}{1253\,D^{4\cdot75}}$
Hazen-Williams: $V = 0\cdot354\,C_H D^{0\cdot63}S_f^{0\cdot54}$	$\dfrac{10\cdot67LQ^{1\cdot85}}{C_H^{1\cdot85}\,D^{4\cdot87}}$	$1\cdot85$	$\dfrac{10\cdot67L}{C_H^{1\cdot85}\,D^{4\cdot87}}$
Manning: $V = \dfrac{0\cdot397}{n}\,D^{2/3}S_f^{1/2}$	$\dfrac{10\cdot31n^2LQ^2}{D^{5\cdot33}}$	2	$\dfrac{10\cdot31n^2L}{D^{5\cdot33}}$

In waterworks practice the Hazen-Williams formula is well favoured and the expression for head loss with the more customary terminology (Q in l/s and d in mm) is

$$h_f = \frac{12\cdot25 \times 10^9 LQ^{1\cdot85}}{C_H^{1\cdot85}d^{4\cdot87}} \qquad (6.21)$$

Charts and nomographs facilitate computation.

When the standard λ-R diagram is consulted in order to assess values of λ likely to pertain in the various components of the system, it often leads to simplification if a single exponential expression is derived which is a good approximation to the resistance law throughout the range of conditions which are applicable. Ex. 6.2 illustrates one way in which this may be done.

Example 6.2

The velocities in a system of 152 mm (6 in.) diameter pipe are expected to vary between 0·6 and 1·5 m/s. The corresponding range of λ is estimated to be 0·022 to 0·018. From these data determine an exponential expression for the head loss per 1000 m of pipe.

$$h_t = \frac{\lambda L Q^2}{12 \cdot 1 \, D^5} = rQ^m \quad \text{so that} \quad \left(\frac{Q_2}{Q_1}\right)^m = \frac{\lambda_2}{\lambda_1}\left(\frac{Q_2}{Q_1}\right)^2$$

Substituting $Q_2/Q_1 = V_2/V_1 = 2\cdot5$ and $\lambda_2/\lambda_1 = 0\cdot818$, we obtain $m = 1\cdot78$, whence $r = 8160$. Thus the head loss per 1000 m of pipe is given by $h_t = 8160 \, Q^{1\cdot78}$.

6.5.4 *Hardy Cross Method of Analysis*

Apart from compliance with the laws of pipe resistance the two basic hydraulic conditions which must be fulfilled in any pipe system are:

(a) For continuity of flow, the algebraic sum of the flow in the pipes meeting at any junction point must be zero, or

$$Q = 0 \tag{6.22}$$

(b) For continuity of pressure, the algebraic sum of the head losses in any closed circuit within the system must be zero, or

$$h_t = 0 \tag{6.23}$$

(*a*) *Head Balance.* Let us suppose that the pipe characteristics and the flows entering and leaving a network are known and that it is required to determine the flow in each of the pipe components. If the pressures throughout the system are also required to be determined, then the pressure head at one point in the network must also be initially known.

The procedure devised by Hardy Cross,[1] which is known as a *head balance*, is as follows:

(i) Give assumed flows, Q_a, to the various pipes in the system such that the condition represented by Eq. (6.22) is satisfied.

(ii) Calculate the value of h_{fa} for each pipe in accordance with Eq. (6.20), $h_{fa} = rQ_a{}^m$, assuming that m is constant.

(iii) Divide the pipe network into a number of closed circuits of sufficient number to ensure that each pipe is included in at least one circuit.

(iv) Determine the algebraic sum of the head losses ($\sum h_{fa}$) in each circuit. Unless it so happens, which is unlikely, that the assumed flows are correct, Eq. (6.23) will not be initially satisfied.

[1] HARDY CROSS (1936) 'Analysis of Flow in Networks of Conduits or Conductors', *Univ. Illinois Bull.* 286.

(v) Compute the value of $\sum (h_{fa}/Q_a)$ for each closed circuit without regard to sign.

(vi) Determine the correction ΔQ to be applied to the assumed flows in each circuit using the expression

$$\Delta Q = -\frac{\sum h_{fa}}{m\sum (h_{fa}/Q_a)} \qquad (6.24)$$

the derivation of which is explained below.

(vii) Revise the flows in accordance with

$$Q = Q_a + \Delta Q \qquad (6.25)$$

paying due regard to sign. As some pipes are common to two or more circuits, more than one correction may have to be applied so that a state of unbalance will normally persist.

(viii) Repeat the procedure until by convergence the desired accuracy of balance is obtained.

The derivation of Eq. (6.24) is as follows:

We have

$$h_f = rQ^m = r(Q_a + \Delta Q)^m = rQ_a{}^m\left(1 + \frac{\Delta Q}{Q_a}\right)^m$$

Expanding by the binomial theorem,

$$h_f = rQ_a{}^m\left[1 + m\frac{\Delta Q}{Q_a} + \frac{m(m-1)}{2!}\left(\frac{\Delta Q}{Q_a}\right)^2 + \cdots\right]$$

Assuming that ΔQ is small compared with Q_a, in other words that the initial assumed values for the flows are nearly correct, the third and succeeding terms may be neglected, with the result that

$$h_f = rQ_a{}^m\left(1 + m\frac{\Delta Q}{Q_a}\right)$$

and for the circuit, $\sum h_f = \sum rQ_a{}^m + m\Delta Q \sum rQ_a^{m-1}$, where ΔQ is outside the numerator sign because it is the same for each pipe in the circuit. Now, one of the basic conditions is $\sum h_f = 0$ so that

$$m\Delta Q \sum rQ_a^{m-1} = -\sum rQ_a{}^m$$

or

$$\Delta Q = -\frac{\sum rQ_a{}^m}{m\sum (rQ_a{}^m/Q_a)} = -\frac{\sum h_{fa}}{m\sum (h_{fa}/Q_a)}$$

which is in the form of Eq. (6.24).

The number of successive approximations which is required is dependent largely on the margins of error in the initial first estimates of flow.

With good judgement, gained from experience, it is generally found that three or four adjustments suffice, although for intricate networks the number is much greater. There is little point in working to fine limits since the results obtained cannot be more accurate than the basic data, which of necessity are often somewhat uncertain. The numerical work involved can be extensive and some orderly and systematic method of tabulation is essential. This is best illustrated by means of the following example.

Example 6.3

Determine the flow in the various pipes of the network shown. Inflow is 90 l/s at A, and outflow is 60 l/s at D and 30 l/s at F. Take $C_H = 100$ in the Hazen-Williams formula.

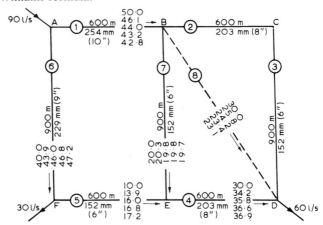

Ex. 6.3

From the Hazen-Williams formula,

$$r = \frac{12 \cdot 25 \times 10^9}{100^{1 \cdot 85}} \left(\frac{L}{d^{4 \cdot 87}} \right) = 2 \cdot 44 \times 10^6 \left(\frac{L}{d^{4 \cdot 87}} \right)$$

Tabulating for r:

Pipe	L(m)	d(mm)	$d^{4 \cdot 87} \times 10^{-9}$	$r \times 10^3$
1	600	254	514·6	2·85
2	600	203	172·8	8·49
3	900	152	42·2	52·10
4	600	203	172·8	8·49
5	600	152	42·2	34·73
6	900	229	310·8	7·08
7	900	152	42·2	52·10

124

1st adjustment / 2nd adjustment

Circuit	Pipe	$r \times 10^3$	Q_{a1}	$Q_{a1}^{1.85}$	h_{fa1}	$\dfrac{h_{fa1}}{Q_{a1}}$	ΔQ_1	$\Delta Q'$	Q_{a2}	$Q_{a2}^{1.85}$	h_{fa2}	$\dfrac{h_{fa2}}{Q_{a2}}$	ΔQ_2	$\Delta Q'_2$	Q (l/s)
I	1	2·85	50	1390	3·96	0·079	−3·9		46·1	1196	3·41	0·074	−2·1		
	7	52·10	20	255	13·29	0·665	−3·9		20·3	262	13·65	0·672	−2·1		
	5	34·73	−10	−71	−2·47	0·247	−3·9		−13·9	−130	−4·51	0·324	−2·1		
	6	7·08	−40	−920	−6·52	0·163	−3·9	+4·2	−43·9	−1093	−7·74	0·176	−2·1	+1·6	
				$\Sigma = +8·26$		$\Sigma = 1·154$				$\Sigma = +4·81$		$\Sigma = 1·246$			
II	8	60·59	30	541	32·78	1·093	−4·2		25·8	409	24·78	0·961	−1·6		
	4	8·49	−30	−541	−4·59	0·153	−4·2		−34·2	−688	−5·84	0·171	−1·6		
	7	52·10	−20	−255	−13·29	0·665	−4·2	+3·9	−20·3	−262	−13·65	0·672	−1·6	+2·1	
				$\Sigma = +14·90$		$\Sigma = 1·911$				$\Sigma = +5·29$		$\Sigma = 1·804$			

3rd adjustment / 4th adjustment

Circuit	Pipe	$r \times 10^3$	Q_{a3}	$Q_{a3}^{1.85}$	h_{fa3}	$\dfrac{h_{fa3}}{Q_{a3}}$	ΔQ_3	$\Delta Q'_3$	Q_{a4}	$Q_{a4}^{1.85}$	h_{fa4}	$\dfrac{h_{fa4}}{Q_{a4}}$	ΔQ_4	$\Delta Q'_4$	Q (l/s)
I	1	2·85	44·0	1098	3·13	0·071	−0·8		43·2	1061	3·02	0·070	−0·4		42·8
	7	52·10	19·8	252	13·13	0·663	−0·8		19·8	252	13·13	0·663	−0·4		19·7
	5	34·73	−16·0	−169	−5·87	0·367	−0·8		−16·8	−185	−6·43	0·383	−0·4		−17·2
	6	7·08	−46·0	−1192	−8·44	0·183	−0·8	+0·8	−46·8	−1230	−8·71	0·186	−0·4	+0·3	−47·2
				$\Sigma = +1·95$		$\Sigma = 1·284$				$\Sigma = +1·01$		$\Sigma = 1·302$			
II	8	60·59	24·2	363	21·99	0·909	−0·8		23·4	341	20·66	0·883	−0·3		23·1
	4	8·49	−35·8	−749	−6·36	0·178	−0·8		−36·6	−781	−6·63	0·181	−0·3		−36·9
	7	52·10	−19·8	−252	−13·13	0·663	−0·8	+0·8	−19·8	−252	−13·13	0·663	−0·3	+0·4	−19·7
				$\Sigma = +2·50$		$\Sigma = 1·750$				$\Sigma = +0·90$		$\Sigma = 1·727$			

Check for head balance A to D: (1)–(8): $2·98+20·18 = 23·16$; (1)–(7)–(4), $2·98+12·92+6·72 = 22·62$; (6)–(5)–(4), $8·84+6·70+6·72 = 22·26$.

Pipes (2) and (3) are supposedly replaced by an equivalent pipe (8) such that $h_{f8} = h_{f2} + h_{f3}$, or $r_8 = r_2 + r_3 = 60.59 \times 10^{-3}$

The pipe discharges comprise six unknowns. In order to include all the pipes, two circuits for head balance are required – these are taken as (1)–(7)–(5)–(6) and (8)–(4)–(7). Tentative discharges are assigned (clockwise positive, anti-clockwise negative) and the Hardy Cross step-by-step procedure is tabulated (p. 125).

(b) *Quantity Balance.* A modified procedure due to R. J. Cornish[1] may be adopted when the pressure head at each entry point is given and it is required to determine the pressure heads and flows throughout the network. The procedure, known as a *quantity balance*, is as follows:

  (i) Make a reasonable assumption as to the pressure head, h_a, at each junction point where the pressure is not initially known.

  (ii) Select one of these junction points and record the value of h_{fa} for each pipe connected to the junction.

  (iii) Calculate the corresponding flow, Q_a, in accordance with Eq. (6.20).

  (iv) Unless the pressure heads have been correctly assessed in the first instance, which is unlikely, the algebraic sum of the flows at the junction point will not be zero. Record the excess or deficiency of inflow $\sum Q_a$.

  (v) Compute the value of $\sum (Q_a/h_{fa})$ for the various pipes at the junction without regard to sign.

  (vi) Determine the correction Δh to be applied to the head at the junction from the expression

$$\Delta h = \frac{m \sum Q_a}{\sum (Q_a/h_{fa})} \qquad (6.26)$$

the derivation of which is explained below.

  (vii) Amend the pressure head at the junction in accordance with

$$h = h_a + \Delta h \qquad (6.27)$$

  (viii) Obtain amended pressure heads for the remaining junction points in the network in turn, taking advantage of all previous amendments.

  (ix) As the head corrections cannot be applied simultaneously, a state of unbalance will persist and the procedure must be repeated until the required accuracy of balance is obtained.

---

[1] CORNISH, R. J. (1939–40) 'The Analysis of Flow in Networks of Pipes', *J. Inst. C.E.*, **13**, 147.

The derivation of Eq. (6.26) is as follows:

Since $h_{\mathrm{fa}} = rQ_{\mathrm{a}}^{m}$, we can write $h_{\mathrm{fa}} + \Delta h = r(Q_{\mathrm{a}} + \Delta Q)^{m}$. Expanding the bracket term as before, $h_{\mathrm{fa}} + \Delta h = rQ_{\mathrm{a}}^{m} + m\Delta Q\, rQ_{\mathrm{a}}^{m}/Q_{\mathrm{a}}$, from which

$$\Delta h = \frac{m\,\Delta Q}{Q_{\mathrm{a}}/h_{\mathrm{fa}}}$$

We do not know the value of ΔQ for each individual pipe, but only for all the pipes connected to a junction, so that

$$\Delta h = \frac{m\sum Q_{\mathrm{a}}}{\sum(Q_{\mathrm{a}}/h_{\mathrm{fa}})}$$

which is in the required form of Eq. (6.26).

As with the head balance, the number of adjustments required depends on the accuracy of the initial assumptions. When both pressures and flows appear in the basic data the analysis may proceed via head or quantity balance whichever is the more convenient.

Example 6.4

The elevation and pressure heads at junction points in the pipe network of Ex. 6.3 are as follows:

Junction	Elevation head (m)	Pressure head (m)
A	155	34
C	150	—
D	146	28
F	152	32

The inflow and draw-off points are at A, D, and F, as previously.

Assuming $C_{\mathrm{H}} = 100$ in the Hazen-Williams formula, determine the flow in the various pipes of the system.

Sketch the pressure contours on the assumption that the pipe gradient between the given elevation heads is uniform.

The sum of the pressure and elevation heads, that is the hydraulic grade line elevation, is known for the junction points A, D, and F, and the respective values are inserted in the layout diagram. Also the flow in pipe (6) may be directly determined since $h_{\mathrm{f6}} = 5$ m and $r_6 = 7\cdot08 \times 10^{-3}$, so that $Q_6 = (5 \times 10^3/7\cdot08)^{1/1\cdot85} = 34\cdot7$ l/s.

There are two effective junction points (B and E) where the grade line elevation is unknown and the Hardy Cross procedure is applied to these points with a view to obtaining the necessary quantity balance. Tentative heads are assigned and tabulation is as follows overleaf.

Junction	Pipe	$r \times 10^3$	1st adjustment h_{a1}	h_{fa1}	Q_{a1}	$\dfrac{Q_{a1}}{h_{fa1}}$	Δh_1	2nd adjustment h_{a2}	h_{fa2}	Q_{a2}	$\dfrac{Q_{a2}}{h_{fa2}}$	Δh_2	h (m)	h_r (m)	Q (l/s)
B	1	2·85	186·0	3·0	43·0	14·33	+1·2	187·2	1·8	32·6	18·11	-0·1	187·1	1·9	33·6
	7	52·10		-7·0	-14·1	2·01			-8·4	-15·6	1·86			-8·3	-15·5
	8	60·59		-12·0	-17·4	1·45			-13·2	-18·4	1·39			-13·1	-18·3
				$\Sigma = +11·5$		$\Sigma = 17·79$			$\Sigma = -1·4$		$\Sigma = 21·36$				
E	7	52·10	179·0	8·2	15·4	1·88	-0·2	178·8	8·3	15·5	1·87	0	178·8	8·3	15·5
	5	34·73		5·0	14·7	2·94			5·2	15·0	2·88			5·2	15·0
	4	8·49		-5·0	-31·4	6·28			-4·8	-30·7	6·40			-4·8	-30·7
				$\Sigma = -1·3$		$\Sigma = 11·10$			$\Sigma = -0·2$		$\Sigma = 11·15$				

N.B.   Flow towards a junction is positive, flow away is negative

128

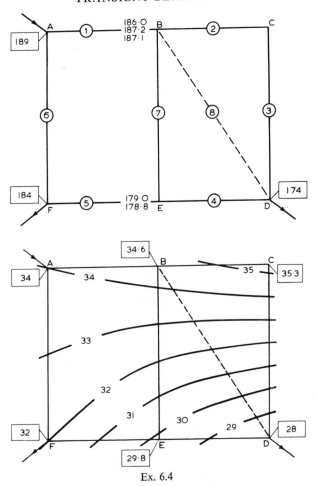

Ex. 6.4

The friction head loss in pipe (2) is $(8.49/60.59) \times 13.1 = 1.8$ m so that the grade line elevation at C is 185·3 m. The pressure head is therefore **35·3** m.

The pressure head at B is $187.1 - 152.5 = \mathbf{34.6}$ m and at E it is $178.8 - 149.0 = \mathbf{29.8}$ m.

Contours of pressure are sketched in by interpolation.

6.6 Transient Behaviour

6.6.1 General Considerations

In those cases of unsteady flow where the rate of change of discharge is appreciable the inertia of the water column can no longer be neglected since significant pressures of a transient nature are developed. The phenomenon is known as *water hammer* and is so called because of the noise and vibration with which it is sometimes accompanied.

Clearly, when designing a pipeline it is important to ensure that it will be capable of withstanding the maximum or minimum (sub-atmospheric) pressures resulting from valve adjustment and other possible actions. It is fairly obvious that water hammer pressures have potentially the greatest magnitude in those cases where a pipeline is long and velocity high. Fortunately, in the design of a water supply distribution system it is rarely necessary to make any special allowance for water hammer. This is because, in general, pipe velocities are relatively low, the numerous branches have a cushioning effect on pressure fluctuation, and the valve operating mechanism precludes abrupt flow adjustment. Consequently, the difference (usually 100 per cent) between working and test pressures is normally regarded as providing an ample margin of safety.

A pipeline supplying a turbine plant and a long pumping main are instances where careful consideration needs to be given to possible transient pressures. Velocities in turbine penstocks are relatively high (3–6 m/s) and it is the function of the turbine gates to regulate the flow so that the output at all times accords with the variable electrical demand. The potentially most dangerous transient condition is when a total rejection of load (e.g. due to transmission failure) suddenly occurs during a period of peak output as a result of which the turbine gates immediately move to the closed position in order to shut off the flow. Unless the pipeline is very short the abrupt arresting of the flow will induce extremely high pressures which it would be quite uneconomical to cater for. A practical solution is to limit the pressure rise to reasonable proportions (usually about 25 per cent) by the provision of an automatic pressure relief valve and diversion conduit at the turbine installation.

In the case of a very long pipeline (and 3–5 km is not unusual) the effective length may be considerably reduced by the provision of a large capacity regulating tank, called a surge tank, located as near as possible to the turbine (Fig. 6.7). On a sudden load rejection, the continuing discharge from the low pressure conduit is prevented from flowing down the penstock and is diverted into the surge tank. The consequent rise in water level within the tank applies the necessary retarding force to the water column, which, after a period of damped oscillation, finally reaches static equilibrium. Likewise, on a sudden starting up of the turbine, water flows temporarily out of the surge tank into the penstock, the resultant lowering of the tank water level imparting a desirable acceleration to the draw-off from the reservoir.

Pressure surges in pumping mains are generally attributable to a sudden power failure. The cessation of flow is not immediate since the inertia of the rotating parts and the water column must first be absorbed. Pressure is lowered during the short period of deceleration and in particularly adverse circumstances (local high points merit special attention) pressures may fall below that of the vapour and result in a parting of the water column. The subsequent reuniting of the column when the flow reverses is conducive to impact pressures of considerable magnitude. With long pipelines it is customary to install a reflux valve of the flap type adjacent to the pump. On flow reversal the valve slams shut

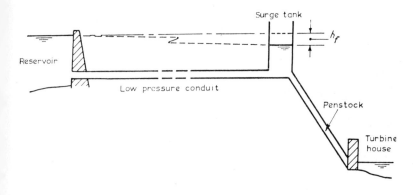

Figure 6.7   Long hydro-electric pipeline with surge tank

thus protecting the pump from the rise of pressure consequent upon the return surge. The provision of relief valves and air chambers serves to limit the maximum pressures to which the pipeline is subjected. Another possibility is to increase the inertia of the rotating parts by the addition of a flywheel.

The determination of water hammer pressures is amenable to mathematical analysis. Owing to the many variables, however, practical problems may be of considerable complexity. In the present text, therefore, our consideration must necessarily be limited to the most elementary cases. The simplest procedure is to regard the water as incompressible, but this assumption leads to appreciable error when flow adjustment is rapid. Elasticity of water and pipeline must then be taken into account.

6.6.2 *Incompressible Theory*

The water column is assumed to behave as a rigid rod so that any change brought about at one end of a pipeline is immediately felt at the other to the same extent. Fig. 6.8 shows a uniform pipeline, length L and sectional area A, connected to a reservoir (or surge tank). Discharge is controlled by a valve at the downstream end, the mass of water in motion being ρAL. During a period of flow adjustment the instantaneous velocity is V and the retardation $-dV/dt$ (negative because $+dV/dt$ is an acceleration). Thus in accordance with Newton's second law the pressure force developed at the valve is given by

$$\Delta p\, A = -\rho AL \frac{dV}{dt}$$

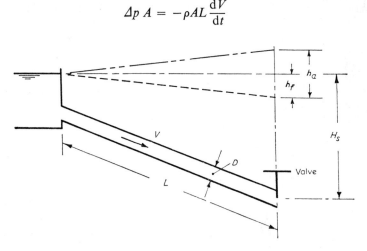

Figure 6.8   Flow adjustment in a uniform pipeline (incompressible theory)

where Δp is the surge pressure that is superimposed on the normal pressure. The dynamic, or acceleration, head h_a ($= \Delta p/\rho g$) at the valve is therefore

$$h_a = -\frac{L}{g}\frac{dV}{dt} \tag{6.28}$$

and diminishes linearly towards the reservoir.

Taking into account pipe friction losses, the instantaneous pressure head h at the valve during the period of flow adjustment is given by

$$h = H_s - \frac{\lambda L V^2}{2gD} - \frac{L}{g}\frac{dV}{dt} \tag{6.29}$$

Provided that the retardation is known, this equation may be solved.

Thus if a steady flow velocity V_0 is reduced to zero at a uniform rate during a period t_c, then the maximum pressure head at the valve occurs at the instant of total closure and is

$$h_{max} = H_s + \frac{LV_0}{gt_c} \qquad (6.30)$$

Actually, to achieve uniform retardation would ordinarily necessitate a very complex sort of valve movement.

From Eq. (6.28) it will be observed that when closure is instantaneous the pressure rise is infinite. Although in practice instantaneous closure is a physical impossibility, nevertheless very rapid closures can be effected and the experimental evidence from these shows that the dynamic pressures developed are much lower than the theoretical. The discrepancy is due to the disregarding of the elastic properties. Some indication of the practical limit of validity of the incompressible theory is given by the fact that when retardation is linear and $t_c > L/60$ (t_c in seconds, L in metres), the calculated pressure rise according to both theories is approximately the same.

A procedure for determining the discharge of a pipeline – the Gibson method – is based on the incompressible theory and entails the autographic recording of the pressure rise at a valve whilst closure is gradually effected. The procedure is described in Appendix B of B.S.353:1962, *Methods of Testing Water Turbine Efficiency*.

When the valve at the end of the pipeline (Fig. 6.8) is opened pressure remains constant at both ends whilst the discharge progressively increases. Neglecting friction, the velocity at any instant is equal to $\sqrt{2gh}$. Substituting in Eq. (6.29),

$$\frac{V^2}{2g} = H_s - \frac{L}{g}\frac{dV}{dt} \qquad (6.31)$$

It will be observed that the maximum dynamic head is when $V = 0$, that is to say at the instant of opening.

From Eq. (6.31),

$$dt = \frac{L}{g}\left(\frac{1}{H_s - V^2/2g}\right)dV$$

$$= \frac{L}{\sqrt{2gH_s}}\left(\frac{1}{\sqrt{2gH_s} + V} + \frac{1}{\sqrt{2gH_s} - V}\right)dV$$

Integrating,

$$t = \frac{L}{\sqrt{2gH_s}}\ln\frac{\sqrt{2gH_s} + V}{\sqrt{2gH_s} - V} \qquad (6.32)$$

Now $\sqrt{2gH_s}$ is the ultimate steady velocity V_0 and therefore the equilibrium condition ($V = V_0$) is approached asymptotically, being only attained at time infinity. In practice, of course, owing to the elastic and viscous properties, the time required for the establishment of steady flow is finite. An approximation is given by the value of t corresponding to $0.95V_0$.

Example 6.5

A pipeline 15 m long and 229 mm (9 in.) diameter terminates in a nozzle 76 mm (3 in.) diameter. The static head at the nozzle is 9 m. If the control valve adjacent to the nozzle is suddenly opened, estimate the length of time for steady flow to be established. Neglect friction and assume $C_v = 1.0$ for the nozzle.

The ultimate velocity of the jet is $\sqrt{19.61 \times 9} = 13.3$ m/s and the corresponding pipe velocity is $13.3/9 = 1.48$ m/s. In Eq. (6.32) the value of H_s must be based on V_0 since the remainder of the head is required for accelerating the flow in the nozzle. Thus

$$t = \frac{15}{1.48} \ln \frac{1.48 + 0.95 \times 1.48}{1.48 - 0.95 \times 1.48} = 37 \text{ s}$$

6.6.3 *Elastic Theory*

Because of the elastic properties of the fluid the pressure and velocity changes produced by some disturbance do not affect all points in a pipeline simultaneously. What happens in fact is that a pressure wave is propagated along the pipeline at an extremely rapid but nevertheless finite rate. Until the wave reaches a particular point the original steady flow conditions at that point are maintained. An analogy may be drawn with a train of loose-coupled trucks. On starting, the trucks come successively under the influence of the draw-bar pull of the engine and there is some time lag before the last truck commences to move. The extensibility of the train serves to reduce the accelerative pull that would otherwise be required.

There are two cases to be considered – a rigid pipeline and an elastic pipeline.

(A) RIGID PIPELINE. The diagram (Fig. 6.9) shows the downstream end of a rigid pipeline. Under conditions of steady flow the velocity is V_0 and the pressure p. A sudden closure of the valve is effected and as a result a pressure wave is propagated upstream at celerity[1] c. Flow is

---

[1] We refer to the celerity of a pressure wave in order to make some distinction with the velocity of pipe flow which is extremely small by comparison. It follows that the relative ($c + V_0$) and absolute (c) wave celerities are virtually the same.

arrested as the pressure wave passes and due to transformation of kinetic energy the pressure and density are increased to $p + \Delta p$ and $\rho + \Delta\rho$ respectively. Thus in one unit of time a mass $\rho A V_0$ is compressed into a mass $\Delta\rho\, Ac$ so that

$$\rho V_0 = \Delta\rho\, c \qquad (6.33)$$

Again, in one unit of time a column of water, volume cA, moving at V_0, is brought to rest, so in accordance with Newton's second law the force produced is given by $\Delta p\, A = \rho AcV_0$, or

$$\Delta p = \rho c V_0 \qquad (6.34)$$

N. E. Joukowsky[1] propounded this relationship and gave a practical demonstration of its validity in 1897. It is known as *Joukowsky's law* and is the fundamental equation in pressure wave analysis.

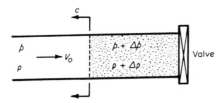

Figure 6.9   Pressure wave in a closed conduit

Eliminating V_0 by combining Eqs. (6.33) and (6.34) we obtain $c^2 = \Delta p/\Delta\rho$. But according to Eq. (2.3) (p. 7), the bulk modulus K of a fluid is represented by $\Delta p \div \Delta\rho/\rho$ so that

$$c = \sqrt{\frac{K}{\rho}} \qquad (6.35)$$

By inserting appropriate values for K and ρ in Eq. (6.35) the celerity of a pressure wave in any given medium may be determined. Since sound is transmitted by means of a pressure wave the expression also gives the speed of sound in the medium. Over the range of conditions normally pertaining, the values of K and ρ for water are 2150 N/mm² and 1000 kg/m³ respectively, so that $c = 1440$ m/s. It is interesting to note that this is approximately four times the speed of sound in air calculated on the same basis.

Substituting in Eq. (6.34) the value of the pressure rise Δp (in kN/m² above the normal) due to the passage of a compression wave in water is

$$\Delta p = 1440V_0 \qquad (6.36)$$

[1] SIMIN, O. (1904) 'Water Hammer, with special Reference to Researches of Prof. Joukowsky', *Proc. Am. Water Works Assoc.*, **24**.

– in other words 1440 kN/m² for every m/s of velocity that is arrested. It will be noted that Δp is independent of the length of pipeline, but as will be shown the latter does have a most important influence on the pressure developed in an actual case.

Owing to the very rapid advance of the pressure wave there is no heat generated during compression so the process may be regarded as adiabatic. Thus another approach is afforded – one which utilises the principle of conservation of energy. Equating, per unit of time, the kinetic energy lost to the strain energy gained, we obtain

$$\frac{\rho A c V_0^2}{2} = \frac{\Delta p^2 A c}{2K}$$

whence as before

$$\Delta p = V_0 \sqrt{K\rho} = \rho c V_0 \tag{6.37}$$

(B) ELASTIC PIPELINE. Elasticity of the pipe material has the effect of reducing the celerity of the compression wave and therefore the pressure rise. As will be demonstrated, these modifications are far from negligible. When applying the principle of energy conservation to the instantaneous closure problem depicted in Fig. 6.9 we must now make allowance for the strain energy gained by the pipe. If we assume that the pipeline is constrained in the longitudinal direction whilst free to expand circumferentially, then the strain energy developed per unit of time is $f_t^2 \pi D \Delta D \, c/2E$, where f_t is the increased circumferential stress, E the Young's modulus, and D and ΔD are the internal diameter and wall thickness, respectively. Since f_t is equal to $\Delta p \, D/2\Delta D$ the strain energy gained is $\Delta p^2 \pi D^3 c/8E\Delta D$.

Now, for each unit of time:

K.E. lost by water = S.E. gained by water + S.E. gained by pipe

or

$$\frac{\rho \pi D^2 c V_0^2}{8} = \frac{\Delta p^2 \pi D^2 c}{8K} + \frac{\Delta p^2 \pi D^3 c}{8E\Delta D}$$

from which

$$\Delta p = \rho V_0 \sqrt{\frac{K}{\rho} \left(\frac{1}{1 + (K/E)(D/\Delta D)} \right)} \tag{6.38}$$

A comparison of Eqs. (6.37) and (6.38) shows that the combined bulk modulus is less than the fluid bulk modulus in accordance with

$$K_p = \frac{K}{1 + (K/E)(D/\Delta D)} \tag{6.39}$$

Similarly, the reduced wave celerity is given by

$$c_p = \frac{c}{\sqrt{1 + (K/E)(D/\Delta D)}} \qquad (6.40)$$

If we consider a typical case – a cast iron ($K/E = 0 \cdot 02$) pipeline with the ratio $D/\Delta D = 100$, we find that $c_p = 0 \cdot 577c = 831$ m/s, indicating that appreciable over-estimation of water hammer pressures results if pipe elasticity is ignored.

We are now in a position to trace the subsequent history of the surge wave produced by instantaneous closure of the valve at the end of a pipeline connected to a reservoir. For simplicity, friction is neglected and the pipeline is assumed horizontal. The sequence of events (Fig. 6.10) is as follows:

$t = 0$ TO L/c_p. During the time that the wave of compression is travelling towards the reservoir, water is still entering the pipeline from the open end and this water serves to make good the loss in volume due to the increase of density. At time $t = L/c_p$ the whole of the water column has been brought to rest and is at a higher than normal pressure whilst the pipeline is distended throughout. This stationary condition is only momentary because pressure at the open end must remain constant but the pipeline pressure is in excess of hydrostatic. As a result, a flow of water back into the reservoir is initiated.

$t = L/c_p$ TO $2L/c_p$. As the wave front travels back at celerity c_p towards the valve, pressure is relieved behind it. On reaching the valve at time $t = 2L/c_p$, pressure is momentarily normal throughout the pipeline, but a state of disequilibrium still exists because the entire water column is moving at velocity V_0 towards the reservoir. Owing to fluid inertia, pressure at the valve instantaneously falls to below normal and a negative wave, sometimes called a wave of rarefaction, is propagated in the direction of the reservoir.

$t = 2L/c_p$ TO $3L/c_p$. When the negative wave reaches the reservoir at time $t = 3L/c_p$ the water column is again entirely at rest, but pressures throughout are subnormal and the pipe diameter has shrunk. Consequently, water starts to flow again into the pipeline.

$t = 3L/c_p$ TO $4L/c_p$. Pressure returns to normal behind the wave front, so that when the latter reaches the valve the entire water column is at normal pressure but moving towards the valve at velocity V_0. This will be recognised as the condition existing at the instant of valve closure. Thus one complete surge cycle occupies a period $t = 4L/c_p$.

Theoretically, the process is continued indefinitely with undiminished vigour, but in practice, of course, frictional damping rapidly reduces the intensity of the phenomenon. If the period of oscillation of a valve on its seating happens to coincide with that of the pipeline, resonance will occur, resulting in a continued and possibly intensified valve 'fluttering' (cf. vibration noises in domestic plumbing systems).

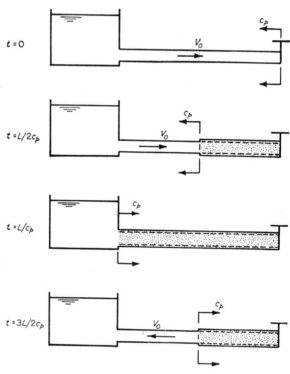

Figure 6.10   (1st portion)

From the sequence described it is evident that the end conditions have a characteristic effect on the reflection of the water hammer waves. At the reservoir end the pressure must remain constant but the velocity is free to change, whereas at the closed end the velocity remains the same (zero) but the pressure can alter. Surge waves reaching the reservoir are reflected negatively and return with the same algebraic sign, whereas waves reaching the closed end are reflected positively, thereby undergoing an instantaneous change of twice the amplitude and returning with reversed sign.

The theoretical pressure-time diagrams (neglecting velocity head) for a point adjacent to the valve and for the mid-point of the pipeline are shown in Fig. 6.11. Provided that valve closure is completed before the first pressure wave generated returns to the valve (i.e. $t_o < 2L/c_p$),

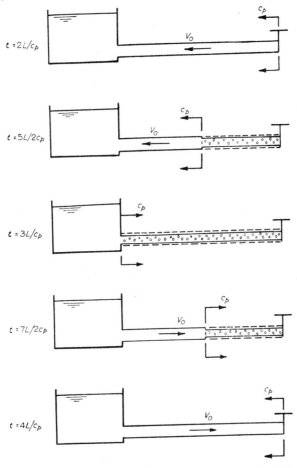

Figure 6.10 (2nd portion) Progress of pressure wave following instantaneous valve closure

the maximum pressure rise is the same as when the valve is closed instantaneously. This fact may be better appreciated if we regard the movement as taking place in a number (n) of instantaneous steps. At each step the velocity is abruptly reduced by ΔV and the pressure increased by $\Delta p/n$. The pressure rise in the first interval $2L/c_p$ is thus

139

the sum of the increments $\Delta p/n$ (Fig. 6.12). Pipeline slope has no effect on the celerity of the wave or the dynamic pressures developed, but it does of course mean that there is a variation in the normal and resultant pressures along the pipeline.

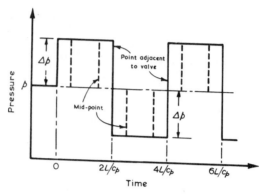

Figure 6.11   Theoretical pressure-time diagram following instantaneous closure

An examination of the theoretical and actual (Fig. 6.13) pressure-time diagrams for a near instantaneous closure would normally reveal little difference between the calculated and recorded maximum pressures. Indeed, the generally striking confirmation of the elastic theory,

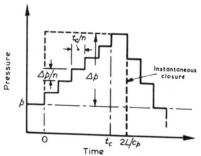

Figure 6.12   Theoretical pressure-time diagram for point adjacent to valve—closure rapid ($t_c < 2L/c_p$) but not instantaneous

afforded by experimental evidence, fully justifies the analytical approach being employed in the solution of practical design problems.

More commonly, emergency closure is effected in a period that is longer than $2L/c_p$ but less than the time for which the fluid behaviour may be reasonably regarded as incompressible. This is because the newly generated positive waves after time $t = 2L/c_p$ are offset by reflected negative waves of earlier origin. A solution of the problem involves a knowledge of valve characteristics and is beyond the scope of the present text. Arithmetical step-by-step methods may be employed, or a graphical procedure developed by L. Bergeron[1] and others. In the latter case the pressure rise at the terminal points is represented by the intersection of the wave and system characteristics. The more complex problems, involving compound and branched pipes, necessitate the aid of computers.[2]

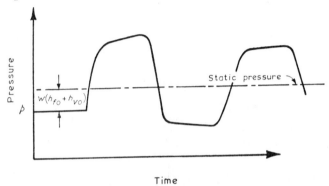

Figure 6.13 Actual pressure-time diagram for point adjacent to valve following sudden closure

Surges in pumping mains and flow establishment after rapid valve opening are other aspects of transient behaviour that are amenable to analysis by the aforementioned methods.

Example 6.6

A steel pipeline 1068 mm (42 in.) diameter conveys 2 cumecs under a head of 430 m.

What must be the thickness of the walls if the pipe is to withstand the pressure rise caused by sudden closure of a valve? Determine also the rise in pressure.

The safe stress for the steel is 0.1 kN/mm^2, $E = 207$ kN/mm^2 and $K = 2.15$ kN/mm^2. To what extent would the pressure rise be modified if the pipe was rigid?

[1] BERGERON, L. (1961) *Water Hammer in Hydraulics and Wave Surges in Electricity.* Wiley.
[2] PAYNTER, H. M. (1953) 'Surge and Water Hammer Problems', *Trans. Am. Soc. C.E.*, **118**, 962.

Under the initial steady conditions the velocity $V_0 = 2/[(\pi/4) \times 1\cdot07^2] =$ 2·23 m/s, and the pressure $p_n = 4220$ kN/m².

The maximum permissible pressure $p_{max} = (2 \times 10^5/1\cdot07)\Delta D$ kN/m². Thus the pressure rise must be limited to

$$\Delta p = 187 \times 10^3 \times \Delta D - 4220 \qquad \text{(i)}$$

Another expression for Δp is obtained from Eq. (6.38):

$$\Delta p = 2\cdot23 \times 1440\sqrt{\frac{1}{1 + 0\cdot0111/\Delta D}} \qquad \text{(ii)}$$

Combining Eqs. (i) and (ii), and solving for ΔD by trial and error or graphical means, we obtain $\Delta D = 0\cdot0378$ m, or 37·8 mm. Hence $\Delta p =$ **2840** kN/m².

For a rigid pipe, utilising Eq. (6.34), $\Delta p = 1440 \times 2\cdot23 =$ **3210** kN/m².

Further Reading

AM. SOC. C.E. and AM. SOC. MECH. E. (1933) *Symposium on Water Hammer.*

BABBITT, H. E., DOLAND, J. J., and CLEASBY, J. L. (1962) *Water Supply Engineering.* McGraw-Hill (6th Edition).

BERGERON, L. (1961) *Water Hammer in Hydraulics and Wave Surges in Electricity.* Wiley.

CAMP, T. R. and LAWLER, J. C. (1969) 'Water Distribution', Sect. 37 of *Handbook of Applied Hydraulics* (Eds. DAVIS, C. V. and SORENSEN, K. E.). McGraw-Hill (3rd Edition).

FAIR, G. M. and GEYER, J. C. (1958) *Elements of Water Supply and Waste-Water Disposal.* Wiley.

PARMAKIAN, J. (1955) *Waterhammer Analysis.* Prentice-Hall.

PICKFORD, J. (1969) *Analysis of Surge.* Macmillan.

RICH, G. R. (1963) *Hydraulic Transients.* Dover (2nd Edition).

TWORT, A. C. (1963) *A Textbook of Water Supply.* Arnold.

CHAPTER SEVEN

Uniform Flow in Channels

7.1 Introduction

Channels have a significant role in the spheres of river control, inland navigation, land drainage, irrigation, water supply, and sanitation. It is not surprising, therefore, that from the earliest days of hydraulics, engineers and scientists have sought to obtain a greater understanding of the basic laws governing channel flow.

The characteristic feature of all channel flow is the presence of a free surface. As used in the present context the word 'channel' has a wide meaning since it is deemed to embrace river channels as well as artificial channels, such as canals and flumes; also included are enclosed conduits (e.g. sewers and culverts) operating partially full. Specifically excluded are pipes operating under pressure, which always flow full. Unlike pipe flow the hydraulic gradient is coincident with the free surface of the water and the sectional area of flow expands or contracts in conformity with the discharge.

In accordance with the definition in Ch. 4, Sect. 4.2(d), the flow is *uniform* when the mean velocity from one section to another is constant. With a free surface this also implies a constant cross-section and depth. The latter is known as the *normal depth* and is dependent on the channel characteristics and discharge. Uniform flow is the result of an exact balance between the gravity and frictional forces.

Because of the free surface, uniform flow in a channel must necessarily be steady flow. If the velocity is increased to a very high value (in excess of about 6 m/s), air entrainment occurs and the flow becomes unsteady and pulsating in character. Thus very rapid flow cannot be uniform. Incidentally, at exceptional velocities (approaching 30 m/s), the increased sectional area of flow, or 'bulking' as it is called, may amount to 50 per cent.[1]

The flow is said to be *non-uniform* when the mean velocity varies throughout the length of channel under consideration. The changes in

---

1 (May 1961) 'Air Entrainment in Open Channels: Report of Task Force', *Proc. Am. Soc. C.E.*, **87**, 73.

velocity may be brought about by a variation in channel section or bed gradient, or by a hydraulic structure, such as a weir or sluice, interposed in the line of flow. All of these influence the flow behaviour for an infinite distance and as every channel must have some such feature, even if it is only at the beginning and end, uniform flow is an ideal state which is never actually attained. In the majority of cases, however, where relatively long straight channels (of constant cross-section and bed slope) are involved, the flow for the most part is so nearly uniform that the assumption of this condition is reasonable, especially as it greatly simplifies the analysis. The inherent limitations of channel formulae and the difficulties of precise discharge measurement provide further justification.

The present chapter is mainly concerned with the relationship between discharge and channel characteristics since this is the primary concern in design. The whole question of flow in channels is of great importance to the hydraulic engineer and merits a more detailed consideration than is possible in a book of this nature.

7.2 Laminar Flow

In laminar flow the viscous forces predominate and there is no eddying or transverse current. This type of flow is associated with very low velocities, small cross-sections, or viscous sluggish liquids. Such conditions are rarely encountered in normal channel flow, and even in river channels, where the velocity is sometimes low, the depth and irregularity of section are generally sufficient to ensure the turbulent state. However, laminar flow is a distinct possibility in the case of small-scale hydraulic models. As the laws of turbulent and laminar flow differ appreciably, it is important to adopt a model scale that is sufficiently large to ensure turbulent flow. Hydraulic models are discussed in Ch. 11.

The sheet run-off on paved or natural earth surfaces is generally laminar in character. Although not strictly speaking in the category of channel flow, it may be regarded as such if a wide shallow channel is envisaged. This is a simple boundary condition that is amenable to theoretical analysis in the following manner.

Fig. 7.1 shows uniform laminar flow, with vertical depth d, down a slope inclined at angle θ to the horizontal. Unless the slope is exceptionally steep (i.e. $S > 1/10$), $\cos \theta \simeq 1$ and the depth normal to the bed may be taken as d. The fluid behaviour may be depicted as a series of elemental layers sliding one over the other, the relative motion being governed by Newton's law of viscosity, namely $\tau = \mu(dv/dy)$, where τ is the shear stress, μ the dynamic viscosity, and dv/dy the velocity gradient.

We will consider the equilibrium of the column of fluid above the plane xx; for convenience the plan area is assumed unity. The component of the gravitational force parallel to the slope is $\rho g(d - y)\sin\theta$ and this is resisted by the shear force τ. Thus

$$\tau = \rho g(d - y)\sin\theta \qquad (7.1)$$

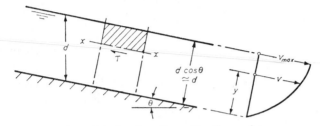

Figure 7.1   Laminar flow with a free surface

Substituting for τ and putting $\sin\theta = S$, the longitudinal slope, we obtain $\mu(dv/dy) = \rho g S(d - y)$, from which $dv = (gS/\nu)(d - y)\,dy$, where ν is the kinematic viscosity (μ/ρ). Integrating, and remembering that $v = 0$ when $y = 0$,

$$v = \frac{gSy}{\nu}\left(d - \frac{y}{2}\right) \qquad (7.2)$$

Thus the velocity distribution is parabolic and the maximum velocity occurs at the surface, being given by

$$v_{max} = \frac{gSd^2}{2\nu} \qquad (7.3)$$

The mean velocity V is obtained from

$$V = \frac{1}{d}\int_0^d v\,dy = \frac{1}{d}\int_0^d \frac{gSy}{\nu}\left(d - \frac{y}{2}\right)dy$$

or

$$V = \frac{gSd^2}{3\nu} = \frac{2}{3}v_{max} \qquad (7.4)$$

A simple calculation shows that the filament of mean velocity is located at $0.42d$ above the bed.

It should be noted that the velocity is independent of the surface roughness and this is, of course, characteristic of laminar flow. The nature of the surface is however important since excessive roughness, being conducive to eddying and turbulence, tends to disrupt the rectilinear nature of the flow. For this reason it seems likely that over natural surfaces both laminar and turbulent flow will be present.

145

7.3 Fundamental Relationships

7.3.1 *Velocity Distribution*

In earlier chapters mention has been made of the velocity gradient that exists in turbulent flow past a solid boundary. In the case of channel flow the distribution of velocity is dependent on a number of factors which include boundary configuration, surface roughness, and discharge.

Fig. 7.2 shows a typical cross-section of a natural river channel with contours of velocity superimposed. These would normally be interpolated from point velocity data obtained by traversing with a current meter. This instrument (Plate 1) is extensively used for purposes of discharge measurement in rivers and artificial channels. Its construction is such that the angular speed of the rotating element, either propeller or mounted cups, is proportional to the water velocity. An electrical

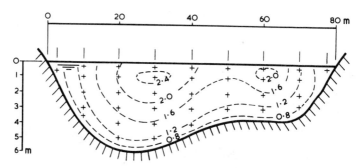

Figure 7.2   Cross-section of a river channel showing distribution of velocity (m/s)

make-and-break circuit transmits signals to a revolution counter or headphones.

In Fig. 7.2 it will be noted that the minimum velocity occurs adjacent to the bed and sides and that there is an increase in velocity towards the free surface. The filament of maximum velocity is depressed slightly below the surface and this can be attributed to the secondary circulatory motion induced by the proximity of the sides. It could therefore be reasoned that with wide shallow channels the maximum velocity would be located at the surface, and this is indeed so in practice.

A knowledge of the normal velocity distribution in a vertical is of some assistance in flow gauging by current meter, since it is more economical to record only one or possibly two representative velocities in a vertical rather than undertake a traverse for the full depth. The mean velocity in a vertical occurs at approximately 0·6 × bed depth, measured from the surface. A closer approximation is the average of the velocities at 0·2 and 0·8 × bed depth, which accords with a parabolic distribution, although there is no theoretical justification for this. As will be seen later, both approximations are nearly in compliance with a logarithmic distribution. The mean velocity varies between 0·8 and 0·95 times the surface velocity, the usual value being about 0·85.

In the transverse direction the distribution is much more irregular, except that in general the velocity adjacent to the sides is nearly always less than that nearer to the centre. Because of the inconsistent nature of the distribution, it is not possible to define the location of any single point of measurement that will yield the mean velocity over the entire cross-section. Even in the case of channels that are of relatively simple geometrical shape (e.g. rectangular), the asymmetry associated with a free surface is such that the velocity distribution is still very complex.

There is one form of channel, however, namely a wide shallow one, where a straightforward application of boundary layer theory does lead to a velocity distribution equation that gives good agreement with the facts. This will now be discussed.

According to the Kármán-Prandtl hypothesis, the equation (Eq. (5.34), p. 94) for the velocity v at distance y from a rough pipe wall is given by

$$v = 5 \cdot 75 v_* \log \frac{33y}{k}$$

where v_* $(= \sqrt{\tau_0/\rho})$ is the shear velocity and k is the effective roughness of the surface excrescences (i.e. equivalent grain size in the Nikuradse pipe experiments). This same equation is applicable to flow past a plane rough boundary such as would normally be the bed condition of a wide shallow channel. The velocity distribution with depth is seen to be logarithmic.

The mean velocity over the full depth d is given by

$$V = \frac{1}{d} \int_0^d v \, \mathrm{d}y = \frac{5 \cdot 75 v_*}{d} \int_0^d \log \frac{33y}{k} \, \mathrm{d}y$$

from which

$$V = 5 \cdot 75 v_* \log \frac{12 \cdot 1 d}{k} \qquad (7.5)$$

and the height y_m above the bed at which the filament velocity has the mean value is obtained from $\log (33 y_m/k) = \log (12 \cdot 1 d/k)$, or

$$y_m = 0 \cdot 37 d \qquad (7.6)$$

which compares very favourably with the conventional $0 \cdot 4d$ (above the bed).

It is interesting to extend the comparison to the average of the velocities at $0\cdot2d$ and $0\cdot8d$.

At $0\cdot2d$:

$$(v)_{0\cdot2d} = 5\cdot75v_* \log \frac{6\cdot6d}{k}$$

and at $0\cdot8d$:

$$(v)_{0\cdot8d} = 5\cdot75v_* \log \frac{26\cdot4d}{k}$$

giving an average velocity

$$V' = 5\cdot75v_* \log \frac{13\cdot2d}{k} \tag{7.7}$$

which compares reasonably well with Eq. (7.5). Thus the extension to channel flow of the Kármán-Prandtl hypothesis, with its sound theoretical basis, clearly gives a very fair representation of conditions in practice.

7.3.2 λ-R *Relationship for Channels*

In the consideration of the fundamentals of pipe flow (Ch. 5) appropriate expressions for the friction factor λ were derived and it was shown that these could be depicted on a single diagram known as the standard λ-R diagram.

Obviously, it would be of advantage if these expressions could be suitably adapted for relevance to channel flow. There is some justification for this approach, since in the case of a closed conduit and a very gradual increase in discharge it is found that the transition from the partially full to the full condition is not abrupt.

Firstly, an appropriate term must be substituted for the diameter D in the Reynolds number (VD/v) for pipes. This term must be dimensionally similar and the ratio sectional area of flow: sectional wetted perimeter, $A:P$, known as the *hydraulic radius* or *hydraulic mean depth*, R, is a logical length characteristic. Moreover, it has the significant merit of being related to the shape. Thus the Reynolds number for channels is VR/v; also since $R = D/4$, the relationship between R for channels and R for pipes is $R_{channel} = R_{pipe}/4$.

The *Darcy-Weisbach equation* ($h_f = \lambda LV^2/2gD$) is modified to

$$\lambda = \frac{8gRS}{V^2} \tag{7.8}$$

The *smooth channel law* (cf. Eq. (5.28), p. 90) becomes

$$\frac{1}{\sqrt{\lambda}} = 2 \log \frac{R\sqrt{\lambda}}{0\cdot627} \tag{7.9}$$

The *rough channel law* (cf. Eq. (5.33), p. 93) becomes

$$\frac{1}{\sqrt{\lambda}} = 2 \log \frac{14\cdot 8R}{k} \qquad (7.10)^1$$

And the composite *transition law* (cf. Eq. (5.38), p. 96) is

$$\frac{1}{\sqrt{\lambda}} = -2 \log \left(\frac{k}{14\cdot 8R} + \frac{0\cdot 627}{R\sqrt{\lambda}} \right) \qquad (7.11)$$

When considering the validity of these logarithmic formulae it is important to remember that their derivation, though logical, is somewhat artificial and thus lacks the sound analytical base of the parent formulae. The presence of a free surface and the pronounced influence of channel shape are complicating factors which make any theoretical approach exceedingly difficult.

Because of the additional variables the λ-R diagram (Fig. 7.3) is not of such utility as is the case in pipe flow, where the diameter or hydraulic radius is constant for a given pipe.

For laminar flow the λ-R relationship (on a log plot) is represented by a straight line. This is because λ is inversely proportional to R and in the simplest case, that of a wide shallow channel, the full equation may be derived mathematically by making use of Eqs. (7.4) and (7.8). Thus

$$S = \frac{3\nu V}{gR^2} = \frac{\lambda V^2}{8gR}$$

so that

$$\lambda = \frac{24}{R} \qquad (7.12)$$

For other channel shapes the numerical term will differ somewhat.

The transition between laminar and turbulent flow in pipes is found to extend over the range R = 2000 to R = 4000. The equivalent range in channel flow would be for R = 500 to R = 1000. Actually, this assessment of the lower limit is reasonable but the upper limit is somewhat low. An arbitrary upper limit of 2000 seems appropriate, but it is by no means clearly defined.

These limiting Reynolds numbers serve to give some useful indication of the kind of situation in which the flow is likely to be laminar or

---

[1] Substituting $\lambda = 8\tau_0/\rho V^2 = 8v_*^2/V^2$ in Eq. (7.10) we obtain $V/\sqrt{8}v_* = 2 \log (14\cdot 8R/k)$, which for a wide shallow channel becomes $V = 5\cdot 66v_*$ log $(14\cdot 8d/k)$. The small discrepancy with Eq. (7.5) arises from the adjustment of the constants in the development of the pipe friction factor formula.

turbulent. Thus in the case of sheet run-off, 2·5 mm deep, the minimum value of λ for laminar flow is 0·048 and the surface slope must not be steeper than $1/79$ $(S = 3\nu^2 R/gR^3)$. Paved surfaces (e.g. the transverse fall on roads) are generally in this category. Again, in the case of flow in a wide shallow channel, say 1 m deep, the flow will be turbulent provided that the velocity exceeds 0·002 m/s. This is an extremely low velocity

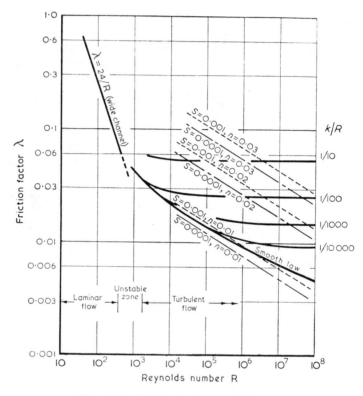

Figure 7.3　λ-R diagram for channel flow

and in any normal watercourse or well-designed channel it will almost invariably be well exceeded.

In the turbulent zone the diagram exhibits the characteristic smooth and rough law features. However, because channel surfaces in general, and the surfaces of channels in natural material in particular, are rougher than the surface of pipe walls, the rough law formula is much more relevant. It will be noted that in this region the lines on the diagram are horizontal indicating that the friction factor is independent of the Reynolds number.

These logarithmic formulae are not amenable to straightforward application in a design office. For rapid evaluation the exponent type of formula ($V = cR^{z}k^{y}S^{z}$) is to be preferred, since it is well suited to representation on charts and nomographs.

7.4 Empirical Formulae

7.4.1

From the earliest days of hydraulic science engineers have recognised the need for a flow formula that would assist them in the design of channels. For such a formula to be acceptable in practice it must be relatively simple to apply and reasonably accurate. The history of development is one of steady progress in fulfilling this requirement.

Early investigators were handicapped by the lack of data, but this was remedied in the late 19th century when more satisfactory methods of discharge measurement resulted in a wide range of detailed information becoming available. A number of empirical formulae have been published and have enjoyed varying periods of favour. Only the three most important are considered here – the Chézy formula, the Kutter formula, and the Manning formula. Of these the Manning formula is the one that is now regarded as the most appropriate.

7.4.2 *Chézy Formula*

This formula was established by Chézy in 1775, and forms the basis of most of those that were put forward later. It may be derived by dimensional analysis (see Ch. 4, Sect. 4.12) or in the following manner.

Consider the uniform flow of water between two cross-sections *aa* and *bb* in a channel with constant section and bed slope. In Fig. 7.4, L is the distance between the cross-sections, θ the inclination of the bed, V the mean velocity, A the sectional area of flow, and P the wetted perimeter.

The weight of the water prism is wAL and the component parallel to the bed $wAL \sin \theta$.

Now, as the resistance force = shear stress × wetted area of contact, we can write

$$\text{resistance force} = \tau_0 PL$$

where τ_0 is the shear stress at the boundary. Since the shear stress is not constant over the surface, τ_0 must be the mean value.

For uniform flow the gravity and resistance forces are in equilibrium, so that

$$\tau_0 PL = wAL \sin \theta \tag{7.13}$$

The shear stress is assumed proportional to the square of the mean velocity, or $\tau_0 = \kappa V^2$, where κ is a constant coefficient. Substituting for τ_0 in Eq. (7.13) we obtain

$$V^2 = \frac{w}{\kappa} \frac{A}{P} \sin \theta \tag{7.14}$$

Since A/P is the hydraulic radius R and $\sin \theta = S$, Eq. (7.14) may be simplified to

$$V = C\sqrt{RS} \tag{7.15}$$

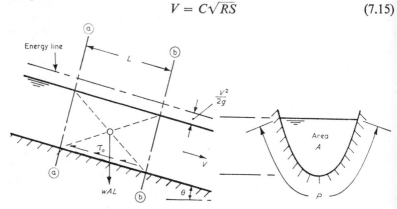

Figure 7.4   Uniform flow in a sloping channel

This is the well-known *Chézy formula*. The coefficient $C\ [= (w/\kappa)^{1/2}]$ is known as the *Chézy coefficient*. It is a pure number but has the dimensions $L^{1/2}T^{-1}$, i.e. (acceleration)$^{1/2}$.

We can obtain a direct relationship between C and λ since from Eq. (7.8), $V = (8g/\lambda)^{1/2}(RS)^{1/2}$, so that

$$C = \sqrt{\frac{8g}{\lambda}} \tag{7.16}$$

After the development of the Chézy formula various attempts were made to express C in terms of the surface roughness only, but the lack of agreement between the measured and the calculated discharges indicated that additional factors were involved. The explanation is to be found in Eq. (7.16) because we know from the previous section that λ for channels may be expressed in terms of the hydraulic radius (shape), the surface roughness, and the Reynolds number. In other words

$$C = \phi(R, k, R) \tag{7.17}$$

152

Under normal conditions of channel flow the Reynolds number is relatively large and the surface comparatively rough. In these circumstances the Reynolds number factor ceases to be significant and may be omitted. Thus

$$C = \phi(R, k) \tag{7.18}$$

From Eqs. (7.10) and (7.16) we obtain $C = 2\sqrt{8g} \log (14 \cdot 8R/k)$,[1] or

$$C = 17 \cdot 7 \log \frac{R}{k} + 20 \cdot 7 \tag{7.19}$$

7.4.3 Ganguillet and Kutter Formula

An exhaustive study of the gauging data obtained by Darcy, Bazin, and other investigators enabled the two Swiss engineers, Ganguillet and Kutter, to publish in 1869 the following formula for the Chézy coefficient:

$$C = 0 \cdot 552 \left[\frac{41 \cdot 6 + 1 \cdot 811/n + 0 \cdot 00281/S}{1 + (n/\sqrt{3 \cdot 28R})(41 \cdot 6 + 0 \cdot 00281/S)} \right] \tag{7.20}$$

where n is a coefficient, known as *Kutter's n*, dependent on the boundary roughness. Values of n appropriate to a limited range of conditions were quoted by the authors. It will be noted that a slope term is included; this only slightly influences the value of C and was introduced so as to give conformity with the Mississippi gaugings carried out by Humphreys and Abbot. As these gaugings were subsequently shown to be inaccurate, there is no longer any justification for its inclusion. The formula was widely used for a great many years, tables and charts being available to facilitate computation. It has been superseded by the much simpler Manning formula.

7.4.4 Manning Formula

Robert Manning, an Irish engineer, studied the various discharge data and formulae that were currently available and in a paper[2] read in 1889 presented a somewhat elaborate formula for the Chézy coefficient which was later simplified to

$$C = \frac{R^{1/6}}{n} \tag{7.21}$$

---

[1] An expression for C also follows from Eq. (7.5), since by substituting $C\sqrt{RS}$ for V and gRS for τ_0/ρ we obtain $C = 5 \cdot 75 \times 3 \cdot 13 \log (12 \cdot 1d/k)$, or

$$C = 18 \cdot 0 \log \frac{d}{k} + 19 \cdot 5$$

The slight discrepancy with Eq. (7.19) arises from the minor adjustments introduced in proceeding from the velocity gradient to the friction factor equations.

[2] MANNING, R. (1891) 'On the Flow of Water in Open Channels and Pipes', *Trans. Inst. C.E. of Ireland*, **20**, 161.

The complete *Manning formula* is thus

$$V = \frac{R^{2/3} S^{1/2}}{n} \tag{7.22}$$

The term n is a coefficient of surface roughness almost identical numerically with Kutter's n. This is of some convenience since it means that roughness values determined in connection with the Kutter formula are equally applicable to the Manning formula. Inherently, n has the dimensions $TL^{-1/3}$, so that a factor 1·486 has to be introduced in the numerator for conversion to British units.

This formula is widely employed today. It has the essential ingredients of shape and roughness. Being of the simple exponent type it is relatively straightforward in application, and charts and nomographs are available to facilitate rapid calculation. Moreover – as Manning himself demonstrated – it compares favourably with earlier formulae and is within the normal limits of accuracy of discharge measurement. In view of its simple form, it is not surprising to find that on the Continent the credit for its origination and introduction is attributed elsewhere, namely to Gauckler (1868) and Strickler (1923).

As was explained in Ch. 5, Sect. 5.8.4, the formula is particularly appropriate to the 'rough turbulent zone', the zone in which most channels in fact operate. In this earlier consideration a direct relationship between n and k was established. It was shown (Fig. 5.10, p. 102) that over the range $k/D = 1/10$ to $k/D = 1/1000$ the value of $k^{1/6}/n$ varied very little and that an average value was 26·0. Although a somewhat wider equivalent range of the ratio k/R is associated with channel flow, the variation of $k^{1/6}/n$ is still not large and it is particularly stable in the important central portion. Thus the curve relating n and k (Fig. 5.11, p. 102) is reasonably applicable to both channel and pipe roughness. Incidentally, it is relevant to point out that Strickler in 1923, before the modern turbulence theories were developed, quoted a value of 24·1 for $k_{50}^{1/6}/n$, where k_{50} is the median grain size (m).[1]

It should not be inferred from the foregoing remarks that the Manning formula, with n as a parameter of roughness, is without limitations. Utilising Eqs. (7.8) and (7.22)

$$\lambda = \frac{8gn^2}{R^{1/3}} \tag{7.23}$$

---

[1] The sixth power relationship of k and n may be questioned. It arises because the dimension of n is $TL^{-1/3}$ or (acceleration)$^{-1/2}$ × (length)$^{1/6}$; the $\sqrt{g}$ term has vanished in the constants.

Table 7.1

Values of *n* Appropriate to Various Channel Conditions

Category	Surface or Condition	n (s/m$^{1/3}$)
Rivers and streams	Earth, free from weed, with straight alignment, stones up to 75 mm av. size	0·02–0·025
	ditto with poor alignment	0·03–0·05
	ditto with weeds and poor alignment	0·05–0·15
	Gravel, free from weed, with straight alignment, stones 75 mm to 150 mm av. size	0·03–0·04
	ditto with poor alignment	0·04–0·08
	Gravel with stones and boulders exceeding 150 mm av. size, mountain rivers	0·04–0·07
Unlined canals and artificial channels	Earth, good alignment	0·018–0·025
	Earth, poor condition, stony bed	0·025–0·04
	Rock	0·025–0·045
Lined channels	Concrete	0·012–0·017
	Hand placed pitching	0·025–0·035
	Dressed stone, jointed	0·013–0·02
	Planed timber flume	0·011–0·013
	Unplaned timber flume	0·012–0·015
Pipes	Cast iron	0·010–0·014
	Concrete	0·011–0·015
	Salt-glazed stoneware	0·011–0·015
	Clay drainage tile	0·012–0·016
	Riveted steel	0·014–0·017
Models	Cement–sand mortar	0·011–0·013
	Smooth hardboard	0·009–0·011
	Perspex	0·009
	Glass	0·009–0·010

Since $R = \nu\mathrm{R}/V$ and $V = \sqrt{8g}\mathrm{R}^{1/2}S^{1/2}/\lambda^{1/2}$ we may eliminate R from Eq. (7.23) and obtain

$$\lambda = 78 \cdot 5 n^{1 \cdot 8} S^{0 \cdot 1} \nu^{-0 \cdot 2} \mathrm{R}^{-0 \cdot 2} \qquad (7.24)$$

This λ-R expression will plot logarithmically as a series of straight inclined lines for various values of n, S, and ν. The superimposed lines in Fig. 7.3 are for a few typical values of n and S, with ν taken as $1 \cdot 14$ mm²/s. Some indication is given of the extent to which n is influenced by S and R.

Values of n for some typical channel conditions are listed in Table 7.1. More detailed information is available in the specialist literature. In the case of lined channels it is generally possible to make a fairly close assessment of n. But channels in natural material present some difficulty, since n must take into account a number of factors; these include seasonal weed growth, differing bed and bank material, irregularity of section, and alignment. Reliance has necessarily to be placed on good judgement based on experience. Photographs of channels with known values of n are of assistance (e.g. V. T. Chow, Fig. 55). It is generally found that the roughness coefficient of the main channel of a river differs appreciably from that of the flood plain; for overspill conditions, instead of attempting to assess a mean roughness it is preferable to divide the cross-section into separate portions in the manner explained in Ex. 7.1.

In the measurement of river flow, advantage is sometimes taken of a straight uniform and stable reach of channel for the establishment of a gauging station. All that is required is a float chamber, connected to the river channel, and a small hut housing an autographic chart recorder which registers the rise and fall of water level. Provided conditions at the site are not influenced by variable weed growth or by the raising and lowering of movable gates, it is reasonable to expect a definite relationship to exist between water level, or stage as it is called, and discharge. If we consider the simplest case, that of a wide shallow channel with vertical sides, then, in accordance with the Manning formula, $Q \propto d^{5/3}$, and this gives some general indication of the form of a typical stage-discharge or rating curve.

Example 7.1

The cross-section of a river channel is to be designed in such a way that normal winter discharges are contained within a main central channel, while exceptional flows spill over into a flood plain on either side, bounded by low banks constructed parallel with the river. The following particulars apply:

(a) Ground level of flood plain 33·50 A.O.D., (b) crest of flood bank 35·00 A.O.D., (c) river channel section to be segmental with a maximum tangential slope of 1:1, (d) freeboard to flood bank to be 0·3 m so as to allow for possible wave action, (e) longitudinal slope of river channel 1/1000, (f) $n = 0·04$ for main channel and 0·05 for flood plain, (g) main channel section to be designed to pass a bank-full discharge of 85 cumecs, (h) flood banks to be sited so as to pass a total peak flood discharge of 230 cumecs.

Determine the top width and lowest bed level of the main river channel, and the minimum distance between flood banks.

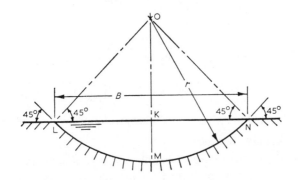

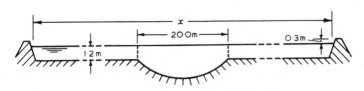

Ex. 7.1

Main channel – bank-full condition ($Q_1 = 85$ cumecs)

Area of sector OLMN $= \pi r^2/4 = 0·785r^2$; area of triangle OLN $= 0·5r^2$. Thus area of segment LKNM (A_1) $= 0·285r^2$. Wetted perimeter LMN (P_1) $= 2\pi r/4 = 1·57r$. Hence $R_1 = A_1/P_1 = r/5·51$.

Substituting in the Manning formula:

$$85 = 0·285r^2 \times \left(\frac{r}{5·51}\right)^{2/3} \times \left(\frac{1}{1000}\right)^{1/2} \bigg/ 0·04$$

from which $r = 14·2$ m. Thus $A_1 = 57·2$ m², $P_1 = 22·2$ m, and top width $B = 1·414r = \mathbf{20·0}$ m. Also, KM $= 0·293r = 4·15$ m, indicating a minimum bed level of **29·35** m A.O.D.

157

Peak flood condition ($Q_2 = 230$ cumecs)

For the purpose of evaluation the cross-section is divided into a deep main channel and a shallow side channel, the latter representing the flood plain. These channels are regarded as operating independently with a differing surface roughness and mean velocity.

(i) Main channel:

$$A'_2 = 57 \cdot 2 + 1 \cdot 2 \times 20 \cdot 0 = 81 \cdot 2 \text{ m}^2 \text{ and } P'_2 = 22 \cdot 2 \text{ m, so that}$$
$$R'_2 = 81 \cdot 2/22 \cdot 2 = 3 \cdot 66 \text{ m. Thus}$$

$$Q'_2 = 81 \cdot 2 \times 3 \cdot 66^{2/3} \times \left(\frac{1}{1000}\right)^{1/2} \bigg/ 0 \cdot 04 = 153 \text{ cumecs}$$

(ii) Side channel:

The required capacity of the flood plain is 77 cumecs. As the section is wide and shallow, $R \simeq$ depth of flow ($1 \cdot 2$ m). Hence

$$77 = 1 \cdot 2(x - 20 \cdot 0) \times 1 \cdot 2^{2/3} \times \left(\frac{1}{1000}\right)^{1/2} \bigg/ 0 \cdot 05$$

giving $x = \mathbf{110}$ m for the approximate width of the flood plain.

7.5 Best Hydraulic Section

A variety of channel shape is possible and it is of some importance to determine the best proportioned section for the usual hydraulic criterion of maximum discharge.

The Manning formula may be expressed in the form

$$Q = \frac{A^{5/3} S^{1/2}}{n \, P^{2/3}}$$

Now Q is a maximum when A^5/P^2 is a maximum. If x is some variable, such as a leading dimension, the condition for maximum discharge is $(\mathrm{d}/\mathrm{d}x)(A^5/P^2) = 0$. Thus

$$5 \frac{A^4}{P^2} \frac{\mathrm{d}A}{\mathrm{d}x} - 2 \frac{A^5}{P^3} \frac{\mathrm{d}P}{\mathrm{d}x} = 0$$

or

$$5P \frac{\mathrm{d}A}{\mathrm{d}x} - 2A \frac{\mathrm{d}P}{\mathrm{d}x} = 0 \qquad (7.25)$$

For a given sectional area A, $\mathrm{d}A/\mathrm{d}x = 0$, and hence the maximum discharge is obtained when

$$\frac{\mathrm{d}P}{\mathrm{d}x} = 0 \qquad (7.26)$$

that is to say when the wetted perimeter is a minimum. This section is known as the *best hydraulic section*. Furthermore, it follows that for a given discharge the sectional area is a minimum, although the excavation volume is not a minimum unless bank-full flow conditions pertain.

Accordingly, the most efficient channel shape is the semi-circle. Normally, however, this is not a practical form either to construct or to maintain. The usual shape for new canals and channels is the rectangular or trapezoidal and the best proportions may be determined as follows:

(a) Rectangular Section

$$A = bd; \quad P = b + 2d$$

With a given value of A, $P = A/d + 2d$. For minimum P, $\mathrm{d}P/\mathrm{d}d = -A/d^2 + 2 = 0$, from which $bd/d^2 = 2$, or

$$b = 2d \qquad (7.27)$$

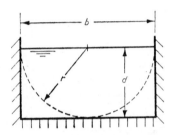

Figure 7.5 'Best' rectangular section

that is to say the 'best' proportions for a rectangular channel are such that the inscribed semi-circle is tangential to the bed and sides (Fig. 7.5). For this condition

$$R = \frac{A}{P} = \frac{d}{2} \qquad (7.28)$$

In the case of a section that is relatively wide in comparison with the depth (width $> 25 \times$ depth) $R \simeq d$. More generally, the value of R is between d and $d/2$.

(b) Trapezoidal Section
If the side slope is s horizontally to 1 vertically, then

$$A = (b + sd)d; \quad P = b + 2d\sqrt{1 + s^2}$$

With a given value of A, $P = (A - sd^2)/d + 2d\sqrt{1 + s^2}$. For minimum P, $\mathrm{d}P/\mathrm{d}d = -A/d^2 - s + 2\sqrt{1 + s^2} = 0$, from which

$$b + 2sd = 2d\sqrt{1 + s^2} \qquad (7.29)$$

159

Thus the most efficient hydraulic section is when the top width is twice the length of a sloping side. A semi-circle radius r may then be inscribed within the cross-section (Fig. 7.6). This follows since

$$\operatorname{cosec} \alpha = \frac{(b + 2sd)}{2d}$$

and also $\operatorname{cosec} \alpha = \sqrt{1 + s^2}$, where α is the side slope angle with the horizontal.

The corresponding value of R is obtained from

$$R = \frac{(b + sd)d}{b + 2d\sqrt{1 + s^2}} = \frac{(b + sd)d}{b + (b + 2sd)} = \frac{d}{2} \qquad (7.30)$$

which is the same value as for a rectangular section.

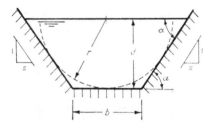

Figure 7.6 'Best' trapezoidal section

In a similar manner it may be shown that the criterion for maximum velocity is the same as that for maximum discharge (i.e. $dP/dx = 0$), and this is somewhat disadvantageous since it means that on the steeper slopes the velocities associated with the best hydraulic section may exceed those that the bed and sides are capable of withstanding without scour. The provision of a suitable lining serves to raise the permissible velocity, but this is expensive and there are other factors, such as bank stability and a sharp increase in excavation cost with depth, that generally militate against the adoption of the 'best section' channel. Unless, therefore, a channel is to be cut in rock, a shallower type of section must normally be provided with side slopes in natural material of the order of $1\frac{1}{2}:1$.

Example 7.2

An artificial drainage channel is of trapezoidal section with bed width 3 m and side slopes $1\frac{1}{2}:1$. What is the flow depth for the 'best hydraulic section'?

Owing to weed growth, the roughness coefficient n is increased from 0·03 in the winter to 0·05 in the summer. For the same discharge, determine the flow depth in summer corresponding to a 1·2 m flow depth in winter.

Substituting in Eq. (7.29), $3 + 3d = 2d\sqrt{1 + 1·5^2}$, from which $d = 4·95$ m. This indicates a relatively deep and narrow section and one that is not likely to be acceptable in practice.

For winter conditions: $\quad Q_w = \dfrac{A_w^{5/3} S^{1/2}}{0·03 P_w^{2/3}}$

For summer conditions: $\quad Q_s = \dfrac{A_s^{5/3} S^{1/2}}{0·05 P_s^{2/3}}$

Now $Q_w = Q_s$, so that

$$\left(\frac{A_s}{A_w}\right)^{5/3} \left(\frac{P_w}{P_s}\right)^{2/3} = 1·667 \tag{i}$$

Substituting $A_s = d_s(3 + 1·5d_s)$, $A_w = 1·2(3 + 1·5 \times 1·2) = 5·76$, $P_s = 3·61d_s + 3$, and $P_w = 7·33$, Eq. (i) becomes

$$\frac{d_s^{5/3}(3 + 1·5d_s)^{5/3}}{5·76^{5/3}} \times \frac{7·33^{2/3}}{(3·61d_s + 3)^{2/3}} \tag{ii}$$

the solution to which by trial and error is $d_s = \mathbf{1·56}$ m.

7.6 Enclosed Conduits

Conduits for the purpose of conveying surface water and sewage are nearly always designed to operate in the partially full condition, so that the formulae for channel flow are applicable. Precast pipes, of earthenware or concrete, have the advantage of good discharge characteristics and structural form, while at the same time being easy to handle and lay. Concrete pipes in standard sizes up to 1524 mm (60 in.) diameter are held in stock by most manufacturers. The roughness coefficient n for a concrete pipe is usually taken as 0·012 or 0·013.

A knowledge of the variation of discharge and mean velocity with depth is of some importance and in the case of a circular conduit may be determined as follows:

Let us suppose (Fig. 7.7) that the angle subtended at the centre of a pipe, diameter D, by the free surface is ϕ. Then

$$A = \frac{\pi D^2}{4} - \frac{D^2 \phi}{8} + \frac{D^2}{4} \sin \frac{\phi}{2} \cos \frac{\phi}{2}$$

$$= \frac{\pi D^2}{4} - \frac{D^2 \phi}{8} + \frac{D^2}{8} \sin \phi = \frac{D^2}{4}\left(\pi - \frac{\phi}{2} + \frac{\sin \phi}{2}\right)$$

and

$$P = \pi D - \frac{D}{2}\phi = D\left(\pi - \frac{\phi}{2}\right)$$

so that

$$R = \frac{D}{4}\left(1 + \frac{\sin\phi}{2\pi - \phi}\right) \tag{7.31}$$

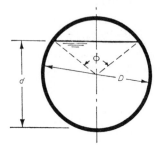

Figure 7.7   Section of a circular
conduit flowing partially full

With a given roughness coefficient[1] and longitudinal slope, the Manning formula may be expressed as $Q = \kappa A R^{2/3}$, where κ is a constant coefficient. Thus

$$Q = \kappa \frac{D^2}{4}\left(\pi - \frac{\phi}{2} + \frac{\sin\phi}{2}\right)\left[\frac{D}{4}\left(1 + \frac{\sin\phi}{2\pi - \phi}\right)\right]^{2/3}$$

$$= \kappa \frac{D^{8/3}}{10\cdot08}\left(\pi - \frac{\phi}{2} + \frac{\sin\phi}{2}\right)\left(1 + \frac{\sin\phi}{2\pi - \phi}\right)^{2/3} \tag{7.32}$$

When the free surface just reaches the pipe soffit, $Q = Q_F$ and $\phi = 0$, so that

$$Q_F = \frac{\kappa\pi D^{8/3}}{10\cdot08} \tag{7.33}$$

For the pipe operating half full:

$$Q_{1/2F} = \frac{\kappa\pi D^{8/3}}{20\cdot16} \tag{7.34}$$

The ratio Q/Q_F is given by

$$\frac{Q}{Q_F} = \frac{1}{\pi}\left(\pi - \frac{\phi}{2} + \frac{\sin\phi}{2}\right)\left(1 + \frac{\sin\phi}{2\pi - \phi}\right)^{2/3} \tag{7.35}$$

[1] Actually, research has shown that for circular sewers n is not constant but tends to decrease with increasing flow depth.

The curve of Q/Q_F against the proportionate flow depth is shown in Fig. 7.8. It will be noted that the maximum discharge occurs when the pipe is slightly less than full. This is because the addition of a relatively small area of flow at the top of the pipe entails a disproportionately large increase in the length of wetted perimeter. By differentiation of Eq. (7.32), the angle ϕ corresponding to Q_{max} is found to be 57° 36′ and Q_{max}/Q_F is then 1·08. The relationship $Q_F : Q_{max} : Q_{1/2F}$ is thus $1 : 1·08 : 0·5$.

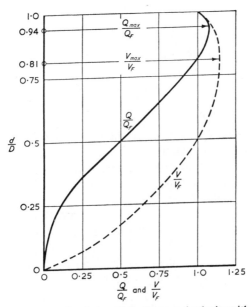

Figure 7.8   Variation of discharge and velocity with flow depth in a circular conduit

The corresponding expression for the ratio V/V_F is

$$\frac{V}{V_F} = \left(1 + \frac{\sin\phi}{2\pi - \phi}\right)^{2/3} \tag{7.36}$$

and it is noteworthy that for the full and half full conditions:

$$V_F = V_{1/2F} = \frac{\kappa D^{2/3}}{2.52} \tag{7.37}$$

The relationship between V/V_F and the proportionate flow depth is shown graphically in Fig. 7.8. The value of V is a maximum when $\phi = 102° 33′$ and it may be shown that $V_F : V_{max} : V_{1/2F}$ is $1 : 1·14 : 1$.

In the customary time-area method of determining the pipe sizes in a sewerage system, it is necessary to estimate the time of flow over each component length. This is normally based on the calculated size of pipe running full, although, owing to the effect of storage and the need to adopt standard pipe sizes, the pipes would in practice generally operate in the partially full condition. The small variation in velocity over the range half full to full shows that the error arising from the pipe full assumption is likely to be insignificant.

Sewers, whether of the foul or surface-water type, usually carry some solid matter in suspension. For self-cleansing, the design velocity should not be less than 0·75 m/s, and in order to avoid abrasion it should not exceed 3 m/s. Tables and charts are available to facilitate design calculations (e.g. see footnotes, p. 99).

Egg-shaped sewers (large end uppermost) have the advantage of a more uniform velocity at all depths, but they are not generally favoured owing to their higher cost and increased headroom.

7.7 Scouring and Silting

One of the most important aims in efficient channel design is to ensure the minimum burden of maintenance. With a lined channel the problem is mainly one of constructional economics since the velocity required to cause erosion is relatively high, usually in excess of 3 m/s. However, in the case of channels in erodible material the problem is at once much more difficult owing to the complex phenomenon of scouring and silting, which may or may not be associated with dune formation, according to the particular conditions. Much research has been carried out on this topic in recent years, but it is still far from being fully understood. Design procedure is largely empirical, as is evidenced in the extensive specialist literature.

The transported material, called *sediment*, is not restricted to fine silt and sand but may include materials as diverse as colloidal clay, pebbles, and boulders. The total quantity of sediment in movement at any point is composed of a *bed load* and a *suspended load*. The former is transported by rolling or saltation along the bed of the channel, while the latter is borne along in the general stream, the density decreasing with the height above the bed. The ability of the flow to carry more suspended load depends on the density of that already existing and on the degree of turbulence; at the saturation point no further increase in suspended load is possible.

The transporting power of a flow stream acting on the bed may be shown to be proportional to the sixth power of the velocity. This relationship is derived from a consideration of the equilibrium of a particle at the moment when transportation along the bed commences.

Thus for a spherical particle radius r, the drag force is equal to $C_D(w/2g)\pi r^2 V^2$, while the frictional force due to contact with the bed is $\mu_f(w_s - w)(4\pi r^3/3)$, where C_D and μ_f are the coefficients of drag and friction respectively, and w_s is the specific weight of the particle. Equating, we obtain $V^2 \propto r$ so that the transporting power is proportional to V^6. The analysis is somewhat oversimplified and several authorities contend that the transporting power is more nearly proportional to V^5.

In the case of a channel carrying a given sediment load it is found that if the velocity is gradually decreased a point is reached when the stream can no longer support the load and deposition occurs. If, later on, the velocity is increased then at another critical velocity the bed commences to scour. Usually, this second critical velocity is higher than the first, and appreciably so if the sediment has had sufficient time to bed down on a well-graded pattern or a natural cement of colloidal silt has filled the interstices. In the intermediate range of velocity the channel will neither silt nor scour, the corresponding shape and longitudinal slope obviously being the best for satisfactory operation. Unfortunately, the sediment load and discharge are rarely constant for any length of time, so that any sectional shape proposed must be in the nature of a compromise. Generally, it is considered preferable that the velocity should be relatively high, since it is easier to take measures against scour than siltation. Also, a relatively fast flow stream is less conducive to weed growth, which is an impediment to flow and expensive to remove.

There are two principal methods of erodible channel design – based on the assumption of (a) maximum permissible velocity, and (b) maximum permissible tractive force. There is also a third approach, based on the so-called 'regime theory', developed by irrigation engineers in India and successfully adopted for the design of constant discharge canals. Indian conditions are somewhat unique and as the formulae are wholly empirical, the method has not found much favour outside the country of origin.

The maximum permissible velocity is the greatest mean velocity that is possible without the occurrence of scour. It is primarily dependent on the boundary material and sediment load. Also, since the boundary velocity is the criterion and the velocity distribution is not uniform, it is also dependent on the depth of flow. Because of the obvious complexities, the value in any particular case is not easy to assess and good engineering judgement based on experience is important. Table 7.2 gives some typical mean velocity values, based mainly on the recommendations of Fortier and Scobey.[1] They are applicable to channels

---

[1] FORTIER, S. and SCOBEY, F. C. (1926) 'Permissible Canal Velocities', *Trans. Am. Soc. C.E.*, 89, 940.

with a flow depth up to approximately 0·9 m; greater flow depths allow of higher mean velocities. Ex. 7.3 illustrates a simple application of this design criterion.

The second method of approach is based on the concept of a critical tractive force. This is the shear stress which requires to be exerted on the boundary surface for transportation to commence. Only in the case of a wide channel is the tractive force uniformly distributed; normally it

Table 7.2

Maximum Permissible Water Velocities and Unit Tractive Force Values for the Design of Stable Channels (Flow Depth < 0·9 m)

Material	Roughness coefficient* n	Clear water		Water transporting colloidal silts	
		Velocity V	Unit tractive force τ_0	Velocity V	Unit tractive force τ_0
	$(s/m^{1/3})$	(m/s)	(N/m^2)	(m/s)	(N/m^2)
Fine sand, colloidal	0·02	0·45	1·5	0·75	4·0
Sandy loam, non-colloidal	0·02	0·55	2·0	0·75	4·0
Alluvial silts, non-colloidal	0·02	0·6	2·5	1·0	7·0
Fine gravel	0·02	0·75	4·0	1·5	15·5
Stiff clay, very colloidal	0·025	1·15	12·5	1·5	22·0
Alluvial silts, colloidal	0·025	1·15	12·5	1·5	22·0
Coarse gravel, non-colloidal	0·025	1·2	14·5	1·8	32·0
Cobbles and shingles	0·035	1·5	43·5	1·7	53·0

* Based on a straight channel of uniform section.

is greater on the bed than on the sides. *Du Boys equation* (1879) for the unit tractive force is obtained from Eq. (7.13) and is

$$\tau_0 = wRS \qquad (7.38)$$

Maximum permissible values of τ_0 may be derived from laboratory and field data. Some typical values, based mainly on U.S. Bureau of Reclamation recommendations, are included in Table 7.2.

Example 7.3

A head-race canal is to be designed to deliver 30 cumecs to a turbine installation. The canal is to be constructed in natural material for which the maximum permissible velocity is estimated to be 0·95 m/s. Assuming a trapezoidal section with side slopes 2:1 and a depth not exceeding 3 m, determine a suitable top width and longitudinal slope. Take $n = 0·025$.

For the maximum velocity, $A = 30/0.95 = 31.6$ m^2. Assuming $d = 3$ m, then $b = A/d - 2d = 4.5$ m. Hence the top width = **16.5** m.

On this basis $R = 31.6/(4.5 + 6\sqrt{5}) = 2.35$ m. Substituting in the Manning formula, $0.95 = 2.35^{2/3} \times S^{1/2}/0.025$, from which $S = 1/\mathbf{5535}$.

The problem may be extended to a determination of the unit tractive force. Thus $\tau_0 = 9810 \times 2.35/5535 = 4.16$ N/m^2, which is probably reasonable. It is interesting to note that if d is taken as 2.5 m, the top width = 17.6 m and $S = 1/3540$, with $\tau_0 = 4.66$ N/m^2.

The section with $d = 3$ m appears the most satisfactory since it approaches nearest to the hydraulic best section.

Further Reading

BAKHMETEFF, B. A. (1932) *Hydraulics of Open Channels*. McGraw-Hill.

BLENCH, T. (1969) *Mobile-bed Fluviology*. Univ. Alberta Press (2nd Edition).

BUREAU OF RECLAMATION, *Canals and Related Structures*, Design Standard No. 3. U.S. Govt. Printing Office.

CHOW, V. T. (1959) *Open-Channel Hydraulics*. McGraw-Hill.

GONCHAROV, V. N. (1964) *The Dynamics of Channel Flow*. Oldbourne Press.

GRAF, W. H. (1971) *Hydraulics of Sediment Transport*. McGraw-Hill.

HENDERSON, F. M. (1966) *Open Channel Flow*. Macmillan.

KING, H. W. and BRATER, E. F. (1963) *Handbook of Hydraulics*, Sect. 7. McGraw-Hill (5th Edition).

LELIAVSKY, S. (1959) *An Introduction to Fluvial Hydraulics*. Constable (Dover reprint, 1966.)

LELIAVSKY, S. (1965) *Irrigation and Hydraulic Design*, vol. 1, Chs. 2–3. Chapman and Hall.

RAUDKIVI, A. J. (1967) *Loose Boundary Hydraulics*. Pergamon Press.

ROZOVSKII, I. L. (1961) *Flow of Water in Bends of Open Channels*. Oldbourne Press.

——, (Mar. 1963) 'Friction Factors in Open Channels: Report of Task Force,' Proc. Am. Soc. C. E., **89**, 97.

SELLIN, R. H. J. (1969) *Flow in Channels*. Macmillan.

WOODWARD, S. M. and POSEY, C. J. (1941) *Hydraulics of Steady Flow in Open Channels*. Wiley.

YALIN, M. S. (1972) *Mechanics of Sediment Transport*. Pergamon.

CHAPTER EIGHT

Non-uniform Flow in Channels

8.1 Introduction

Non-uniform flow is characterised by a surface slope that is not parallel to the bed. Study of the subject is primarily concerned with the analysis of surface profiles and energy gradients. An ability to predict these features is of obvious advantage when designing channel improvement works.

For purposes of analytical treatment non-uniform flow may be conveniently classified as (a) gradually varied flow, and (b) rapidly varied flow. There is no definite dividing line.

In gradually varied flow the degree of non-uniformity is slight, the change in flow conditions normally extending over a considerable distance. Boundary friction, of course, has to be taken into account. Owing to the irregularity of section and bed slope, as well as the influence of control structures, the flow in river channels is mostly of this type.

In rapidly varied flow there is an abrupt or very rapid change in the sectional area of flow within a short distance. Boundary friction losses are relatively insignificant in comparison with losses due to turbulent eddying. The flow at channel transitions usually falls within this category.

Generally, it is permissible to regard the velocity distribution as uniform. This simplifies the analysis, particularly as the depth and therefore the precise distribution are often initially unknown. The resulting error is usually quite small in relation to the overall limits of accuracy that are applicable. However, it must be remembered that the distribution is less uniform than in pipe flow, and in circumstances of exceptionally irregular distribution, the correction coefficients α or β should appear in some suitable form in the relevant energy or momentum equations.

8.2 Specific Energy and Critical Depth

8.2.1. *Definition of Specific Energy*

The energy at a cross-section of a channel, referred to the base of the channel, is termed the *specific energy* or *specific head*. The concept of

168

specific energy was introduced by Bakhmeteff[1] and it is the basis of present-day theory of non-uniform flow. It is to be distinguished from *total head* (p. 33) which is a measure of the energy referred to a certain datum.

Fig. 8.1 shows a prismatic channel (i.e. channel with constant section and alignment) with exaggerated bed and surface slopes. At section *aa*, the pressure head against the bed is $d \cos^2 \theta$ (obtained from a consideration of the normal component of the weight of water acting on unit area of the bed – see inset). Thus the specific energy is given by

$$E_s = d \cos^2 \theta + \frac{V^2}{2g} \tag{8.1}$$

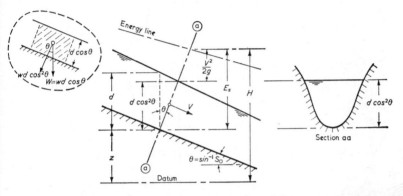

Figure 8.1   Specific energy in channel flow

Now, unless the slope is unusually steep (i.e. $\sin \theta > 1/10$), $\cos^2 \theta \simeq 1$. In this case, with negligible error, d may be taken as the depth of flow so that

$$E_s = d + \frac{V^2}{2g} \tag{8.2}$$

Under the same conditions the total head is given by

$$H = z + d + \frac{V^2}{2g} \tag{8.3}$$

These energy head expressions are based on the assumption that the piezometric pressure at any depth complies with the hydrostatic law. This is valid provided that the stream lines are parallel or near parallel;

[1] BAKHMETEFF, B. A. (1912) *Varied Flow in Open Channels*. St. Petersburg, Russia.

if they have an appreciable curvature or are sharply divergent, then accelerative forces in the plane of the cross-section are operative and serve to modify the normal pressure head-depth relationship (see Ch. 4, Sect. 4.9.2). It is therefore a basic requirement that the flow be uniform or gradually varied. As exceptionally steep slopes ($\sin \theta > 1/10$) produce a high velocity, unsteady type of flow, quite incompatible with either of these conditions, the simplified form of specific energy expression (Eq. (8.2)) is almost always applicable.

In uniform flow the specific energy is constant and the energy grade line is parallel to the bed. In non-uniform flow, however, whereas the energy grade line must always slope downwards in the direction of flow, the specific energy may increase or decrease according to the particular channel and flow conditions.

8.2.2. *Variation of Specific Energy with Depth of Flow*

We will now consider the relationship between the specific energy and the depth of flow when the latter is varied over a wide range in a channel of arbitrary section (Fig. 8.2). The discharge Q is deemed to remain

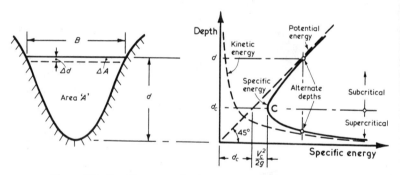

Figure 8.2   Specific energy-depth relationship (constant discharge)

constant and the variation of flow depth may be brought about by amending the roughness, longitudinal slope, or conditions up and downstream. For each depth of flow there is a corresponding specific energy. The flow depth-specific energy curve (Fig. 8.2) has as its components the potential energy line and the kinetic energy curve. The former, for equal scales, is inclined at 45 degrees; the latter is asymptotic to the two axes.

It will be observed that the specific energy declines to a minimum at the point C and then increases again. The depth and velocity at this significant point are called the *critical depth*, d_c, and the *critical velocity*,

V_c. The latter should not be confused with the transition from laminar to turbulent flow where the value is much lower.

The critical depth represents the least possible specific energy with which the fixed discharge Q is able to flow in the channel of given shape. For every value of the specific energy, other than the minimum, there are two possible depths of flow, one above and the other below the critical value, and known as *high stage* and *low stage* respectively. These corresponding depths are referred to as *alternate*.

It is evident that a loss of specific energy at depths above the critical is associated with a reduction in depth, whereas at depths below the critical there is an increase in depth. The high and low stage surface profiles associated with a local rise in bed level are a good example (Fig. 8.3).

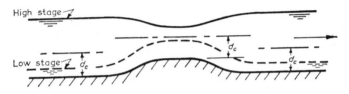

Figure 8.3   Surface profiles at a local rise in bed level

At depths near the critical a very small change in specific energy brings about a relatively much greater variation in depth; this is a feature that is borne out in practice, an unstable and wave-like motion being evident. Obviously, it is desirable to avoid designing a channel such that it is likely to operate near the critical depth.

The critical depth is clearly an important criterion in defining the nature of the flow. When the depth is above the critical the velocity and flow are referred to as *subcritical*. Correspondingly, for depths below the critical the term *supercritical* is applied. Sometimes the flow is described as tranquil or streaming instead of subcritical, and rapid or shooting instead of supercritical. Whilst these terms are more graphic, the previous ones are to be preferred both on the grounds of convention and also to avoid any risk of mis-association with turbulent and laminar flow.

8.2.3. *Criterion for Critical Depth* (Q *constant*)

Since the critical depth corresponds to minimum specific energy, the conditions for $d = d_c$ may be determined by differentiating the specific energy expression and equating to zero.

Firstly, it is necessary to write Eq. (8.2) in the form

$$E_s = d + \frac{1}{2g}\frac{Q^2}{A^2}$$

in which the variable V has been replaced by the constant Q and the sectional area A (a function of d).

For minimum specific energy:

$$\frac{dE_s}{dd} = 1 + \frac{Q^2}{2g}\frac{d}{dA}\left(\frac{1}{A^2}\right)\frac{dA}{dd} = 0$$

Now it is evident from Fig. 8.2 that in the limit $dA = B\,dd$, or $dA/dd = B$, where B is the surface width, so that

$$1 - \frac{Q^2}{2g}\frac{2}{A^3}B = 0$$

or

$$\frac{Q^2 B}{gA^3} = 1 \qquad (8.4)$$

This is the general expression which, for the depth to be critical, must be satisfied irrespective of channel shape. Since both A and B will have the values corresponding to d_c we can write

$$\frac{Q^2 B_c}{gA_c{}^3} = 1 \qquad (8.5)$$

The critical velocity V_c is given by

$$V_c = \frac{Q}{A_c} = \sqrt{\frac{gA_c}{B_c}} \qquad (8.6)$$

It is convenient to introduce a term called the *mean* or *hydraulic depth*, d_m, such that $d_m = A/B$. Eq. (8.6) then becomes

$$V_c = \sqrt{gd_{mc}} \qquad (8.7)$$

where d_{mc} is the mean depth of the section corresponding to the critical depth d_c. This expression may be written in the dimensionless manner

$$\frac{V_c}{\sqrt{gd_{mc}}} = 1 \qquad (8.8)$$

The term $V/\sqrt{gd_m}$ is a form of Froude number F (see Ch. 4, Sect. 4.12(b)) and thus if the Froude number for a certain flow condition is greater than unity the flow is supercritical; conversely, if it is less than unity then the flow is subcritical.

The use of the Froude number in this way is helpful in checking on the nature of the flow. More significant, however, is the fact that $\sqrt{gd}$ is the speed of propagation of a small gravity wave in shallow water (see Sect. 8.10.3, p. 208). The Froude number for given channel conditions is thus the ratio of the water velocity to that of the wave velocity. It follows that in subcritical flow waves caused by disturbances and obstructions can progress upstream whereas in supercritical flow they can only travel downstream.

8.2.4 *Criterion for Maximum Discharge* (E_s constant)

Let us now suppose that the specific energy is held constant whilst the discharge is varied. Eq. (8.2) may be rearranged to give

$$Q = \sqrt{2g}A(E_s - d)^{1/2} \qquad (8.9)$$

The graph of discharge against depth is shown in Fig. 8.4. It will be observed that there are two values of d for each value of Q, except at the

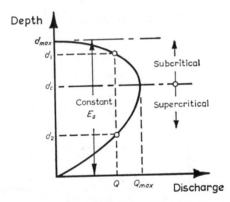

Figure 8.4   Discharge-depth relationship (constant specific energy)

maximum. The criterion for maximum discharge may be obtained by differentiation as before. Thus

$$\frac{dQ}{dd} = \sqrt{2g}\left[A\frac{(E_s - d)^{-1/2}}{2}(-1) + \frac{dA}{dd}(E_s - d)^{1/2}\right] = 0$$

Hence, remembering $dA/dd = B$, we obtain

$$E_s - d = \frac{A}{2B}$$

173

and utilising Eq. (8.9),

$$\frac{Q^2 B}{g A^3} = 1 \tag{8.10}$$

which is identical with Eq. (8.4). In other words, for constant specific energy the maximum discharge occurs at the critical depth, or

$$\frac{Q_{max}^2 B_c}{g A_c^3} = 1 \tag{8.11}$$

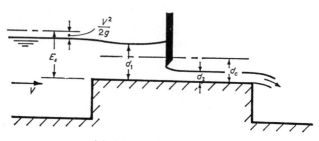

(a) Gate partially raised

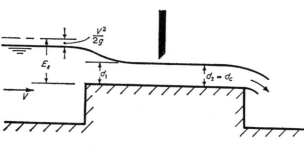

(b) Gate fully raised

Figure 8.5   Sluice control at a place of raised bed level

It may also be noted that by combining Eqs. (8.9) and (8.11) so as to eliminate Q_{max} we obtain for the critical depth condition

$$d_c + \frac{d_{mc}}{2} = E_s \tag{8.12}$$

The concept of constant specific energy can usefully be illustrated by considering the flow behaviour at a sluice gate situated near the mid-point of a raised portion of channel bed (Fig. 8.5). The upstream specific energy head E_s (referred to the raised bed level) is assumed to remain constant.

When the gate is closed the upstream depth d_1 is equal to E_s while the downstream depth d_2 is zero. If the gate is now raised by an amount less than d_c the surface profile is as indicated with $d_1 > d_c$ and $d_2 < d_c$. For a complete raising of the gate, upstream and downstream levels must be the same, which can only be so when $d_1 = d_2 = d_c$ and the discharge is a maximum. We have now created the flow conditions of a simple broad-crested weir. This type of hydraulic structure is discussed in Ch. 9, Sect. 9.4.3. It will be noted that for a constant width-depth ratio Eq. (8.12) becomes $d_c = \frac{2}{3}E_s$ (cf. Eq. (9.12), p. 223) and the discharge per unit width is given by $q_{max} = \sqrt{g}(\frac{2}{3})^{3/2}E_s^{3/2}$ (cf. Eq. (9.13), p. 223).

8.2.5 *Computation of Critical Depth*

For channel sections where there is a simple relationship between surface width and depth Eq. (8.4) may be solved directly. Thus the critical depth for a rectangular section of width b is given by

$$d_c = \sqrt[3]{\frac{Q^2}{gb^2}} \tag{8.13}$$

Replacing Q/b by q, the discharge per unit width,

$$d_c = \sqrt[3]{\frac{q^2}{g}} \tag{8.14}$$

Also

$$V_c = \sqrt{gd_c} \tag{8.15}$$

and

$$d_c = \frac{2}{3}E_{sc} \tag{8.16}$$

For other sections commonly encountered in practice, but not so readily evaluated, tables and charts[1] are available to facilitate a direct or rapid solution. In any event a solution can always be effected by either

(a) Plotting the curve of A^3/B against the depth, the intersection of the curve with the vertical line Q^2/g yielding the critical depth,

or

(b) Trial and error substitution in Eq. (8.4) (or Eq. (8.8)) for various depths until the equation is satisfied.

---

[1] For example: KING, H. W. and BRATER, E. F. (1963) *Handbook of Hydraulics*, Tables 8-4 to 8-12. McGraw-Hill (5th Edition).

With either method a useful first estimate of the critical depth is usually obtained by calculating the critical depth of the approximately equivalent rectangular section.

Example 8.1

A trapezoidal channel with bed width 6 m and side slopes 2:1 carries a discharge of 10 cumecs. Determine the critical depth, critical velocity, and minimum specific energy.

It is assumed that the approximately equivalent rectangular section has a width of 7 m. Then from Eq. (8.13), $d_c = [10^2/(9 \cdot 81 \times 7^2)]^{1/3} = 0 \cdot 593$ m. The first estimate of d_c is therefore taken as $0 \cdot 59$ m. With this value, $B = 8 \cdot 36$ m, $A = 4 \cdot 24$ m^2, so that $d_m = 0 \cdot 507$ m and $F = V/\sqrt{gd_m} = 1 \cdot 06$, indicating supercritical flow. Hence $d_c > 0 \cdot 59$ m.

By subsequent trial and error the value of d_c satisfying the condition of $F = 1 \cdot 0$ is found to be $d_c = \mathbf{0 \cdot 61}$ m. Hence $V_c = \mathbf{2 \cdot 26}$ m/s and $E_{sc} = 0 \cdot 61 + 2 \cdot 26^2/19 \cdot 6 = \mathbf{0 \cdot 87}$ m.

It is to be noted that the value of d_c/E_{sc} is $0 \cdot 68$.

8.2.6 *Significance of Bed Slope*

The bed slope required to produce uniform flow in a channel operating at the critical depth is called the *critical slope*, S_c. An expression for the critical slope may be derived from the Manning formula, $V = R^{2/3}S_0^{1/2}/n$, where S_0 is the bed slope (uniform flow). At the critical depth $V = \sqrt{gd_{mc}}$, $R = R_c$, and $S_0 = S_c$, so that

$$S_c = \frac{gd_{mc}n^2}{R_c^{4/3}} \tag{8.17}$$

The critical slope is seen to depend on the discharge and on the boundary roughness, as indeed it would have been logical to infer. In the case of a wide shallow channel $d_{mc} \simeq d_c \simeq R_c$, so that

$$S_c = \frac{gn^2}{d_c^{1/3}} \tag{8.18}$$

If uniform flow occurs in a channel with bed slope less than the critical ($S_0 < S_c$), the nature of the flow must be subcritical, the bed slope being termed *subcritical* or, more commonly, *mild*. Likewise with a bed slope greater than the critical ($S_0 > S_c$), the flow is supercritical and the bed slope is termed *supercritical* or *steep*.

8.2.7

From the foregoing it is apparent that the critical depth and specific energy criteria have a most significant role in the analysis of channel

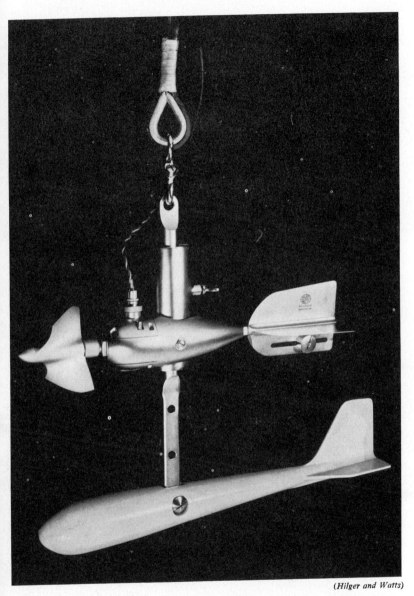

(Hilger and Watts)

1   Helix Current Meter

(*from "The Severn Bore" by F. W. Rowbotham* David and Charl

2 The Severn bore

(*Glenfield and Kenne*

3 Vertical lift gates at Tongland spillway, Scotland

flow. In particular, the unique relationship that exists between velocity and depth for the critical condition means that there is useful scope for practical application in flow measurement and in the establishment of control points.

Finally, it must be emphasised that the above relationships are based on the assumptions that:

(a) The increase of pressure with depth follows the hydrostatic law (i.e. flow is uniform or gradually varied).
(b) The bed slope is not excessive ($S_0 < 1/10$).
(c) The velocity distribution is uniform.

Consequently, any application of the specific energy concept to flow through control structures, where the stream lines are sharply curved, must be expected to yield results that are only approximate. Thus in the case of the sluice structure (Fig. 8.5), cited by way of illustration only a cautious qualitative prediction of flow behaviour is strictly speaking justified.

8.3 Transition Through Critical Depth

8.3.1 *Subcritical to Supercritical Flow*

Let us consider the flow behaviour in a uniform channel whose bed slope gradually increases from $S_0 < S_c$ to $S_0 > S_c$ (Fig. 8.6(a)). The discharge is assumed constant. Since the section is uniform the critical

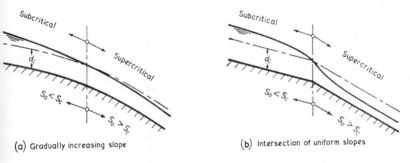

(a) Gradually increasing slope   (b) Intersection of uniform slopes

Figure 8.6   Transition from subcritical to supercritical flow

depth line is parallel to the bed. The depth of flow steadily decreases with increasing slope and passes through the critical depth at the point where the slope is critical ($S_0 = S_c$). This transition from subcritical to supercritical flow is comparatively smooth and is accompanied by little loss of energy or turbulence. Specific energy is lost in moving to the critical depth, but afterwards there is a gain brought about by the external addition of energy derived from the steepening slope. Points of equal specific energy on the energy diagram (Fig. 8.2) are approximately in conformity with the actual depths recorded in the subcritical and supercritical flow regions.

In the case of the intersection of two uniform slopes, a mild slope preceding a steep one (Fig. 8.6(b)), the general effect is very similar, although it is likely that the surface will be rather more disturbed in the transition zone. It should be noted that upstream of the intersection the depth cannot (theoretically, at least) become less than the critical depth, since to do so would require the supply of energy from an external source and this is not available until the steeper slope is encountered.

8.3.2 *Supercritical to Subcritical Flow*

Fig. 8.7 shows the longitudinal section of a uniform channel at a transition between two constant bed slopes, a steep slope changing to

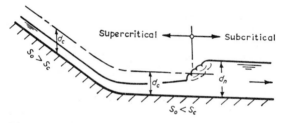

Figure 8.7   Transition from supercritical to subcritical flow

a mild. This could represent conditions at the toe of a spillway chute. The flow upstream is supercritical whereas a short distance downstream it is subcritical at the normal depth. At some intermediate point the transition from supercritical to subcritical flow occurs.

Unlike the transition from subcritical to supercritical flow, the reverse process is far from smooth. What in fact happens is that on entering the mild slope the high velocity of the flow is retarded by frictional resistance, resulting in an increase of depth and a loss of specific energy. Then at some distance upstream of the hypothetical intersection point of the rising surface and the critical depth line, the water abruptly behaves in a highly turbulent manner with much frothing and boiling – the visible indications of air entrainment. Because the mixture has a lesser density than that of normal water, there is appreciable 'bulking' with the surface standing higher than would otherwise be the case. Following a steep but uneven rise of the disturbed surface to a depth approximately equal to the normal depth d_n, the turbulent eddying dies out fairly rapidly and the water flows away freely in the subcritical state. This phenomenon is known as a *hydraulic jump* or *standing wave*.

Although more generally used in connection with the measuring flume (Ch. 9, Sect. 9.6), the name 'standing wave' is very apt for two reasons. Firstly, the water particles do have a wave-like gyratory motion (Fig. 8.8(a)) under the surface roller that develops. Secondly, the surface roller is stationary, the lower end ceaselessly dashing itself against the oncoming current, but without progressing upstream – this is in fulfilment of Eq. (8.8) which states that the Froude number at the critical depth is unity, or in other words that the upstream speed of travel of the shallow-water wave is equal to the downstream velocity of flow.

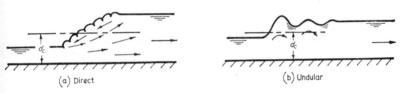

(a) Direct   (b) Undular

Figure 8.8   Types of hydraulic jump

The physical explanation for the jump phenomenon lies in the fact that whilst it would be conceivable for the depth to increase steadily to the critical value, it cannot proceed beyond this to the normal depth without the addition of specific energy from an external source, which in view of the uniform slope is not available. Instead, we find that at a point where the specific energy is still in excess of that represented by the uniform flow condition, the jump discontinuity occurs and the water surface rises rapidly to the normal depth. The turbulent expansion and retardation of the high velocity jet are associated with appreciable energy loss (ultimately dissipated as heat) and the final specific energy is that appropriate to the normal depth.

Normally, the jump takes the *direct* form, described above and illustrated in Fig. 8.8(a). When the downstream depth is only slightly above the critical, however, the jump is usually *undular* in character, the waves rising and falling with a damped oscillatory motion until steady conditions finally obtain (Fig. 8.8(b)). There are, of course, many intermediate forms. The loss of energy increases with the height of jump and it is therefore least with the undular type.

179

In addition to its very considerable merit as an energy dissipator (discussed more fully in the next chapter) it serves other useful practical purposes as follows:

(a) Efficient fluid mixing—due to the highly turbulent nature of the phenomenon. This attribute may be of particular advantage where pollution is involved.

(b) The enhancement of the discharge at a sluice by preventing the tailwater from backing up against the gate. The effective head promoting flow through the sluice is thereby increased.

(c) The recovery of head downstream of a measuring flume (see Ch. 9, Sect. 9.6).

8.3.3 *Analysis of the Hydraulic Jump*

Owing to the initially unknown but appreciable energy loss associated with the hydraulic jump, the application of Bernoulli's equation to the

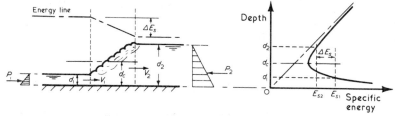

Figure 8.9   Energy loss at a hydraulic jump

channel sections before and after the jump does not provide an adequate means of analysis. On the other hand, the momentum equation is well suited. This is because of the pronounced variation in mean velocity between the two ends of the jump and the fact that a knowledge of internal energy changes is not required. The general agreement between theoretical and experimental results confirms the soundness of the momentum approach.

Let us consider the occurrence of a hydraulic jump in a prismatic channel with horizontal bed carrying a discharge Q (Fig. 8.9). The depth before the jump d_1 is called the *initial depth* and the depth after the jump d_2 the *sequent depth*. This terminology avoids any confusion with the alternate depths and also conveys the correct impression of non-reversibility. The corresponding sectional areas and velocities are A_1, A_2 and V_1, V_2.

The external forces acting on the mass of water between the two sections are the upstream and downstream hydrostatic pressures and the frictional resistance along the boundaries. In the short length of the jump the energy loss due to skin friction is relatively insignificant in comparison with the major internal losses and it may therefore be neglected. Utilising the expression (Eq. (3.6)) for the hydrostatic pressure force on a plane surface, the net force acting parallel to the bed is

$$P_2 - P_1 = w(A_2 \bar{y}_2 - A_1 \bar{y}_1) \qquad (8.19)$$

where $\bar{y}_1$, $\bar{y}_2$ are the respective depths from the surface to the centroids of area A_1, A_2.

The change of momentum $M_1 - M_2$ per sec is given by $(wQ/g) \times (V_1 - V_2)$, or

$$M_1 - M_2 = \frac{wQ^2}{gA_1A_2}(A_2 - A_1) \qquad (8.20)$$

Equating, in accordance with the momentum equation, we obtain

$$P_2 - P_1 = M_1 - M_2 \qquad (8.21)$$

or

$$A_2 \bar{y}_2 - A_1 \bar{y}_1 = \frac{Q^2}{gA_1A_2}(A_2 - A_1) \qquad (8.22)$$

This equation enables the height of the jump to be evaluated provided that either upstream or downstream conditions are known. Ex. 8.2 illustrates the procedure.

The energy dissipated in the jump is represented by the loss of specific energy, or

$$E_{s1} - E_{s2} = (d_1 - d_2) + \frac{(V_1^2 - V_2^2)}{2g}$$

from which

$$E_{s1} - E_{s2} = (d_1 - d_2) + \frac{Q^2}{2gA_1^2A_2^2}(A_2^2 - A_1^2) \qquad (8.23)$$

In design problems concerning the hydraulic jump it is often necessary to ascertain the jump heights appropriate to a range of upstream and downstream conditions. For a given discharge the compilation of a force + momentum $(P + M)$ diagram avoids the need for tedious repetitive calculation. It will be noted from the momentum equation (Eq. (8.21)) that $P_1 + M_1 = P_2 + M_2$. In other words, for a given

181

$P + M$ there are two depths, constituting the initial and sequent depths, d_1 and d_2, satisfying the equation

$$P + M = wA\bar{y} + \frac{wQ^2}{gA} \qquad (8.24)$$

Thus for a constant discharge

$$P + M = \phi(d) \qquad (8.25)$$

The shape of the resulting curve (Fig. 8.10) is rather similar to that for the specific energy. If one depth is known the other may be obtained by a simple vertical projection between the upper and lower limbs. At the minimum value of $P + M$ the two depths coincide giving the critical depth.

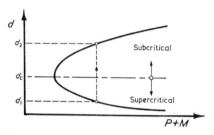

Figure 8.10   Pressure force plus momentum
diagram

Some simplified expressions for the jump result when a rectangular channel, the most common case, is considered. Thus substituting $A_1 = bd_1$, $\bar{y}_1 = d_1/2$, $A_2 = bd_2$, and $\bar{y}_2 = d_2/2$ in Eq. (8.22), we obtain

$$\frac{b}{2}(d_2{}^2 - d_1{}^2) = \frac{Q^2}{gbd_1d_2}(d_2 - d_1)$$

or

$$d_1{}^2 + d_1d_2 - \frac{2q^2}{gd_2} = 0 \qquad (8.26)$$

where q is the discharge per unit width. Utilising the real root of the quadratic,

$$d_1 = \sqrt{\frac{d_2{}^2}{4} + \frac{2q^2}{gd_2}} - \frac{d_2}{2} \quad \text{or} \quad d_2 = \sqrt{\frac{d_1{}^2}{4} + \frac{2q^2}{gd_1}} - \frac{d_1}{2} \qquad (8.27)$$

It is worth noting that from Eq. (8.27) the ratio of the two depths may be expressed in the dimensionless form

$$\frac{d_2}{d_1} = \sqrt{\tfrac{1}{4} + 2F_1{}^2} - \tfrac{1}{2} \qquad (8.28)$$

where F_1 $(= V_1/\sqrt{gd_1})$ is the upstream Froude number. For the jump to occur $d_2 > d_1$, or from the above expression $F_1 > 1$; that is to say the upstream flow must be supercritical.

The energy lost in the jump is

$$E_{s1} - E_{s2} = (d_1 - d_2) + \frac{q^2}{2gd_1{}^2d_2{}^2}(d_2{}^2 - d_1{}^2)$$

Utilising Eq. (8.26) so as to eliminate q, the expression may be simplified to

$$E_{s1} - E_{s2} = \frac{(d_2 - d_1)^3}{4d_1d_2} \tag{8.29}$$

The length of the jump is a subject that has received considerable attention from investigators without, as yet, any entirely satisfactory formula being evolved. This is no doubt due to the fact that the problem is not amenable to theoretical analysis. In addition, practical complications arise from the general instability of the phenomenon and the difficulty of defining the terminal points. As a general statement it may be said that for a rectangular channel the length varies between 5 and 7 times the height of the jump. For the purpose of ensuring a reasonable factor of safety in design, the bed protection works should be based on the upper limit of length. The length of the jump is greater in a trapezoidal channel and there is also some asymmetry due to the non-uniform velocity distribution. For these reasons a rectangular section is generally preferable in the transition region.

The discussion so far has referred to a channel with horizontal bed. In the case of a sloping channel a rigorous analysis would need to take into account the weight component of the water mass comprising the jump. The problem has been studied in detail,[1] but space does not permit an account to be given here. Suffice it to say that provided the slope is not steeper than about 1/10, the error arising from the assumption of a horizontal bed is relatively insignificant in comparison with the overall limits of accuracy.

Example 8.2

A trapezoidal channel with bed width 6 m and side slopes $1\frac{1}{2}:1$ conveys 8·5 cumecs. If the normal depth downstream of the jump of 1·2 m, determine the height of the jump. Also determine the height of the jump for the rectangular channel having the same downstream sectional area of flow and depth.

[1] ELEVATORSKI, E. A. (1959) *Hydraulic Energy Dissipators*, Ch. 6. McGraw-Hill.

Trapezoidal channel

$$A_1 = d_1(6 + 1 \cdot 5d_1); \qquad A_2 = 9 \cdot 36 \text{ m}^2$$

$$A_1 \bar{y}_1 = \Sigma \, y \, dA_1 = \int_0^{d_1} [6 + 3(d_1 - y)]y \, dy = 3d_1^2 + 0 \cdot 5d_1^3$$

$$A_2 \bar{y}_2 = 5 \cdot 18 \text{ m}^3$$

Substituting in Eq. (8.22),

$$5 \cdot 18 - 3d_1^2 - 0 \cdot 5d_1^3 = \frac{8 \cdot 5^2[9 \cdot 36 - d_1(6 + 1 \cdot 5d_1)]}{9 \cdot 81 \times 9 \cdot 36d_1(6 + 1 \cdot 5d_1)}$$

or

$$5 \cdot 97 - 3d_1^2 - 0 \cdot 5d_1^3 = \frac{7 \cdot 36}{d_1(6 + 1 \cdot 5d_1)} \qquad \text{(i)}$$

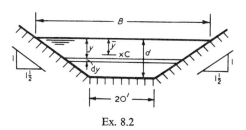

Ex. 8.2

Assuming an average width at the critical depth of 7 m, the approximate value of d_c for the equivalent rectangular channel is given by

$$d_c = \sqrt[3]{\frac{8 \cdot 5^2}{(9 \cdot 81 \times 7^2)}} = 0 \cdot 53 \text{ m}$$

Clearly, the value of d_1 is appreciably less, probably not more than 0·3 m. In view of this relatively low value a close estimate may be obtained by neglecting second order powers and above in Eq. (i). Then $5 \cdot 97 = 7 \cdot 36/(6d_{a1})$, or $d_{a1} = 0 \cdot 21$ m. Hence, by trial and error, $d_1 = 0 \cdot 20$ m and $d_2 - d_1 = \mathbf{1 \cdot 00}$ m.

Rectangular channel

The equivalent width is $b' = 9 \cdot 36/1 \cdot 2 = 7 \cdot 8$ m, so that $q' = 1 \cdot 09$ cumecs/m width. Substituting in Eq. (8.27),

$$d_1' = \sqrt{\frac{1 \cdot 44}{4} + \frac{2 \times 1 \cdot 09^2}{9 \cdot 81 \times 1 \cdot 2}} - \frac{1 \cdot 2}{2} = 0 \cdot 15 \text{ m}$$

so that $d_2 - d_1' = \mathbf{1 \cdot 05}$ m.

8.4 General Equation of Gradually Varied Flow

An expression for the surface slope in gradually varied flow may be derived from a consideration of the rate of change of energy heads along a channel (Fig. 8.11).

The total energy head is given by

$$H = z + d + \frac{V^2}{2g}$$

Differentiating with respect to L, the distance along the channel in the direction of flow,

$$\frac{dH}{dL} = \frac{dz}{dL} + \frac{dd}{dL} + \frac{d}{dL}\left(\frac{V^2}{2g}\right)$$

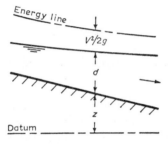

Figure 8.11   Non-uniform flow profile

Now, $-dH/dL$ is the energy or friction slope S_f (negative because energy is decreasing in the direction of flow), $-dz/dL$ is the bed slope S_0, and dd/dL is the surface slope relative to the bed. Then $-dz/dL - dd/dL$ is the surface slope relative to the horizontal datum plane, denoted by S_w; it is of course more commonly referred to as the hydraulic gradient.

Substituting, we obtain

$$\frac{dd}{dL} = S_0 - S_f - \frac{d}{dL}\left(\frac{V^2}{2g}\right) \qquad (8.30)$$

Now

$$\frac{d}{dL}\left(\frac{V^2}{2g}\right) = \frac{dA}{dL}\frac{d}{dA}\left(\frac{Q^2}{2gA^2}\right) = -\frac{dA}{dd}\frac{dd}{dL}\frac{Q^2}{gA^3} = -\frac{dd}{dL}\frac{Q^2B}{gA^3}$$

so that Eq. (8.30) becomes

$$\frac{dd}{dL} = \frac{S_0 - S_f}{1 - Q^2B/gA^3} \qquad (8.31)$$

185

This is the *general equation of gradually varied flow* and it is of fundamental importance in profile analysis. Positive values of dd/dL indicate
increasing depths and negative values decreasing depths.

Provided there is compliance with the basic requirement that the
flow be only gradually varied, it is reasonable to assume that the energy
slope may be represented by one of the uniform flow formulae for the
same discharge and depth. There are obvious reasons why this assumption cannot be reasonably extended to cases of rapidly varied flow, not
least of which is the fact that with an expanding flow stream there are
additional losses due to turbulent eddying.

Adopting the Chézy formula,

$$S_f = \frac{V^2}{C^2 R}$$

(8.32)

and adopting the Manning formula,

$$S_f = \frac{V^2 n^2}{R^{4/3}}$$

(8.33)

Two important surface slope conditions may be identified from an
inspection of Eq. (8.31):

(a) $dd/dL = 0$. In this case the surface is parallel to the bed so that the
flow must be uniform. It is an implicit requirement for uniform flow that
the energy and bed slopes be the same, and indeed the equation is
satisfied when $S_f = S_0$.

(b) $dd/dL = \infty$. The equation yields this value when $Q^2 B/gA^3 = 1$,
which from Eq. (8.4) is seen to be the condition for flow at the critical
depth – in other words, the surface slope is infinity at the critical
depth. This apparently anomalous situation is not so far from reality
as might at first be inferred, since, in the transition from supercritical to
subcritical flow, a hydraulic jump occurs, with the consequence that at
the critical depth the surface slope is indeed steeply inclined. The
discrepancy with theory may be accounted for by the fact that, before
this stage is reached, one of the basic conditions – namely that the flow
shall be gradually varied – is violated, so that the equation is no longer
truly valid. It does nevertheless give a useful qualitative indication.

The surface slope is also infinity at the point where a free overfall
terminates a channel of mild slope (Fig. 8.12). In this case there is a
progressive lowering of the surface as specific energy is lost until at the
brink the specific energy attains its minimum value and the depth is

critical. It cannot decrease further without the addition of external energy and this is not available until the brink is reached. In practice it is found that the depth at the brink is less than the critical depth, the

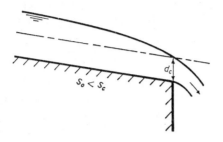

Figure 8.12   Free overfall

latter occurring at a distance of approximately $4d_c$ upstream. The discrepancy with theory at this point is due to the sharp curvature of the stream lines (rapidly varied flow) and to the fact that the pressure throughout the flow stream is atmospheric instead of conforming to a hydrostatic distribution.

8.5 Classification of Surface Profiles

The surface slope equation (Eq. (8.31)) is utilised in establishing the various forms of surface profile that may be encountered in non-uniform flow. Both numerator and denominator are influenced by the channel characteristics and discharge, so in order to present the analysis in the simplest terms, our consideration will be limited to the case of a wide and relatively shallow rectangular channel carrying a constant discharge. Channels of other sectional shape commonly met with in practice will have profile characteristics differing only in degree.

Eq. (8.31) may be rearranged in the form

$$\frac{dd}{dL} = \frac{S_0(1 - S_f/S_0)}{1 - Q^2B/gA^3} \qquad (8.34)$$

According to the Manning formula the energy slope for a wide and shallow channel is given by

$$S_f = \frac{V^2n^2}{d^{4/3}} = \frac{Q^2n^2}{b^2d^{10/3}}$$

187

Also, we know that

$$S_0 = \frac{V^2 n^2}{d_n^{4/3}} = \frac{Q^2 n^2}{b^2 d_n^{10/3}}$$

where d_n is the normal depth in uniform flow.

Thus in Eq. (8.34), $S_t/S_0 = (d_n/d)^{10/3}$; also, since $d_c = \sqrt[3]{Q^2/gB^2}$ and $d = A/B$, we may replace $Q^2 B/gA^3$ by $(d_c/d)^3$. Hence

$$\frac{\mathrm{d}d}{\mathrm{d}L} = S_0 \frac{1 - (d_n/d)^{10/3}}{1 - (d_c/d)^3} \tag{8.35}$$

Earlier in this chapter, bed slopes were classified as mild ($S_0 < S_c$), steep ($S_0 > S_c$), and critical ($S_0 = S_c$). These slopes will now be denoted by M, S, and C. To complete the possible range of slopes two further categories must be included, horizontal slopes (H), and adverse slopes (A). This terminology is attributed to Bakhmeteff.[1]

Utilising Eq. (8.35) we are now in a position to sketch in outline the profile curves for the various types of slope and the known relationships of d, d_n, and d_c. The reader can investigate for himself whether the numerator, denominator, and the resulting quotient will be positive or negative. The profiles are depicted in Fig. 8.13. An exaggerated vertical scale has necessarily been adopted. Accompanying each profile is a typical site condition with which it could be associated. It will be observed that there are three possible profiles for slope types M and S and two possible profiles each for the remainder, making twelve in all.

The critical and normal depth lines and the bed form the boundaries of three zones. Each zone has its own distinctive curve which is valid only within the limits of that zone. All the curves in the upper zone have a positive surface slope and are referred to as *backwater curves*, whilst all those in the centre zone have a negative slope and are called *drawdown curves*.

The zone boundaries are approached in the following characteristic manners:

(a) Upper limits of depth – the curves tend to become asymptotic to a horizontal water line.
(b) Normal depth line – approached asymptotically.
(c) Critical depth line – intersected at right angles.
(d) Bed – intersected at right angles.

With regard to case (a), as the depth increases the velocity decreases (for a given discharge), until at depth infinity the velocity is zero and

[1] BAKHMETEFF, B. A. (1932) *Hydraulics of Open Channels*. McGraw-Hill.

the water surface is horizontal. At the critical depth discontinuity (case (c)) the basic equation is not strictly speaking applicable, for the reasons given earlier. Consequently in this region the curves are shown only in dotted outline. For the bed slope type C the zone boundary conditions are somewhat anomalous, but this is due to the fact that at the critical depth the surface slope is indeterminate. When the depth is exceedingly small (case (d)) the velocity is exceptionally high, and this as we know is associated with flow of a pulsating unsteady nature. The curves are again shown dotted, indicating that a qualitative interpretation only is justified.

The following are some explanatory comments on the various profile types:

Type M. The M1 profile is extremely common. Control structures, such as weirs and sluices, and natural features, such as narrows and bends, may produce a backwater effect in a river channel extending for several km upstream, the practical limit being generally taken as the point where the depth is within about 0·01 m of the normal depth. The M2 profile occurs where the depth is reduced, at say an enlargement in section, or at an approach to a free overfall. The M3 profile may be found downstream of a change in slope from steep to mild or at the exit from a sluice gate. It is caused by upstream conditions and is normally terminated by a hydraulic jump. Both M2 and M3 profiles are very short in comparison with the M1 profile.

Type S. The S1 profile is produced by a control structure such as a dam or sluice situated on a steep slope. It commences with a hydraulic jump and terminates at the obstruction. The S2 profile is generally very short and is commonly found at the entrance to a steep channel or at a change of grade from mild to steep. An S3 profile may be produced downstream of a sluice gate situated on a steep slope or downstream of the intersection of a steep changing to a less steep slope.

Type C. As the normal and critical depths coincide there are only two profiles. These are nearly horizontal, but of course the general instability of the critical depth condition manifests itself in appreciable surface undulation.

Type H. This is the lower limit of mild slope. The normal depth is infinity, so there are only two profiles.

Type A. The adverse bed slope (S_0 negative) is rare. When the depth is infinite, $\mathrm{d}d/\mathrm{d}L = 1/S_0$, signifying a profile asymptotic to the horizontal. The profiles are extremely short.

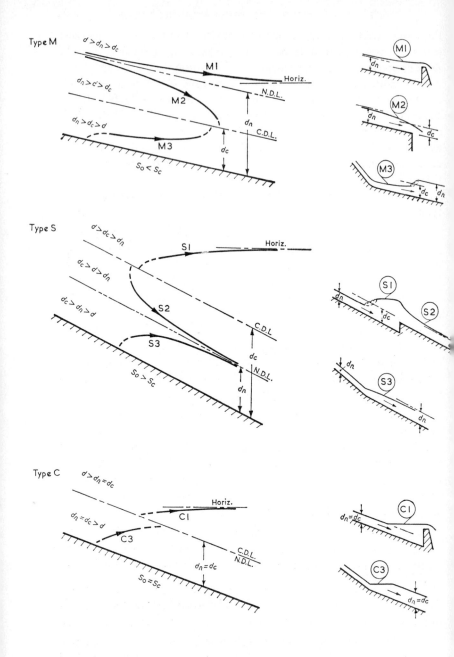

Type M

$d > d_n > d_c$

M1

Horiz.

$d_n > d' > d_c$

M2

N.D.L.

$d_n > d_c > d$

M3

d_n

C.D.L.

d_c

$S_o < S_c$

(M1) d_n

(M2) d_n d_c

(M3) d_c d_n

Type S

$d > d_c > d_n$

S1

Horiz.

$d_c > d > d_n$

S2

$d_c > d_n > d$

S3

C.D.L.

d_c

N.D.L.

d_n

$S_o > S_c$

(S1) d_n d_c

(S2)

(S3) d_n d_n

Type C

$d > d_n = d_c$

C1

Horiz.

$d_n = d_c > d$

C3

C.D.L.
N.D.L.

$d_n = d_c$

$S_o = S_c$

(C1) $d_n = d_c$

(C3) $d_n = d_c$

Figure 8.13 Types of profile curve

190

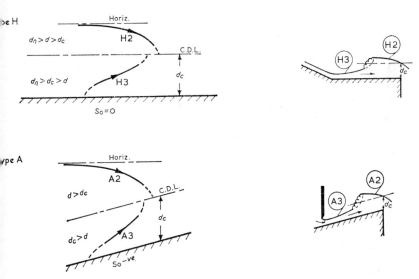

Figure 8.13 (*cont.*)   Types of profile curve

In every case of non-uniform channel flow one of these profile types is applicable. The hydraulic engineer should familiarise himself with this classification since reasonable proficiency in identification under a variety of conditions is a valuable asset in practical design.

8.6 Control Points

Before surface profiles can be evaluated it is necessary to establish the locations of the various *control points*, that is to say points where there is a definite relationship between discharge and depth. Profile plotting proceeds upstream or downstream from these points according to whether the flow is subcritical or supercritical. This follows from an appreciation of the fact that in subcritical flow it is downstream conditions that govern the profile ($V < \sqrt{gd_m}$), whereas in supercritical flow it is upstream conditions that dictate ($V > \sqrt{gd_m}$).

Obvious examples of control points are dams, weirs, and sluices, since the discharge is related to the head by the particular rating curve. As the critical depth depends only on the discharge and channel shape, any well-defined intersection of the profile and critical depth lines also constitutes a control point. However, it is only at the transition from subcritical to supercritical flow that this control is effective, since in

191

the reverse transition a hydraulic jump occurs, the location and height of which are, in general, initially unknown.

Control points also exist at the entry to and exit from a channel. A knowledge of profile characteristics in these regions is important, and to illustrate we will consider the case of a long prismatic channel connecting

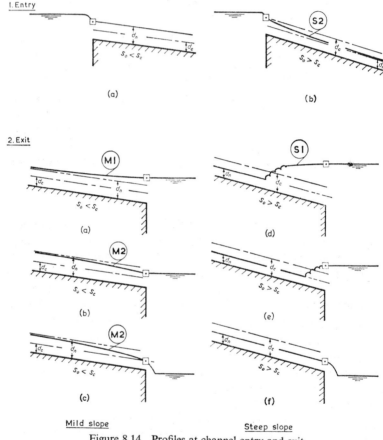

Figure 8.14   Profiles at channel entry and exit

two reservoirs. In view of the considerable channel length a state of uniform flow is attained. The diagrams (Fig. 8.14) show the profiles at entry and exit for both a mild and a steep slope and for a variety of reservoir level at exit. Control points are indicated by square plotted points.

At entry to a mild slope (diagram 1(a)) the uniform flow condition

192

is seen to commence very close to the upstream limit. A sharp drop in the water surface occurs at entry, caused by the conversion of potential to kinetic energy and also by losses due to eddying turbulence. The latter may be appreciable where entry conditions are poor. At entry to a steep slope (diagram 1(b)) the profile drops to the critical depth and then proceeds to the normal depth line by way of a short S2 profile. In both cases the reservoir level and discharge are of course directly related.

Turning now to exit conditions, a high reservoir level at the end of a mild slope (diagram 2(a)) results in an M1 backwater curve, merging at the downstream end with the horizontal pool surface. Theoretically, there should be a slight rise in surface level at the exit point equal to the velocity head; in practice there is no head recovery, the entire velocity energy being dissipated in turbulent eddying. A reservoir level below that of normal depth in the channel (diagram 2(b)) produces an M2 drawdown curve which in the limit has a depth that is critical at the exit point. Any further lowering of the reservoir level (diagram 2(c)) has no influence on the surface profile.

A high reservoir level at the end of a steep slope (diagram 2(d)) produces an S1 profile, commencing upstream at a hydraulic jump located where the depth is sequent to the normal depth. At exit the profile merges with the horizontal pool surface. A reservoir level slightly less high (diagram 2(e)) produces only the jump, which for very small level differences may be imperfect in form. Finally, when the reservoir level is below the normal depth (diagram 2(f)), uniform flow persists to the exit point. Except for the S1 profile the flow is supercritical throughout and, characteristically, upstream conditions are unaffected by downstream reservoir levels.

It is important to add that any hydraulic improvement of the entry and exit geometry (e.g. flaring) may modify the profiles as portrayed.

8.7 Outlining of Surface Profiles

We are now in a position to sketch the profile outlines for a long channel of uniform section having a variety of bed slope. The two channels, shown in Fig. 8.15, each with a sluice gate near the downstream end, will be used to illustrate the procedure. The discharge and sluice gate opening are assumed constant.

The first step is to draw the lines of critical and normal depth. These are parallel to the bed and as the channels are of uniform section the critical depth is the same throughout. Next, the control points are inserted at the appropriate entry and exit sections, at the change of slope from mild to steep, and at the sluice. The latter is a control point in both directions, since the upstream and downstream depths are governed by the sluice gate rating curve.

Referring to each channel in turn:

Channel 1. Proceeding in a downstream direction from the control point *a* because the flow is supercritical, the S2, S3, and M3 profiles may

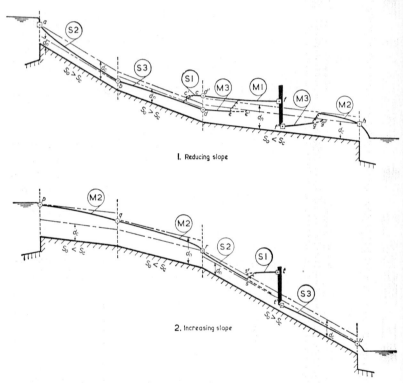

Figure 8.15   Curve identification and control points

be traced as far as *e′*, the intersection with the critical depth line. Similarly, proceeding upstream from the control point *f* because the flow is subcritical, the M1 and S1 profiles extend to the critical depth line at *c′*. At some point between *c′* and *e′* the upper profile has a depth sequent to the lower, and the hydraulic jump transition occurs. The actual position is determined later, when the detailed computation is made, by drawing the locus of sequent depth to the lower profile, making due allowance for the length of the jump.

Downstream of the sluice the position of the hydraulic jump between the M3 and M2 profiles is determined in a similar manner.

Channel 2. Providing that the depth at the sluice is not greatly in excess of the critical, a control point is to be found at r, the transition from subcritical to supercritical flow. A hydraulic jump thus links the S1 and S2 profiles at a point s' where the upper profile is at sequent depth.

Proceeding in an upstream direction from r, because the flow is subcritical, there are two distinct M2 profiles linking with the reservoir at p, the surface level of which must be appropriate to the discharge, due allowance being made for a small drop in level at entry to the channel.

Downstream of the sluice there is an S3 profile terminating in a free overfall to the lower reservoir.

The above treatment is typical, but of course the range of possible channel conditions is so large that there is considerable scope for detailed variation. After identifying the profiles in this manner the actual surface levels may be evaluated by one of the methods described in the next section.

8.8 Profile Evaluation

The general equation of gradually varied flow (Eq. (8.31)) is the basis of all methods of surface profile analysis. These may be described as follows:

(a) *Direct Integration.* From the earliest days hydraulicians have sought to obtain a solution by mathematical means. For the general case of a channel of any given shape the problem is an intractable one. However, by making various simplifying assumptions, a number of investigators have succeeded in achieving a direct integration. Perhaps the best known method is that devised by Bresse.[1] Utilising the Chézy formula he obtained a direct solution for the case of a wide rectangular channel. Tables of the 'Bresse function' facilitate rapid computation of backwater profiles.

In view of the inherent approximations and somewhat restricted application, none of the methods proposed has gained universal acceptance in practice. For this reason a more detailed mention would not be justified here.

---

[1] BRESSE, J. A. CH. (1860) *Cours de Mécanique Appliquée*. Paris.

(b) *Graphical Integration.* Inverting Eq. (8.31) we obtain

$$\frac{dL}{dd} = \frac{1 - Q^2 B / g A^3}{S_0 - S_f} \qquad (8.36)$$

For a constant discharge and bed slope the expression on the right-hand side is some function of the depth. Hence we can write $dL/dd = \phi(d)$. Integrating in order to determine the length L in which the depth changes from d_1 to d_2,

$$\int_0^L dL = L = \int_{d_1}^{d_2} \phi(d)\, dd \qquad (8.37)$$

Thus the length along the channel bounded by the depths d_1 and d_2 is equal to the area under the $\phi(d)$ curve. Fig. 8.16 shows in sketch outline

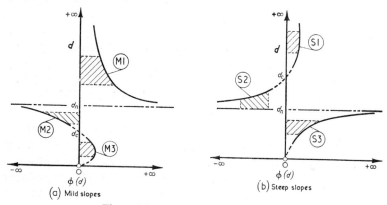

Figure 8.16   Curves relating ϕ(d) and d

the $\phi(d)$ curves corresponding to the principal profile types. The areas shaded represent to scale the length L.

The method of procedure is as follows:

(i) Calculate the value of $\phi(d)$ for depths d_1 and d_2.
(ii) Calculate values of $\phi(d)$ for as many intermediate depths as may be necessary to define the $\phi(d)$ curve with reasonable accuracy. Owing to the sharp curvature, particular care is needed in the region of the critical depth.
(iii) Determine the area under the $\phi(d)$ curve by planimeter or Simpson's rule.
(iv) When a sufficient number of lengths corresponding to various depths have been ascertained a complete profile may be plotted to scale.

The Manning formula is normally employed in the calculation of the energy slope S_f. In view of the asymptotic approach to the normal depth it is generally advisable to halt the integration within 0·01 m of this depth.

A solution can be obtained for any channel shape and the method has wide application. For a given profile, the discharge may be determined by a trial and error procedure. However, when L is given and it is required to find d the method tends to be somewhat laborious. This case is fairly common, particularly in backwater studies, and method (c) is preferable.

(c) *Step-by-step Method*. This method has wide application. It is suitable for the analysis of surface profiles in non-prismatic channels (e.g. rivers), as well as prismatic channels.

From information obtained by hydrographic survey the length to be profiled is divided into a number of reaches, so apportioned that the sectional shape, bed slope, and roughness coefficient are approximately uniform within each reach. Depths are determined successively, proceeding upstream in the case of subcritical flow and downstream for supercritical flow.

In the great majority of river channels the flow is subcritical and the depth not very different to the normal. Profiles upstream are therefore dictated by conditions downstream. Thus it is advantageous, though not essential, to arrange for the downstream end of the length to be investigated to coincide with a control point, such as a sluice, weir, or junction with a large river; the water level will then be known for any given discharge condition. If termination at a control point is not possible, then a tentative water level must be assumed and the profile traced upstream. On repeating the procedure over a range of initial levels it will normally be found, due to the decreasing curvature of the backwater curves, that the effect of small variations in level will fairly quickly be lost.

The basis of computation is as follows:

Fig. 8.17 shows a short reach of channel, length ΔL, where the surface elevation changes from Z_1 to Z_2. As the flow is gradually varied it is reasonable to regard the surface and energy grade lines as straight, the respective slopes being the average of those at each end.

The total heads at sections 1 and 2 respectively are given by

$$H_1 = Z_1 + \frac{V_1^2}{2g}; \qquad H_2 = Z_2 + \frac{V_2^2}{2g} \qquad (8.38)$$

and the difference in total head is

$$H_1 - H_2 = \bar{S}_f \, \Delta L \qquad (8.39)$$

where $\bar{S}_f$ is the average slope of the energy line at the two sections.

The actual procedure is best explained by working through two or three steps in a typical case. The tabulated data below refers to the

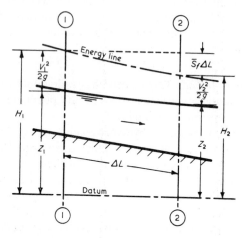

Figure 8.17   Short reach of channel

backwater profile in the tributary of a large river just upstream of the confluence. A bank-full discharge of 4·95 cumecs is investigated.

(1)	(2)	(3)	(4)	(5)	(6)	(7)	(8)	(9)	(10)	(11)	(12)	(13)	(14)	(15)
Dis-tance	ΔL	Z	n	A	V	$\dfrac{V^2}{2g}$	H	R	$R^{4/3}$	$\dfrac{S_f}{\times 10^3}$	$\dfrac{\bar{S}_f}{\times 10^3}$	$\bar{S}_f \Delta L$	H	Re-marks
0	—	78·10	0·03	6·9	0·72	0·03	78·13	0·707	0·630	0·741	—	—	—	Main river
170	170	78·24	0·03	6·1	0·81	0·03	78·27	0·675	0·591	0·998	0·869	0·15	78·28	
270	100	78·36	0·035	6·0	0·83	0·04	78·40	0·665	0·581	1·454	1·226	0·12	78·39	X-bridge

m-s units apply throughout

The various reaches are marked off in the manner previously described. Cols. 1 and 2 are obtained directly from the plans. The value of n (Col. 4) is assessed for each end section, taking into account general channel conditions. At the starting point the water level is fixed by the level in the main river; thus in Col. 3, $Z = 78·10$. Working across, Cols. 5 and 9 are completed by mensuration of the cross-sections. As the discharge is known, Col. 6 follows directly. The value of H (Col. 8) is obtained from Eq. (8.38) and S_f (Col. 11) from Eq. (8.33).

Sometimes it is advisable to insert an additional column to make provision for the eddy loss h_e arising from an expanding flow stream. The value of h_e is based on the variation of mean velocity within the reach and experienced judgement is required in making an assessment. Alternatively, the roughness coefficient is given a value which takes some account of eddy loss.

When dealing with a prismatic channel the step-by-step procedure may be simplified slightly. This is because the calculations can now be based on depths of flow rather than surface elevations and specific energies instead of total heads. Then

$$E_{s2} - E_{s1} = (S_0 - S_f)\varDelta L \tag{8.40}$$

An illustration of the modified procedure is given in Ex. 8.5.

Example 8.3

A sluice set in a rectangular channel 3 m wide with bed slope 1/1000 is adjusted so that the minimum depth in the channel is 0·3 m at a point just downstream of the sluice. Further downstream, the depth increases by means of a hydraulic jump to the normal depth of 1·07 m. Estimate the distance of the jump from the sluice. Assume $n = 0·0135$.

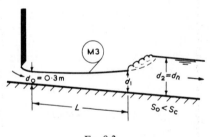

Ex. 8.3

Downstream of the sluice there is a short convergence to the vena contracta, followed by an M3-type surface profile in which the velocity is retarded by boundary friction. The jump occurs at the point where the normal depth is sequent to the lower profile. In order to determine the distance L the two end depths d_0 and d_1 must be known.

For uniform flow, $d_n = 1·07$ m, $A_n = 3·21$ m², $P_n = 5·14$ m, so that $R_n = 0·62$ m. Utilising the Manning formula, $Q = 3·21 \times 0·727/(31·62 \times 0·0135) = 5·46$ cumecs, or $q = 1·82$ cumecs/m width.

Substituting in Eq. (8.27), and bearing in mind that $d_2 = d_n$, we have

$$d_1 = \sqrt{\frac{1\cdot07^2}{4} + \frac{2 \times 1\cdot82^2}{9\cdot81 \times 1\cdot07}} - \frac{1\cdot07}{2} = 0\cdot423 \text{ m}$$

Now, from Eq. (8.33), $S_f = 0\cdot182 \times 10^{-3} V^2/R^{4/3}$, so that

$$\phi(d) = \frac{1 - V^2/gd}{S_0 - S_f} = \frac{10^3(1 - V^2/9\cdot81d)}{1 - 0\cdot182 V^2/R^{4/3}}$$

Tabulating for $\phi(d)$:

d	V	V^2	$\dfrac{V^2}{9\cdot81d}$	$1-\dfrac{V^2}{9\cdot81d}$	R	$R^{4/3}$	$\dfrac{0\cdot182 V^2}{R^{4/3}}$	$1-\dfrac{0\cdot182 V^2}{R^{4/3}}$	$\phi(d)$
0·300	6·07	36·8	12·50	−11·50	0·250	0·158	42·3	−41·3	278
0·362	5·03	25·3	7·13	−6·13	0·292	0·194	23·7	−22·7	270
0·423	4·30	18·5	4·46	−3·46	0·330	0·228	14·8	−13·8	251

Thus

$$L = \int_{0\cdot3}^{0\cdot423} \phi(d) \, \mathrm{d}d = \frac{0\cdot123}{6} (278 + 4 \times 270 + 251) = 330 \text{ m}$$

The vena contracta is very close to the sluice so that the distance from the sluice gate to the jump is approximately **34** m. Incidentally, the critical depth d_c is $\sqrt[3]{1\cdot82^2/9\cdot81} = 0\cdot70$ m.

Example 8.4

A road embankment 60 m wide, with sheet-piled vertical retaining walls, runs parallel to a large river and protects low-lying land from inundation. Owing to an exceptional flood the river level is 0·6 m above that of the road.

Estimate the quantity of water (per ft run of road embankment) discharging into the low-lying area. The transverse surface profile of the road and verges may be regarded as horizontal with the mean value of n equal to 0·04.

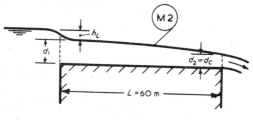

Ex. 8.4

The profile is clearly of the M2 type. At entry the depth is reduced by conversion of potential head to velocity head, while at exit the depth is critical.

200

Eq. (8.36) for a rectangular channel and zero bed slope becomes $dL/dd = -(1 - q^2/gd^3)/S_f$. Substituting $S_f = q^2n^2/d^{10/3} = q^2/(625d^{10/3})$, we obtain $dL = -(625d^{10/3}/q^2 - 63 \cdot 7d^{1/3})\, dd = \phi(d)\, dd$. Integrating,

$$L = 60 = \int_{d_1}^{d_2} \phi(d)\, dd$$

Solution of this equation is by trial and error for various values of q Some idea as to the order of discharge may be gained from an application of the discharge equation for a broad-crested weir (Eq. (9.13), p. 223), which gives $q = 1 \cdot 70 \times 0 \cdot 6^{3/2} = 0 \cdot 79$ cumecs. Owing to frictional effects, the actual discharge will be much less, say $0 \cdot 4$ cumecs.

Tabulating for $q = 0 \cdot 4$ cumecs:

$$h_L = 0 \cdot 05 \text{ m (say)}, \quad d_1 = 0 \cdot 550 \text{ m}, \quad d_2 = d_c = 0 \cdot 254 \text{ m}$$

d	$d^{10/3}$	$\dfrac{625d^{10/3}}{q^2}$	$d^{1/3}$	$63 \cdot 7d^{1/3}$	$\phi(d)$
0·550	0·1360	531·0	0·819	52·1	−478·9
0·451	0·0704	275·0	0·767	48·8	−226·2
0·353	0·0311	121·5	0·707	45·0	−76·5
0·254	0·0103	40·2	0·633	40·2	—

The area under the $\phi(d)$ curve gives $L = 53$ m. A lower value of q is then investigated, and the final q which satisfies $L = 60$ m is obtained by drawing the q-L curve (based on two or three plotted points). In this way q is found to be $0 \cdot 38$ cumecs/m width. A more rigorous analysis would take into account the fact that the critical depth occurs about 1 m upstream of the brink.

Example 8.5

A weir, equipped with crest gates, spans the entire width of a rectangular channel 9 m wide conveying 11 cumecs. The crest is $2 \cdot 5$ m above the bed of the channel and the discharge coefficient is $2 \cdot 0$. The bed slope of the channel is $1/2500$ and $n = 0 \cdot 025$. Determine the profile depth at $\frac{1}{2}$ km intervals for the first 3 km upstream of the control structure.

The normal depth d_n is obtained from the Manning formula. Thus

$$11 = \frac{9d_n}{0 \cdot 025} \times \left(\frac{9d_n}{9 + 2d_n}\right)^{2/3} \times \frac{1}{50}$$

giving $d_n = 1 \cdot 44$ m. Clearly, the surface profile is of the M1 backwater type, but as a check the Froude number at the normal depth is $V_n/\sqrt{gd_n} = 0 \cdot 85/\sqrt{9 \cdot 81 \times 1 \cdot 44} = 0 \cdot 23$, which is < 1. Subcritical flow at all stages is thus confirmed.

From $Q = 2 \cdot 0 L h^{3/2}$ (see Sect. 9.4.2, p. 222) we obtain $h = 0 \cdot 72$ m, so that the depth at the downstream end is 3·22 m. This is the control point in the profile analysis.

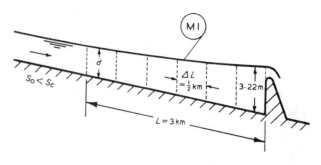

Ex. 8.5

Proceeding upstream and tabulating for sections $\Delta L = \frac{1}{2}$ km apart:

Dis-tance	d	A	V	$\dfrac{V^2}{2g}$	E_s	R	$R^{4/3}$	$\dfrac{V^2}{1 \cdot 60 R^{4/3}}$ $(S_f \times 10^3)$	$\bar{S}_f$ $\times 10^3$	$(S_0 - \bar{S}_f)$ $\times 10^3$	$(S_0 - \bar{S}_f)\Delta L$	E_s	Re-marks
0	3·22	29·0	0·380	0·01	3·23	1·88	2·32	0·0389	—	—	—	—	Weir
500	3·04	27·4	0·402	0·01	3·05	1·81	2·21	0·0457	0·042	0·358	0·18	3·05	
1000	2·86	25·7	0·427	0·01	2·87	1·75	2·11	0·540	0·050	0·350	0·18	2·87	
1500	2·69	24·2	0·455	0·01	2·70	1·68	2·00	0·647	0·059	0·341	0·17	2·70	
2000	2·53	22·8	0·482	0·01	2·54	1·62	1·90	0·765	0·071	0·329	0·16	2·54	
2500	2·38	21·4	0·514	0·01	2·39	1·56	1·80	0·918	0·084	0·316	0·16	2·38	
3000	2·22	20·0	0·550	0·02	2·24	1·49	1·70	1·112	0·101	0·299	0·15	2·24	

m-s units apply throughout

The method of procedure is self-explanatory. At each step a cross-check is made of the specific energy and, if necessary, the depth d is adjusted to produce agreement.

As the depth at 3 km upstream is still 0·78 m above normal it is apparent that the backwater effect extends for a much greater distance upstream.

8.9 Bridge Piers

The pattern of flow around an isolated bridge pier and the associated drag have been discussed in Ch. 4, Sect. 4.11.7. We have now to consider the general effect on channel flow caused by the restricted waterway between bridge piers.

Let us assume that a bridge with piers is erected near to mid-length of a long channel in which the flow is initially uniform. If the flow is supercritical the effect of the piers will be to split the water, leaving a disturbed wake downstream but having no effect upstream other than to create spray, the extent of which will depend primarily on the pier shape.

In the more usual case of subcritical flow, the surface profile is as depicted in Fig. 8.18. The constriction produces an M1-type backwater

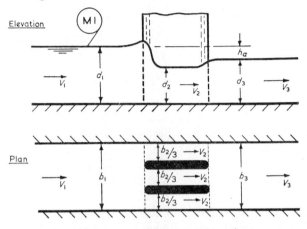

Figure 8.18 Flow between bridge piers

upstream, and an accelerated flow with reduced depth between the piers. Downstream of the bridge, beyond any turbulent wake created by the piers, the flow conditions are near to uniform. The recovery of potential head in the region of expanding flow is often quite small, being dependent on the discharge and boundary geometry. As the backwater effect may raise the water level for a considerable distance upstream, with consequent risk of flooding, it is important to establish the afflux (heading-up)-discharge relationship.

In Fig. 8.18 the bed profile is assumed horizontal. Applying Bernoulli's equation to sections upstream, between the piers, and downstream, we obtain

$$d_1 + \frac{V_1^2}{2g} = d_2 + \frac{V_2^2}{2g} + (h_L)_{1-2}$$

$$= d_3 + \frac{V_3^2}{2g} + (h_L)_{1-2} + (h_L)_{2-3} \qquad (8.41)$$

where $(h_L)_{1-2}$ and $(h_L)_{2-3}$ are the head losses due to the contraction and expansion respectively.

Assuming that the recovery of potential head is negligible (i.e. $(h_L)_{2-3} = (V_2^2 - V_3^2)/2g$), then $d_2 = d_3$ and the afflux h_a is given by

$$h_a = d_1 - d_2 = \frac{V_2^2}{2g} - \frac{V_1^2}{2g} + (h_L)_{1-2}$$

Putting $V_2 = Q/b_2 d_3$, where b_2 is the overall width between the piers, the expression for the discharge becomes

$$Q = b_2 d_3 \sqrt{2g[h_a - (h_L)_{1-2}] + V_1^2}$$

and introducing an empirical discharge coefficient C_b to take account of $(h_L)_{1-2}$ and inaccuracies arising from the simplifying assumptions,

$$Q = C_b b_2 d_3 \sqrt{2gh_a + V_1^2} \qquad (8.42)$$

This formula was developed by d'Aubuisson[1] and has been widely used. Values of C_b for various shapes of bridge pier and waterway geometry are normally based on Yarnell's[2] extensive experimental data. Usually, C_b is between 0·90 and 1·05. For a given discharge and channel characteristics, d_3 (i.e. d_n) may be calculated by a method previously described. The value of h_a follows by simple trial and error substitution in Eq. (8.42).

Another much favoured empirical formula is that proposed by Nagler.[3] It is rather more complicated than d'Aubuisson's, but as there is some provision for recovery of downstream head, a superior accuracy under conditions of low turbulence is claimed.

In important and unusual cases, recourse to hydraulic model tests may be necessary.

8.10 Translatory Waves in Channels

8.10.1

Our consideration of channel flow has so far been limited to steady conditions. Attention will now be directed to the unsteady state. This is represented in practice by such phenomena as surges in the head- and tail-race canals of turbine plants, and flood waves in natural channels. It will be appreciated that the introduction of the time variable inevitably complicates the analysis, so in order to keep within the scope of the present text the simplest cases only will be discussed.

[1] D'AUBUISSON, J. F. (1840) *Traité d'Hydraulique*. Paris.

[2] YARNELL, D. L. (1934) 'Bridge Piers as Channel Obstructions', *U.S. Dept. Ag. Tech. Bull.* 442.

[3] NAGLER, F. A. (1918) 'Obstruction of Bridge Piers to the Flow of Water', *Trans. Am. Soc. C.E.*, **82**, 334.

8.10.2 *Types of Surge*

A sudden increase or decrease of discharge in a channel will produce a *translatory* or *surge wave*. There are four possible basic types of surge wave according to the direction of propagation and the change in depth–upstream or downstream, increase or decrease, respectively. A

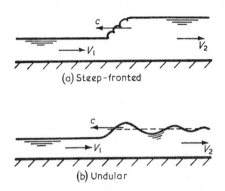

Figure 8.19   Positive backwater surge

wave travelling upstream is known as a backwater surge while its counterpart travelling downstream is known as a flood surge. A wave producing an increase in depth is positive and one producing a decrease is negative.

Positive waves are inherently stable and may be steep-fronted as in Fig. 8.19(a), or, in the case of low amplitudes, undular as in Fig. 8.19(b). Negative waves are always unstable and a steep wave front cannot be

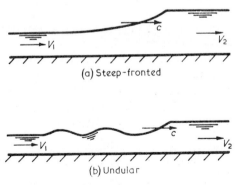

Figure 8.20   Negative flood surge

205

sustained; the reason for this is that the upper particles of water, being in water of a greater depth, travel faster and thus tend to cause a more gradual decrease in depth along the channel. The alternative profiles of negative surge waves are shown in Fig. 8.20.

8.10.3 *Surge Celerity*[1]

Fig. 8.21 shows a horizontal prismatic channel with terminal sluice gate. As the result of a sudden lowering of the gate a positive backwater surge is propagated upstream with celerity c.

In one unit of time the volume of water $A_1 V_1$ is distributed partly to the storage volume $c(A_2 - A_1)$ and partly to the flow out of the section $A_2 V_2$, the suffices 1 and 2 referring to the respective conditions

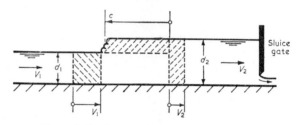

Figure 8.21   Surge wave propagated upstream

preceding and following the passage of the surge wave. Therefore, for continuity, $A_1 V_1 = A_2 V_2 + c(A_2 - A_1)$, from which

$$V_2 = \frac{A_1 V_1 - c(A_2 - A_1)}{A_2} \tag{8.43}$$

A change of momentum occurs as the wave passes. The mass of water whose momentum is changed per unit of time is wA_1/g multiplied by the vectorial sum of the wave celerity and the upstream water velocity $(c + V_1)$. The reduction in velocity is $V_1 - V_2$ so that the change in momentum is given by

$$M_1 - M_2 = \frac{w}{g} A_1(c + V_1)(V_1 - V_2) \tag{8.44}$$

Neglecting the effects of frictional resistance on the bed and sides over the short distance concerned, we can equate the change of momentum to the unbalanced hydrostatic force, so that

$$wA_2 \bar{y}_2 - wA_1 \bar{y}_1 = \frac{wA_1}{g} (c + V_1)(V_1 - V_2) \tag{8.45}$$

---

[1] The speed of travel of surge waves is generally much faster than water velocities and in order to obtain some distinction the word 'celerity' is adopted.

where $\bar{y}_1$ and $\bar{y}_2$ are the respective depths of the centroids of area below the surface. Substituting for V_2 from Eq. (8.43) and simplifying,

$$A_2\bar{y}_2 - A_1\bar{y}_1 = \frac{A_1}{g}\left(\frac{A_2 - A_1}{A_2}\right)(c + V_1)^2$$

from which

$$c = \left[\frac{gA_2(A_2\bar{y}_2 - A_1\bar{y}_1)}{A_1(A_2 - A_1)}\right]^{1/2} - V_1 \qquad (8.46)$$

It should be noted that if the surge wave remains stationary ($c = 0$), Eq. (8.46) becomes

$$V_1 = \left[\frac{gA_2(A_2\bar{y}_2 - A_1\bar{y}_1)}{A_1(A_2 - A_1)}\right]^{1/2} \qquad (8.47)$$

which will be recognised as being identical with Eq. (8.22), the general equation for the hydraulic jump. The latter may thus be described as a stationary surge wave.

Also, it follows from Eq. (8.46) that if d_1 is greater than the critical depth, the surge wave may travel up or downstream according to whether the right-hand side of the equation is positive or negative. If, however, d_1 is less than the critical depth, V_1 is always the greater so that the wave must travel downstream.

For a rectangular channel, $A = bd$ and $\bar{y} = d/2$, so that Eq. (8.46) becomes

$$c = \left[\frac{gd_2(d_1 + d_2)}{2d_1}\right]^{1/2} - V_1 \qquad (8.48)$$

If the surge height is insignificant in relation to the depth ($d_1 \simeq d_2$), this equation may be simplified to

$$c = \sqrt{gd_1} - V_1 \qquad (8.49)$$

For $c = 0$, the water velocity is seen to be the same as the critical velocity in a rectangular channel (Eq. (8.15)).

In practice it is found that the surge wave suffers some distortion in a non-rectangular section due to the influence of the sides. It will be noted that the celerity increases with the wave height and depth; this accounts for the profile of negative waves, also for the fact that even with more gradual changes in flow there is a tendency for a series of small waves to be propagated which, because of the successively deepening water and consequent higher celerity, tend to coalesce into larger waves.

In the case of short channels, the simplifying assumptions of frictionless flow and a horizontal bed are justified on the grounds that the theoretical approach gives reasonably realistic results. But in the case of long channels, the dampening of a surge in its course of travel is significant, so that a step-by-step analysis, taking into account these factors, is normally essential if the passage of the surge is to be traced over any distance.

Incidentally, for still water ($V_1 = 0$), Eq. (8.49) becomes $c = \sqrt{gd}$. This expression is identical with that for the celerity of small gravity waves (Fig. 8.22) whose wave length λ_w is appreciable relative to the depth ($\lambda_w > 25d$).[1] These waves are translatory in character and are often observed in the vicinity of a shallow coastline. The tides are also in this category, the wave length being related to the period T (12·4

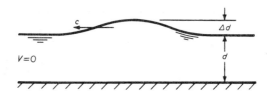

Figure 8.22   Propagation of a small gravity wave

hours for a semi-diurnal tide) by the expression $c = \lambda_w/T$. A simple calculation shows that the wave length is many times greater than even the maximum ocean depth (about 11 km).

Example 8.6

A rectangular head-race canal, 12 m wide, supplying a turbine installation has a bed gradient 1/1800 and a roughness coefficient n assessed at 0·02. Under full load conditions the canal supplies 40 cumecs to the turbines and the flow in the canal is uniform. If, due to a major rejection of load, the turbine gates close rapidly so that only 3 cumecs is accepted, determine the *initial* celerity with which the surge wave is propagated upstream.

The normal depth in uniform flow is obtained from the Manning formula and is found to be $d_n = 1\cdot93$ m. Then $V_n = 1\cdot73$ m/s.

Conditions immediately following instantaneous partial closure are shown in inset. The velocity after the surge has passed is $V_0 = 3/(12d_0) = 0\cdot25/d_0$.

---

[1] If we bring the wave to rest by superimposing an equal and opposite velocity, then for constant specific energy, $d + (c + \Delta c)^2/2g = d + \Delta d + c^2/2g$. Neglecting 2nd order powers of small quantities, $c\,\Delta c = g\,\Delta d$. Also, for continuity, $(c + \Delta c)d = c(d + \Delta d)$, or $d\,\Delta c = c\,\Delta d$. Thus $c = \sqrt{gd}$.

For a rectangular channel the continuity relationship (Eq. (8.43)) gives $0\cdot25/d_0 = [d_n V_n - c(d_0 - d_n)]/d_0$, from which

$$c = \frac{3\cdot08}{d_0 - 1\cdot93} \qquad \text{(i)}$$

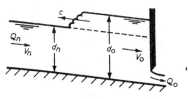

Ex. 8.6

and the momentum relationship (Eq. (8.48)) gives

$$c = \left(\frac{9\cdot81 d_0(1\cdot93 + d_0)}{2 \times 1\cdot93}\right)^{1/2} - 1\cdot73 \qquad \text{(ii)}$$

Equating (i) and (ii), and solving for d_0 by trial and error, we obtain $d_0 = 2\cdot71$ m, so that $c = \mathbf{3\cdot93}$ m/s.

8.10.4 *Bore in a Tidal River*

The characteristics of certain tidal rivers and estuaries are favourable to the passage of a surge wave, usually called a *bore* or *eagre*, which is propagated upstream with the incoming tide. A bore generally attains its maximum celerity in the narrowest part of an estuary and may penetrate a considerable distance up a tidal river channel before its forward momentum is finally absorbed by friction.

The wave front of most bores is undular. Steep fronts are associated only with those bores which have a considerable wave height relative to the water depth. Reference has been made earlier to the tendency for small waves to coalesce into larger ones due to the faster speed of travel in deeper water. This accounts for the well-defined nature of a bore even though the sea level in the mouth of the estuary may be rising quite gradually. The well-known bore of the R. Severn is an excellent example (Plate 2). It first appears in the lower reaches of the estuary and the celerity ranges up to 6 m/s with a wave height of up to $1\cdot5$ m. The bore (eagre) that occurs in the tidal reaches of the R. Trent is also noteworthy.

The factors favouring the phenomenon are (a) a large tidal range, (b) an estuarine channel converging towards the mouth of a tidal river, and (c) shallow water. Both the factors (a) and (c) attain their maximum favourability when the tide ebbs to the greatest extent, which is during the period of spring tides. This is when the bores in the R. Severn and R. Trent may be observed. It is indeed fortunate that bores always occur at times when the water levels are low. Otherwise, specially raised flood banks would be required to contain them.

Fig. 8.23 outlines a bore moving upstream. It is supposed that the bore is of such a magnitude that the velocity V_2 is in the upstream direction, in which case the sign of V_2 in Eqs. (8.43) and (8.44) is reversed. The net result is that Eq. (8.46), giving the celerity of the bore, remains unchanged. In the passage of lesser bores the water velocity may remain in a downstream direction.

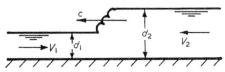

Figure 8.23   Bore moving upstream

Of course, the remarks made earlier on the limitations of the theory when applied to long channels are also relevant to the progress of a bore. In fact, one additional factor, that of change of sectional shape, must be taken into account.

Example 8.7

At low water the downstream velocity of a wide tidal river is 0·6 m/s and the average depth is 1·2 m. If a bore is observed to pass upstream with a celerity of 4·5 m/s, determine its height and the velocity of flow after it has passed.

Substituting in Eq. (8.48),

$$4 \cdot 5 = \left[\frac{9 \cdot 81 d_2 (1 \cdot 2 + d_2)}{2 \times 1 \cdot 2} \right]^{1/2} - 0 \cdot 6$$

from which $d_2 = 1 \cdot 99$ m, and $d_2 - d_1 = \mathbf{0 \cdot 79}$ m.

The velocity V_2 after passage of the bore is obtained from Eq. (8.43), or $V_2 = [4 \cdot 5(1 \cdot 99 - 1 \cdot 2) - 1 \cdot 2 \times 0 \cdot 6]/1 \cdot 99 = \mathbf{1 \cdot 42}$ m/s in an upstream direction.

Further Reading

BAKHMETEFF, B. A. (1932) *Hydraulics of Open Channels*. McGraw-Hill.

CHOW, V. T. (1959) *Open-Channel Hydraulics*. McGraw-Hill.

ELEVATORSKI, E. A. (1959) *Hydraulic Energy Dissipators*. McGraw-Hill.

HENDERSON, F. M. (1966) *Open Channel Flow*. Macmillan.

KING, H. W. and BRATER, E. F. (1963) *Handbook of Hydraulics*, Sect. 8. McGraw-Hill (5th Edition)

LELIAVSKY, S. (1965) *Irrigation and Hydraulic Design*, vol. 1, Chs. 2–3. Chapman and Hall.

SELLIN, R. H. J. (1969) *Flow in Channels*. Macmillan.

WOODWARD, S. M. and POSEY, C. J. (1941) *Hydraulics of Steady Flow in Open Channels*. Wiley.

CHAPTER NINE

Hydraulic Structures

9.1 Introduction

Control of water level and regulation of discharge are necessary for purposes of irrigation, water conservation, flood alleviation, and inland navigation. A wide variety of hydraulic structure is available to suit the particular need; the range is from the weirs or sluices encountered on small watercourses to the overflow spillways of large dams.

Many control structures have a useful secondary function in that they may be utilised for the measurement of discharge, although the movable gates that are usually incorporated result in a complicated head-discharge relationship and some loss of accuracy.

Sharp-crested weirs, long-base weirs and throated flumes are in a rather special category since their sole function is discharge measurement. Parts 4A (1965), 4B (1969) and 4C (1971) of B.S. 3680, *Measurement of Liquid Flow in Open Channels*, set forth detailed proportions with appropriate calibrations for the more customary forms of these structures.

The present chapter is concerned with the hydraulic characteristics of the various types of installation. In the final section the effect of reservoirs on the hydrographs of natural watercourses is examined.

9.2 Sluices and Gates

Water issues at relatively high velocity from the opening caused by the raising of a sluice (Fig. 9.1) or other type of movable gate (Fig. 9.2). The flow behaviour resembles that of a jet issuing from an orifice. Thus a contraction of the jet takes place downstream of the opening and the velocity of efflux is dependent on the square root of the head on the opening. There is a difference, however, in that the presence of the bed prevents any contraction of the jet on the underside, and this also means that the pressure distribution a short distance from the opening is approximately hydrostatic instead of uniformly atmospheric. Although the high velocity of the jet is useful in preventing siltation, it tends to cause scour of an erodible bed, and for this reason some form of protective apron is generally necessary.

In Figs. 9.1 and 9.2 an application of the Bernoulli equation to the upstream and downstream sections gives $h_0 + h_v = h_1 + V_1^2/2g$, where h_v is the upstream velocity head $(\alpha_0 V_0^2/2g)$. Thus $V_1 = \{2g[(h_0 - h_1) + h_v]\}^{1/2}$, so that the discharge Q is given by

$$Q = C_c A \sqrt{2g[(h_0 - h_1) + h_v]} \tag{9.1}$$

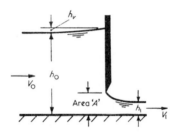

Figure 9.1   Flow under a vertical sluice gate

where A is the area of the opening and C_c is the coefficient of contraction. This expression may be simplified to

$$Q = CA\sqrt{h_0} \tag{9.2}$$

where C is an overall coefficient of discharge which incorporates the coefficient of contraction and the effects of downstream head, velocity of approach, and energy loss. The value of C varies with the head and

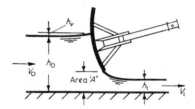

Figure 9.2   Flow under a radial gate

gate opening and is evidently governed primarily by the geometry of the opening and the downstream conditions. If the discharge is free, that is to say the depth h_1 is less than the critical depth for the particular channel section, the value of C is greater than when the jet is submerged (Fig. 9.3), since under the latter condition a lesser effective head is available.

    The energy losses in the converging flow at a sluice gate are relatively small so that there is reasonable similarity to the behaviour of an ideal fluid. A flow net analysis may therefore be used in order to determine an approximate value for C. For the simplest case, that of a vertical sharp-edged gate spanning the full channel width, with free discharge and $h_0/h_1 \to \infty$, it may be shown that C_c in Eq. (9.1) is equal to 0·61. A more general procedure for effecting the calibration is to conduct field tests on the actual structure, measuring both head and discharge, or to investigate the performance of a scale model in a hydraulics laboratory.

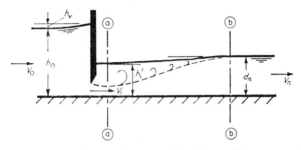

Figure 9.3 Sluice gate with submerged jet

The value of h_1' in Fig. 9.3 may be estimated by applying the momentum equation to sections aa and bb on either side of the expanding jet. Thus, assuming a channel of rectangular section with the sluice opening spanning the entire width, and neglecting friction, we have

$$\frac{w d_n^{\,2}}{2} - \frac{w h_1'^{\,2}}{2} = \rho q (V_1 - V_n) \qquad (9.3)$$

where q is the discharge per unit width and the suffix n denotes the conditions for uniform flow. Knowing q, V_1, and V_n, it is possible to obtain h_1', which is clearly somewhat less than d_n.

The momentum equation may also be utilised to determine the horizontal force acting on a gate. This is demonstrated in the following example.

Example 9.1

The sluice gate in Fig. 9.1 spans a wide rectangular channel and is raised 0·25 m above the channel floor. The upstream depth is 3 m. Estimate the horizontal force per metre run exerted on the gate. Take $C = 2\cdot5$.

The discharge per unit width is given by $q = 2\cdot5 \times 0\cdot25 \times 3^{1/2} = 1\cdot08$ cumecs, so that $V_0 = 1\cdot08/3 = 0\cdot36$ m/s and $V_0^2/2g = 0\cdot01$ m.
Application of Bernoulli's equation, and the assumption that $\alpha_0 = 1$,

leads to $3 + 0\cdot01 = h_1 + 1\cdot08^2/19\cdot6h_1^2$, the solution to which is $h_1 = 0\cdot14$ m. Thus $V_1 = 1\cdot08/0\cdot14 = 7\cdot71$ m/s.

The corresponding expression for the momentum equation is $wh_0^2/2 - wh_1^2/2 - F = \rho q(V_1 - V_0)$, where F is the horizontal force per metre run on the gate. Thus $F = (9\cdot81/2)(9 - 0\cdot02) - 1\cdot08 \times 7\cdot35 = \mathbf{36\cdot1}$ kN.

It should be noted that this method of evaluation gives no information concerning the line of action. However, since the value of F is generally so close to that applicable to hydrostatic conditions (in this case $F = 9\cdot81 \times 2\cdot75^2/2 = 37\cdot1$ kN), a very fair approximation to the actual location is readily obtained.

9.3 Sharp-crested Weirs

9.3.1 *General Characteristics*

Certain installation requirements must be complied with if these structures are to satisfactorily fulfil their role as accurate measuring devices. Thus the upstream face must be vertical and the crest plates, usually of brass or stainless steel, must be provided with an accurately finished square upstream edge, a crest width not exceeding 3 mm and a bevel on the downstream side. This results in a structure that is necessarily somewhat slender and lacking in stability, and therefore too vulnerable to handle the debris-laden flood discharges that occur periodically in most rivers. A further disadvantage is that under continuous operating conditions there is a tendency for the crest to become rounded and this adversely affects the calibration. The field of application is thus generally limited to laboratories, small artificial channels, and streams.

The rectangular weir and the triangular weir (V-notch) are the commonest and most efficient types of sharp-crested structure. Other forms include the trapezoidal weir and the compound weir.

9.3.2 *Rectangular Weir*

Fig. 9.4 shows the pattern of stream lines in two-dimensional flow over a rectangular weir. The water is accelerated as it approaches the weir, the increase in velocity being derived from a reduction in potential head which manifests itself in a downward slope of the surface. A continued acceleration under gravity takes place downstream of the crest, the overflowing sheet of water being called the *nappe*. When the weir is operating correctly the nappe springs clear of the crest and this condition is called *free*. The upward component of velocity adjacent to the upstream face of the weir produces an initial sharp curvature and slight rise in level of the underside of the nappe. The falling nappe has an approximately parabolic trajectory in accordance with the relationships set forth in Ch. 4, Sect. 4.6.5. Both the upper and lower surfaces of the nappe are exposed to the atmosphere so that the pressure distribution throughout is near to atmospheric.

The backwater beneath the nappe tends to rise above the level downstream since a static force must be exerted in order to balance the momentum force created by the deflection of the jet. Referring to the inset diagram of Fig. 9.4, an approximate estimate of the backwater depth d_1 is obtained by substitution in the momentum equation:

$$\frac{wd_1{}^2}{2} - \frac{wd_2{}^2}{2} = \rho q V_2 (1 - \cos \phi) \tag{9.4}$$

where q is the discharge per unit width and ϕ is the angle of inclination of the jet on striking the channel floor.

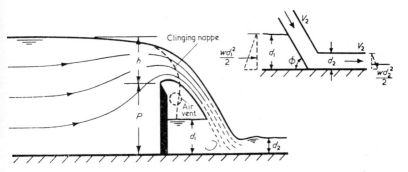

Figure 9.4   Flow over a rectangular weir

Some of the air on the underside of the nappe becomes entrained in the falling jet. Unless this air is replenished, a vacuum pressure is created which draws the backwater upwards and causes the nappe to adhere to the downstream face of the weir. A *clinging* nappe, shown dashed in Fig. 9.4, is associated with an increase in discharge and some instability in the flow behaviour. This condition is quite incompatible with accurate discharge measurement. In cases where the length of the weir is less than the channel width, the underside of the nappe is naturally aerated, but where the weir spans the entire width it is essential to provide air vents. The full span or *suppressed* condition is the more satisfactory since there are no end contractions. Their presence reduces the effective crest length from L to L_e (Fig. 9.5).

In order to determine the head on the weir it is necessary to measure it beyond the point where the downward acceleration commences; the recommended (B.S. 3680: Part 4A: 1965) distance from the weir is

215

between three and four times the maximum head. A hook or float gauge is normally employed and the disturbance caused by surface ripples is avoided if the instrument is located in a small stilling well connected to the channel. A straight uniform approach to the weir ensures that there is reasonable uniformity in velocity distribution and an absence of eddying pulsations.

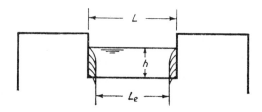

Figure 9.5   Rectangular weir with end contractions

An expression for the discharge may be derived from a consideration of the flow over a hypothetical weir where there is no drawdown or contraction of the nappe. Thus in Fig. 9.6 the upper and lower surfaces of the nappe are horizontal. Upstream, the pressure distribution with depth is hydrostatic while at the crest it is uniformly atmospheric.

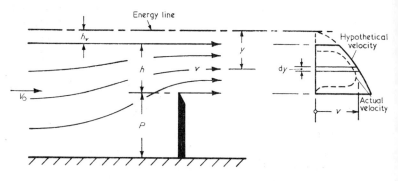

Figure 9.6   Hypothetical flow pattern for a rectangular weir (no drawdown)

Assuming an ideal fluid, the total head is constant throughout the flow stream, so that $P + h + h_v = P + (h + h_v - y) + v^2/2g$, where P is the height of the weir, h the measured head, h_v the upstream velocity head, and v the filament velocity at vertical distance y below the energy line. Hence $v = \sqrt{2gy}$, indicating a parabolic velocity distribution. The actual velocity distribution with normal nappe behaviour resembles that shown in dashed outline.

216

The discharge through an elemental horizontal strip of the nappe of depth dy and unit length is $dq = v\,dy = \sqrt{2gy}\,dy$. Integrating,

$$q = \sqrt{2g} \int_{h_v}^{h+h_v} y^{1/2}\,dy = \tfrac{2}{3}\sqrt{2g}[(h + h_v)^{3/2} - h_v^{3/2}]$$

Thus the total theoretical discharge Q for a suppressed weir, crest length L, is given by

$$Q = \tfrac{2}{3}\sqrt{2g}\,L[(h + h_v)^{3/2} - h_v^{3/2}] \qquad (9.5)$$

The actual discharge is much less, owing principally to the marked contraction of the nappe and the curvature of the stream lines. Introducing a coefficient of discharge C_d and an effective head h_e, the discharge expression becomes

$$Q = \tfrac{2}{3}\sqrt{2g}\,C_d L h_e^{3/2} \qquad (9.6)$$

If the *Rehbock formula*, as quoted in B.S. 3680: Part 4A: 1965 is employed, then $C_d = 0\cdot602 + 0\cdot083h/P$ and $h_e = h + 0\cdot0012$ m.

For a weir with end contractions, either C_d or L must be amended so as to allow for the reduction in discharge. The *Hamilton Smith formula*, in which $C_d = 0\cdot616(1 - 0\cdot1h/L)$ and $h_e = h$, is quoted in the British Standard.

Provided that there is compliance with certain specified design conditions, the values of C_d in these formulae are stated to have an accuracy of 1 per cent. Allowing for errors of head measurement, this means that it should normally be possible to determine the discharge within 2 per cent.

Another approach to obtaining a discharge expression is by means of dimensional analysis. The discharge per unit length q is dependent on the measured head h, the weir geometry represented by P, and the physical quantities ρ, g, μ, and σ (surface tension). Thus $q = \phi(h, P, \rho, g, \mu, \sigma)$. By dimensional analysis these may be combined in the form

$$q = g^{1/2}h^{3/2}\,\psi\left(\frac{h}{P}, \frac{\rho g^{1/2}h^{3/2}}{\mu}, \frac{\rho g h^2}{\sigma}\right)$$

Since $(gh)^{1/2}$ represents a velocity, the second and third terms in the brackets are forms of the Reynolds and Weber numbers respectively. Thus

$$q = g^{1/2}h^{3/2}\,\psi\left(\frac{h}{P}, R, W\right) \qquad (9.7)$$

or

$$Q = CLh^{3/2} \qquad (9.8)$$

where C is an overall coefficient which takes into account the various dependent factors, including any end contractions. This is the *general weir equation*, C being known as the *weir coefficient*.

Clearly, surface tension has a pronounced effect at very low heads, tending to bring about a clinging of the nappe, but is of little consequence under normal operating conditions. If we wish to ascertain the general influence of viscosity we may, from experimental data (h/P constant), express graphically the relationship between the dimensionless term $q/g^{1/2}h^{3/2}$ and R. Except at low Reynolds numbers (i.e. very small heads), the graph is a near horizontal line indicating that viscous effects are relatively insignificant. This confirms an earlier intimation that the pattern of stream lines for an ideal fluid, as derived in Ex. 4.2 (p. 53), gives a very fair representation of actual conditions.

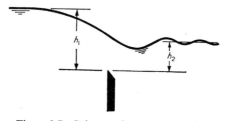

Figure 9.7　Submerged rectangular weir

Thus one of the chief merits of this type of flow measuring device is that the value of C (or C_d) is reasonably stable over a wide range of operating conditions. Very low heads ($h < 0.025$ m), outside the specified range, are an exception since the variation in C is then large and somewhat unpredictable.

In the interests of accurate measurement it is important that the tail-water level should be low enough for there to be no interference with the ventilation or the free discharge of the overflowing jet. If the downstream level is above the crest level (Fig. 9.7) the weir is said to be operating submerged, the ratio of heads ($h_2:h_1$) being called the *submergence ratio*. The effect of the downstream head is to reduce the discharge below that represented by the free condition at the particular upstream head. In these circumstances *Villemonte's formula* is applicable, namely

$$Q = Q_1\left[1 - \left(\frac{h_2}{h_1}\right)^{3/2}\right]^{0.385} \tag{9.9}$$

where Q_1 is the free discharge at head h_1. It will be noted that two head measurements are entailed. The degree of accuracy of the discharge

218

measurement inevitably suffers and the undulatory condition of the downstream surface is an additional hazard. However, on the occasion of exceptionally high flows, some approximate measure of the discharge is very often better than none.

9.3.3 *Triangular Weir or V-notch*

The triangular weir is superior to the rectangular type when very small discharges require to be measured. This is because the small sectional area of the nappe leads to a relatively greater variation in head and thus a more sensitive discharge measurement. On the other hand, the excessive head at the higher flows places an upper limit on the range of application.

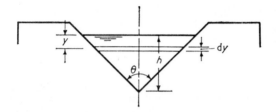

Figure 9.8   Triangular weir

A similar analytical procedure to that adopted for the rectangular weir, that is to say the integration of elemental discharges through horizontal strips of the nappe (Fig. 9.8), leads to the discharge expression

$$Q = \frac{8}{15}\sqrt{2g}\, \tan\frac{\theta}{2}\, h^{5/2}$$

where θ is the apex angle. The velocity head is neglected since it is quite insignificant in comparison with the measured head. Again, the actual discharge is much less, and introducing the necessary discharge coefficient,

$$Q = \frac{8}{15}\sqrt{2g}\, C_{\mathrm{d}} \tan\frac{\theta}{2}\, h^{5/2} \tag{9.10}$$

There are three principal sizes of V-notch, namely $\theta = 90°$, $53°\ 8'$, and $28°\ 4'$, corresponding to $\tan(\theta/2) = 1\cdot0, 0\cdot5$, and $0\cdot25$, respectively. Values of C_{d} for the three sizes over the range $h = 0\cdot05$ m to $0\cdot38$ m are tabulated in B.S. 3680: Part 4A: 1965. The standard of accuracy

219

is stated to be within 1 per cent. Again, there is the merit of very stable values; for instance with the 90° V-notch, C_d is approximately 0·585 for all heads in excess of 0·16 m.

Example 9.2

The measured head on a 90° V-notch is 0·32 m ($C_d = 0·585$). To what extent does a possible error of ±1·5 mm in measuring the head affect the degree of accuracy of the discharge measurement? What is the corresponding effect on the accuracy if a rectangular weir, crest length 1 m, with end contractions suppressed, is employed to handle the same discharge? The crest is 0·6 m above the stream bed.

V-notch

Substituting in Eq. (9.10), $Q = 1·38 \times 0·32^{5/2} = 0·080$ cumecs. The discharge formula may be expressed in the form: $\log Q = \log 1·38 + 2·5 \log h$. Differentiating, $dQ/Q = 2·5(dh/h)$; and introducing finite elements, $\Delta Q/Q = 2·5 \times 0·0015/0·32 = 0·0117$. Thus the possible error in Q is **±1·2** per cent.

Rectangular weir

In Eq. (9.6), $Q = 0·080$ cumecs, $C_d = 0·602 + 0·083h/0·6$, $h_e = h + 0·0012$ m, so that a simple trial and error solution is necessary. It is found that $h = 0·126$ m. In this case, with close approximation, $\Delta Q/Q = 1·5\Delta h/h = 1·5 \times 0·0015/0·126 = 0·0179$, or the possible error is **±1·8** per cent.

Thus it is evident that under the particular conditions quoted the V-notch is the more accurate measuring device.

9.3.4 *Trapezoidal Weir*

The trapezoidal or Cipolletti (the name of its originator) weir, shown in Fig. 9.9, has the merit that no correction for end contractions is

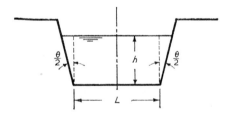

Figure 9.9   Cipolletti weir

required. This is because the sloping sides compensate for the reduction of discharge brought about by the end contractions.

The weir may be regarded as a combination of a rectangular weir and a V-notch, and analysis shows that the requirements are met when $\theta/2 \simeq 14°$. However, the value of C (about 1·9) in Eq. (9·8) is found to vary appreciably with the head and approach conditions so that this form of weir is not suitable for precise discharge measurement.

9.3.5 *Compound Weirs*

In the case of natural watercourses the range of flow is often such that a V-notch is desirable for measuring very low discharges, whilst a rectangular weir is required at the higher stages. A compound weir meets this situation and a typical crest profile is illustrated in Fig. 9.10.

Compound weirs may be calibrated by a division of the nappe into rectangular and triangular portions, utilising the relevant formulae in order to determine the overall head-discharge relationship. A more

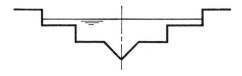

Figure 9.10 Compound weir

accurate calibration results from a site measurement of discharge (volumetric, chemical, or current meter methods) or from model tests in a hydraulics laboratory.

9.4 Solid Weirs

9.4.1

Solid weirs are robust structures capable of handling quite large discharges. Generally, their primary function is the control of water level for purposes of irrigation, navigation, or flood protection. For close and flexible control of upstream water level, the provision of movable crest gates is essential.

Another useful purpose served by this type of weir is the lessening of the bed gradient of a river. The detritus which is brought down accumulates at the weir and must be periodically removed. A larger version is the check dam that is sometimes installed on steep mountain rivers for the interception of boulders.

9.4.2 *General Type*

Solid weirs have a variety of shapes—triangular, trapezoidal, and round-crested forms are typical. The round-crested or ogee weir (Fig. 9.11) is efficient since it has good structural and hydraulic characteristics.

Its surface profile is designed to conform to the underside of a free-falling nappe at or near to the maximum discharge.

The discharge is related to the head by the general weir equation, $Q = CLh^{3/2}$, the head h being measured above the highest point of the crest. The value of the weir coefficient C is not constant but increases with the head; it should also take account of the effect of piers and end contractions.

Standard weir shapes[1] have been developed for which calibration curves are available. A weir of non-standard design must be calibrated in the field or by means of a model in the hydraulics laboratory. Most types of solid weir have an advantage over the sharp-crested weir in that the discharge is unaffected by downstream conditions even when the

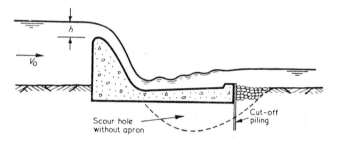

Figure 9.11　Round-crested weir

weir is operating with an appreciable degree of submergence. This is an important consideration in flat country where fall is limited.

Even with low weirs it is important to take measures to prevent downstream scour, which in soft bed material can result in under-mining of the weir structure. A free overfall requires some form of stilling pool where energy can be harmlessly dissipated, whilst a jet which adheres to the downstream face should be deflected on to a concrete apron in some such manner as illustrated in Fig. 9.11. The purpose of the outer lip is to assist in confining turbulence to the apron.

9.4.3 *Broad-crested Weir*

This type of solid weir has a rectangular section and normally spans the full channel width. The behaviour of the nappe is illustrated in Fig. 9.12. It will be noted that a drawdown of the nappe occurs near the

---

[1] For example, the standard weirs of the U.S. Geological Survey and U.S. Dept. of Agriculture.

upstream end of the weir, but subsequently the upper surface runs almost parallel to the crest. If the upstream edge is sharp, separation is likely to occur at this point, but otherwise the underside of the nappe adheres to the crest throughout.

Provided that the crest is of sufficient width ($> 3h_{max}$), the stream lines are aligned for the greater part nearly parallel with the crest, although there may be some surface undulation. Near to the downstream end of the weir the depth is critical, and the velocity at this point is governed by the relationship $V_0 = (gd_o)^{1/2}$, (Eq. (8.15), p. 175), so that

$$Q = Ld_o(gd_o)^{1/2} = g^{1/2}Ld_o{}^{3/2} \qquad (9.11)$$

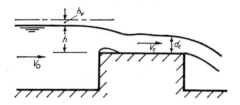

Figure 9.12   Broad-crested weir

In order to derive an expression relating Q and h we must apply Bernoulli's equation to the critical depth section on the crest and to a section just upstream of the weir. We then have

$$\tfrac{3}{2}d_o = h + h_v \qquad (9.12)$$

Substituting for d_o in Eq. (9.11),

$$Q = (\tfrac{2}{3})^{3/2}g^{1/2}L(h + h_v)^{3/2} = 1{\cdot}71L(h + h_v)^{3/2} \qquad (9.13)$$

By introducing a weir coefficient C to take account of energy losses, non-parallelism of stream lines, and the velocity of approach, we obtain a discharge expression in the conventional form $Q = CLh^{3/2}$. The value of C varies with the weir geometry and the discharge; it is usually about 1·5 (see B.S. 3680: Part 4B). This value is lower than that for a sharp-crested weir, signifying that the broad-crested weir causes greater heading-up. However, there are some advantages in that the degree of aeration of the overfall is immaterial and the simple head-discharge relationship continues to apply until the submergence ratio exceeds approximately 0·66, which is of course when the downstream head exceeds the critical depth at the crest.

223

9.5 Special Types of Weir

9.5.1 *Extended Weirs*

An increased discharge for the same head may be obtained if the weir crest, instead of being set in plan at right angles to the channel, is elongated in one of the typical forms illustrated in Fig. 9.13.

Weirs type (a), (c), and (d) direct part of the flow towards the bank so that some form of revetment is required. Types (e) and (f) are favourable

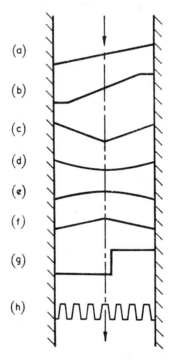

Figure 9.13   Types of extended weir

to the banks but tend to cause scour in the centre of the channel. The labyrinthine type (h) is very effective at low heads, but is expensive to construct.

Discharge characteristics must be determined by field measurement or by model tests for the complete weir. The general weir formula (Eq. (9.8)) is applicable, with L as the actual crest length; the weir coefficient C has a lower value than that for the equivalent transverse weir. At times of extreme flood, with an exceptionally high head on the weir, the enhanced discharge normally associated with crest lengthening, is much less pronounced. This is particularly the case with the weir type (h).

9.5.2 *Side Weirs*

The function of a side weir is to act as an overspill or diversion device in a channel or conduit. Unlike the normal weir there is a flow component parallel to the crest and the head varies along it. The discharge is appreciably less than that of the same length weir placed across the line of flow. Fig. 9.14 shows a typical surface profile at a side weir located on a channel of mild bed slope.

Side weirs have been widely employed as storm overflows on sewerage systems, where it is customary to divert flow in excess of 6 × D.W.F. (dry weather flow) to a river or stream, flows within this limit being passed forward to the sewage purification works. Space requirements are reduced by placing the weirs on both sides of a through channel.

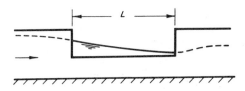

Figure 9.14   Side weir

To some extent, side weirs on sewers have fallen into disrepute in recent years, owing to their tendency to cause unnecessary pollution. Some of the other types of overflow, such as siphons and circular vortex chambers, are superior in this respect.[1]

Detailed research[2] has been carried out with a view to establishing the discharge characteristics of side weirs. The flow behaviour is extremely complex so that no simple empirical formula can be regarded as entirely satisfactory. The *Coleman and Smith formula* for the length of a single weir is

$$L = 1 \cdot 16 b V h_1^{0 \cdot 13} \left(\frac{1}{h_2^{1/2}} - \frac{1}{h_1^{1/2}} \right) \tag{9.14}$$

where b is the width of channel (rectangular) at the weir, V is the average velocity in the channel, h_1 and h_2 are the respective heads at the

[1] (1963) 'Interim Report of the Technical Committee on Storm Overflows and the Disposal of Storm Sewage', pp. 11–12. H.M.S.O.

[2] (1957) 'Symposium of Four Papers on Side Spillways', *Proc. Inst. C.E.*, **6**, 250.

upstream and downstream ends of the weir. The units are m-s and h_2 is usually taken as 0·015 m.

9.6 Throated Flumes

A throated flume is essentially an artificial constriction in a channel which, by producing a change in the velocity and depth, facilitates the

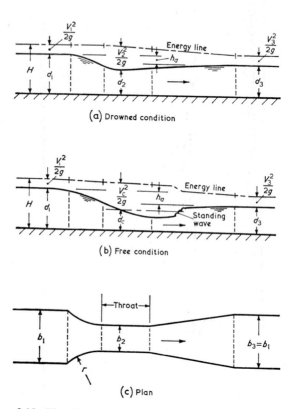

(a) Drowned condition

(b) Free condition

(c) Plan

Figure 9.15　Flow through a level-bed throated flume (not to scale)

measurement of discharge. It is normally constructed in concrete, timber, or steel plate, and the geometry of the design is most important. In the conventional form (Fig. 9.15) it consists of a bell-mouthed entry, parallel throat, and downstream diverging portion (cf. Venturi tube, Fig. 4.12, p. 37).

The afflux or ponding back (h_a) caused by the constriction is relatively small, and this is a considerable asset when the fall available is strictly limited, such as for example on irrigation canals in flat country. Other favourable characteristics are robustness of construction and a relatively clear passage for the water. Maintenance is simplified, since floating debris and suspended matter are unlikely to cause damage or be deposited; also, only limited protection against scour is required as the regime of flow is not greatly disturbed.

In addition to irrigation canals the throated flume is well suited to the measurement of flow in purification plants, streams, and small rivers. The discharge capacity ranges from a small fraction of a cumec up to about 30 cumecs; flumes have been designed for much higher discharges, but the constructional cost tends to be relatively expensive.

Assuming an ideal fluid, the total head at the upstream and throat sections is given by $H = d_1 + V_1^2/2g = d_2 + V_2^2/2g$. Now, as $Q = b_1 d_1 V_1 = b_2 d_2 V_2$, we have

$$d_1 + \frac{Q^2}{2g b_1^2 d_1^2} = d_2 + \frac{Q^2}{2g b_2^2 d_2^2}$$

from which

$$Q = \frac{\sqrt{2g}\, b_2 d_2}{[1 - (b_2 d_2/b_1 d_1)^2]^{1/2}} \sqrt{d_1 - d_2}$$

Substituting $m = b_2 d_2/b_1 d_1$, and introducing a coefficient of discharge C_d to take account of non-uniform velocity distribution and small loss of energy head, we obtain

$$Q = \frac{\sqrt{2g}\, C_d b_2 d_2}{(1 - m^2)^{1/2}} \sqrt{d_1 - d_2} \qquad (9.15)$$

This is the *general discharge equation* and is valid for all conditions of flow. C_d has a value usually between 0·96 and 0·99 but is somewhat unstable. The expression is very similar to that for a pipe constriction (cf. Eq. (4.7), p. 37) – with the important exception that, owing to the free surface, m is now a variable.

If the flow is subcritical throughout, as in Fig. 9.15(a), the flume is said to be operating in the *drowned condition*. Although this operating condition is sometimes unavoidable (e.g. exceptionally high flow), it is difficult to cater for satisfactorily since a complex equation and two water level measurements are involved. Moreover, the surface at the throat tends to be unstable so that d_2, and even more important the small difference $d_1 - d_2$, cannot be accurately determined.

For the flume to be a convenient and efficient measuring device, we wish to establish a direct relationship between Q and a single depth which can be recorded by automatic float equipment. This is possible if the flow in some portion is supercritical or *free*, that is to say it is unaffected by conditions downstream, in which case it is only necessary to measure the upstream depth d_1.

A steeply inclined apron, known as a glacis, suffices, since a hydraulic jump is produced at the foot of the slope where the subcritical flow is resumed. However, with this form of design we are deprived of the principal advantage of the throated flume which is the small head loss. It is therefore more usual for the bed to be horizontal, the proportions being such that a hydraulic jump occurs just downstream of the throat, as shown in Fig. 9.15(b). This is the normal manner in which a throated flume is designed to operate and hence the description *standing-wave flume*.

The transition from subcritical to supercritical flow occurs at the throat, and at the critical depth we know that $d_c = V_c^2/g$ (Eq. (8.15), p. 175). Thus the total head at the upstream and critical depth sections is $H = d_1 + V_1^2/2g = 3d_c/2$, from which $d_c = 2H/3$ and $V_c = \sqrt{2gH}/\sqrt{3}$. As $Q = b_2 d_c V_c$, it follows that

$$Q = \frac{2\sqrt{2g}}{3\sqrt{3}} b_2 H^{3/2} = 1 \cdot 71 b_2 H^{3/2} \qquad (9.16)$$

This is the same form of expression as was established for the flow over a broad-crested weir (Eq. (9.13)). Of course, the two phenomena are essentially similar.

For a real fluid and practical application, allowance must be made for frictional and turbulence losses and for the fact that it is the upstream water level which is measured. Thus

$$Q = 1 \cdot 71 C_d C_v b_2 d_1^{3/2} \qquad (9.17)$$

where C_d and C_v are coefficients of discharge and velocity (approach correction), respectively. The tabular data given in B.S. 3680: Part 4A: 1965 enable C_d and C_v to be determined for the particular conditions. In general the values are such that the overall coefficient C in $Q = Cb_2 d_1^{3/2}$ is of the order of $1 \cdot 65$.

If the afflux is diminished, as for example by lowering a sluice gate downstream, the hydraulic jump is made to advance upstream towards the throat and as the minimum depth approaches the critical it becomes undular in profile. Finally, with still further diminution of the afflux, the drowned profile shown in Fig. 9.15(a) is assumed and the head upstream is no longer unaffected by the conditions downstream.

The energy dissipated in the standing wave is relatively large so that the *submergence ratio* $(d_3:d_1)$ must be appreciably less than unity if the standing-wave flume is to operate satisfactorily in its intended form. The upper limiting value of this ratio is called the *modular limit*. It is normally taken as about 0·75, which incorporates a small safety factor so as to avoid the undular transitional zone.

It is sometimes found impossible to design a level-bed flume which will operate within the modular limit at the lower discharges whilst having an afflux which is not excessive at the higher. The problem is overcome by introducing a streamlined hump in the bed instead of, or more commonly in addition to, a contraction in width (Fig. 9.16). The hump behaves in exactly the same way as a broad-crested weir, but has the advantage of greater accuracy. As the upstream total head is reduced from H to H', a lesser discharge is obtained for almost the same submergence ratio $(d'_3:d'_1)$.

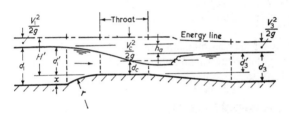

Figure 9.16   Standing-wave flume with raised invert

The Crump weir is of the 'hump' type without throated sides. It is triangular-shaped in elevation with bed slope 1 in 2 upstream and 1 in 5 downstream. The principal points in its favour are a reasonably high modular limit (about 0·8) and inexpensive constructional cost.

The desirable hydraulic characteristics to aim at when designing a standing-wave flume are generally: (a) sufficient float range for accurate measurement, (b) minimum afflux, and (c) a high modular limit. Unfortunately, there is some conflict between these requirements so that a compromise must often be made. Extensive research has been carried out with a view to establishing the most efficient designs in the various circumstances. An accuracy of discharge measurement within ±3 per cent is a reasonable expectation.

Less disturbance is caused to the regime of flow if the cross-section of a flume is made to conform as far as possible with that of the channel in which it is situated. Thus the floor of a flume in a length of sewer pipe is generally dished or semi-circular, whilst in the case of a trapezoidal channel the wing walls are battered. The necessary calibration is obtained by a field method (e.g. current meter traverse) or by means of a hydraulic model. In the case of a simple trapezoidal channel, analytically derived calibration charts are available.[1]

A good example of specific design is the standing-wave flume on the R. Derwent, Derbyshire (Plate 4). It is of the compound trapezoidal type and has been calibrated to measure discharges up to 170 cumecs while operating in the free condition.

The Parshall flume[2] is a type that is well favoured in the United States. Standard proportions and the omission of curved surfaces simplify construction as well as eliminating the need for calibration.

Example 9.3

A standing-wave flume is to be installed in a long rectangular channel. The channel width, bed slope, and roughness coefficient n are 1·5 m, 1/720, and 0·012, respectively. The flume is to be designed to pass 1·4 cumecs with an afflux not exceeding 0·3 m and a modular limit of 0·75. Assume $C_d = 0·98$ and $C_v = 1·10$.

Determine (a) the maximum and minimum permissible throat widths, and (b) the approximate equivalent afflux with a sharp-crested weir spanning the entire width of the channel.

Substituting in the Manning formula,

$$1·4 = \frac{1·5d_n}{0·012}\left(\frac{1·5d_n}{1·5 + 2d_n}\right)^{2/3} \times \left(\frac{1}{720}\right)^{1/2}$$

from which $d_n = 0·620$ m. This may be taken as the depth d_3 downstream of the flume.

The minimum permissible depth upstream is $d_1 = 0·620/0·75 = 0·827$ m. Substituting in Eq. (9.17), we obtain $b_2 = 1·4/(1·70 \times 0·98 \times 1·10 \times 0·827^{3/2}) = 1·015$ m, which is the *maximum* permissible throat width.

For an afflux of 0·3 m, $d_1 = 0·920$ m. Again, from Eq. (9.17), we find $b_2 = 0·867$ m, which is the *minimum* permissible throat width.

With a suppressed sharp-crested rectangular weir the equivalent head is obtained from Eq. (9.6). Assuming $C_d = 0·65$, we have $h_e = [1·4/(\frac{2}{3} \times \sqrt{2g} \times 0·65 \times 1·5)]^{2/3} = 0·619$ m. About 0·15 m must be allowed for aerating the nappe so that the minimum afflux required is approximately **0·77 m**.

[1] ACKERS, P. and HARRISON, A. J. M. (1963) 'Critical-depth Flumes for Flow Measurement in Open Channels', *Hydraulics Research Paper No. 5.* H.M.S.O.
[2] PARSHALL, R. L. (1926) 'The Improved Venturi Flume', *Trans. Am. Soc. C.E.*, **89**, 841.

Example 9.4

The depth of water in the channel downstream of the flume (throat width 1 m) in Ex. 9.3 is prevented from falling below 0·3 m by means of a regulating sluice. If the flume is to be used for measuring flow in the range 0·15 to 1·5 cumecs, determine the minimum height of the hump which must be inserted at the throat, assuming that the same modular limit is applicable. What is the afflux for 1·5 cumecs discharge and normal depth downstream? Assume the same values for the coefficients.[1]

It is necessary to provide a hump at the throat otherwise the flume will be 'drowned' at the lower flows.

Minimum flow (0·15 *cumecs*)

From Eq. (9.17), $d_1' = [0·15/(1·70 \times 0·98 \times 1·10 \times 1·0)]^{2/3} = 0·189$ m. At the modular limit, $d_1' = d_3'/0·75$, so that $d_3' = 0·142$ m and $x = \mathbf{0·158}$ m.

Maximum flow (1·5 *cumecs*)

Substituting again in Eq. (9.17), $d_1' = 0·875$ m. Thus the afflux is $d_1' - (d_n - x) = 0·875 - (0·620 - 0·158) = \mathbf{0·413}$ m.

9.7 Spillways

9.7.1

A spillway is the overflow device for an impounded body of water. Its function is to dispose of excess water without harming the impounding structure or causing undesirable scour downstream. There are three principal spillway types—(a) open, (b) shaft, and (c) siphon. They are usually incorporated in a dam but are sometimes separate structures. Type (a) is the one most frequently encountered.

Spillways are an extremely important feature of any dam project. Because of the great variety of site conditions, both with regard to the spillway itself and to the tailwater, it is generally necessary to have recourse to hydraulic model tests (see Ch. 11, Sect. 11.4.3) in order to ensure that the best possible design is provided.

9.7.2 *Open Spillways*

(a) *Upper Portion.* It is the normal practice for the top of a dam to be utilised as a spillway crest. In the case of an arch dam the overflowing jet generally springs clear after leaving the crest, its kinetic energy being dissipated in a deep stilling pool at the toe. In the much more common

---

[1] Since these coefficients are dependent on the head and weir geometry, a trial and error solution is in fact necessary, utilising the data given in B.S. 3680: Part 4A: 1964.

case of a gravity dam, it is usual to arrange for the jet to adhere to the downstream face. Side channels are sometimes provided just below the crest and running parallel to it; their function is to collect the over-flowing water and direct it in the most efficient manner to the river channel below. Water must never be allowed to spill freely over an earth embankment dam except at a point where there is a properly constructed concrete spillway.

In order to create additional storage at the end of a wet season, the crest height is sometimes raised by the provision of temporary structures such as flash-boards or stop-logs. For permanent control of excess storage and flood release, movable gates are required. These may be in the form of radial, drum, or vertical lift gates (see Ch. 3, Sect. 3.4). Plate 3 shows a flood discharge passing under vertical lift gates at Tongland spillway, Scotland.

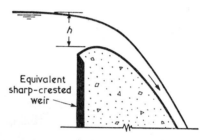

Figure 9.17   Profile of a spillway crest

Since the velocity of water on open spillways is relatively high it is important that the crest profile should guide the flow as smoothly as possible and with the minimum of turbulence. A profile which is excessively convex tends to reduce the pressure of the water flowing over it and this is conducive to separation and possibly cavitation. Separation in itself is not harmful but it is often associated with an intermittent making and breaking of contact which can cause troublesome vibration. If the jet is held on the spillway surface by the sub-atmospheric pressure and this pressure falls to that exerted by the vapour, cavitation will ensue, causing structural damage in the form of pitting and honey-combing of the surface.

In order to minimise these effects a profile is adopted (Fig. 9.17) that conforms to the underside of the aerated nappe for the equivalent sharp-crested weir. As the profile geometry varies with the head, true conformity is only obtained for one particular condition, referred to as the design head. At lesser heads the pressure distribution on the spillway is nearly hydrostatic. At greater heads the pressure at the spillway surface is sub-atmospheric, but experiments[1] have shown that provided the maximum head is not more than about 50 per cent in excess of the design head, the cavitation stage will not be reached. The sub-atmospheric pressure serves a useful purpose in increasing the effective discharge head.

As a result of extensive model and prototype studies the U.S. Waterways Experiment Station[2] has evolved a comprehensive range of standard crest shapes and these form a useful basis for design. The general weir equation is applicable, namely

$$Q = CLh^{3/2} \qquad (9.18)$$

and the value of the *spillway coefficient C* is usually between 1·6 and 2·2, increasing with the head. In most cases there is little difficulty in reconciling the hydraulic best shape with structural requirements.

(b) *Lower Portion.* Water traversing the steep slope of the downstream face of a gravity dam is greatly accelerated and by the time it reaches the foot of the spillway it has attained a very high velocity, well above the critical value. On being deflected by a suitable reverse curve into the nearly horizontal tailwater channel this high velocity stream is only gradually retarded by friction and is thus capable of causing severe scour and erosion for a considerable distance downstream. A dangerous situation may develop if this results in undermining of the dam structure.

Scour can be prevented by concrete lining, but this is an expensive remedy and obviously the better course is to aim at a hydraulic design that will dissipate the excess energy (ultimately as heat) in as short a distance from the toe of the spillway as possible. It has been pointed out earlier (Ch. 8, Sect. 8.3.3) that there is a considerable loss of energy associated with the hydraulic jump, and this is therefore an efficient agency to employ. Fortunately, the essential conditions for the jump phenomenon are nearly always present since the flow upstream is supercritical whilst downstream it is nearly always subcritical. However, the wide variations of the spillway discharge and downstream water level mean that it is not always easy to confine the jump and its associated turbulence to a relatively short distance from the spillway toe.

[1] ROUSE, H. and REID, L. (1935) 'Model Research on Spillway Crests', *Civil Engineering* (U.S.A.), **5**, 10.
[2] *Corps of Engineers Hydraulic Design Criteria*, U.S. Army Engineer Waterways Expt. Sta., Vicksburg, Miss.

Fig. 9.18 shows the flow of water over a dam spillway. A steep uniform slope produces a drawdown curve of the S2 type (Ch. 8, Sect. 8.5) and the flow depth at any point can be determined by step-by-step procedure. A simplified analysis is possible if the effect of friction is neglected. Thus, applying Bernoulli's equation to the crest and toe of the spillway (and neglecting the velocity of approach), we obtain $h_D + h = d_1 + V_1^2/2g$. Now, the spillway rating curve defines the head-discharge relationship (Eq. 9.18) and $V_1 = q/d_1$, where q is the discharge per unit width, so that

$$h_D + h = d_1 + \frac{q^2}{2gd_1^2} \qquad (9.19)$$

Given the discharge, this equation can be solved for d_1 by trial and error procedure.

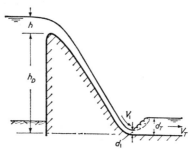

Figure 9.18   Flow over a dam spillway

It has been shown in Ch. 8, Sect. 8.3.3 that the depths upstream and downstream of a hydraulic jump are related by the equation (Eq. (8.27))

$$d_1 = \sqrt{\frac{d_2^2}{4} + \frac{2q^2}{gd_2}} - \frac{d_2}{2} \qquad (9.20)$$

If then it were possible to arrange for the tailwater depth d_T to be always the same as d_2, the jump would never depart from the foot of the spillway and a short concrete apron would give adequate protection. Of course, this is impracticable, since the tailwater depth, being sub-critical, is governed by downstream conditions. In actual fact, the assumption of uniform flow and the application of one of the open-channel formulae (e.g. Manning) will generally furnish a reasonable estimate of this depth.

It is now necessary to consider the flow behaviour when d_T differs from d_2.

If d_T is less than d_2, the jump cannot take place until the depth d_1

has increased to d_1', the sequent depth to which is d_T. The retarding action of friction on the mild slope brings this about, the surface profile being of the M3 type (Fig. 9.19).

Again, if d_T is greater than d_2, the jump is drowned and the high velocity stream plunges into the tailwater (Fig. 9.20). There is little

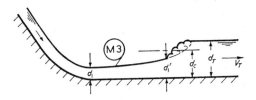

Figure 9.19   Effect of low tailwater level

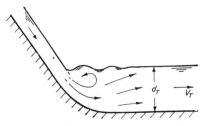

Figure 9.20   Effect of high tailwater level

immediate dissipation of energy so that the flow remains violently disturbed for some distance downstream.

The measures which may be taken to restrict the extent of the turbulence are as follows:

(i) LOW TAILWATER LEVEL. This condition is likely to be associated with a moderately steep downstream channel. The jump may be

Figure 9.21   Stilling basin and check weir

made to occur near the foot of the spillway by lowering the apron to below the level of the bed of the tailwater channel, the pool so formed being called a stilling basin (Fig. 9.21). The solid weir at the end of the

235

apron assists in maintaining the requisite depth of water in the basin as well as regulating the flow and confining the turbulence.

If the spillway velocity is not particularly high an upward sloping sill and baffle pier type of design is quite effective. The function of the concrete baffles, situated at the upstream and downstream end of a stilling basin, is to break up the flow to the greatest possible extent. Careful design is needed, since, with the high velocities normally encountered, there is the possibility of cavitation attack in the vicinity of sharp corners.

(ii) HIGH TAILWATER LEVEL. By sloping the apron, as in Fig. 9.22, the jump is made to occur at some point on the slope. The location of the jump is largely fixed by the intersection of the tailwater surface and

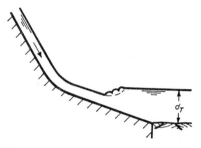

Figure 9.22   Spillway with a sloping apron

the high velocity stream; no satisfactory theoretical analysis is available, but experiment indicates that provided the slope is not greater than 1:6 a modified form of jump will occur.

Figure 9.23   Spillway with a roller-bucket toe

A bucket-shaped toe (Fig. 9.23) is another means of limiting the extent of the downstream turbulence. It was successfully adopted for the spillway of the Grand Coulee Dam (167 m high). By making the radius fairly large and ensuring that the bucket is kept well submerged, a circulatory motion is induced which is very effective in dissipating energy; furthermore, the presence of a reverse current on the bed adjacent to the toe greatly assists in preventing scour at a potentially dangerous point.

The ski-jump is an energy dissipator of a rather different type since its effectiveness rests on a combination of jet dispersion and air resistance. André Coyne, a French engineer, was the originator of this economical form of spillway. As depicted in Fig. 9.24, the toe of the spillway is provided with a concave upward profile (hence 'ski-jump') designed to project the water into the air well above the tailwater level. This produces a wide dispersion of the jet, its energy being dissipated through interaction with the air. Finally, the partially disintegrated jet

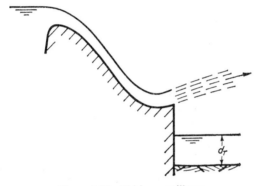

Figure 9.24   Ski-jump spillway

strikes the tailwater stilling pool at a safe distance from the dam. The force exerted on the guiding surface is considerable and can be estimated by a simple application of the momentum equation.

Example 9.5

A small stilling basin is to be constructed at the foot of a dam spillway, the coefficient in the head-discharge relationship being 1·9. The crest of the spillway is 9 m above the floor of the stilling basin and the discharge is 2 cumecs per metre width for both spillway and tailwater channel. The latter has a bed slope of 1 in 2000 with $n = 0.03$, and is wide compared to the depth.

If the energy dissipation is to be confined as far as possible to the stilling basin, determine the approximate height of the vertical 'step' which is required. Also estimate the horizontal force on the step.

The head h on the spillway crest is obtained from $q = 1.9h^{3/2}$, so that $h = (2/1.9)^{2/3} = 1.035$ m.

Substituting in Eq. (9.19), $9 + 1\cdot035 = d_1 + 2^2/19\cdot61d_1^2$. As a first approximation, $10\cdot035 = 4/19\cdot61d_1^2$, or $d_1 = 0\cdot142$ m. Subsequent trial and error leads to $d_1 = 0\cdot144$ m, so that $V_1 = q/d_1 = 13\cdot89$ m/s.

Assuming uniform flow in the tailwater channel and substituting in the Manning formula, $2 = d_T^{5/3}/(0\cdot03 \times 2000^{1/2})$, whence $d_T = 1\cdot808$ m, and $V_T = 1\cdot11$ m/s.

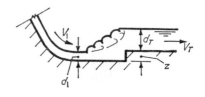

Ex. 9.5

Applying the momentum equation to the foot of the spillway and to a point just downstream of the step, and assuming that the pressure distribution on the step is hydrostatic, we obtain $w[(d_T + z)^2/2 - d_1^2/2] = (wq/g) \times (V_1 - V_T)$, or $(1\cdot808 + z)^2 - 0\cdot020 = (4/9\cdot81)(13\cdot89 - 1\cdot11)$, from which $z = \mathbf{0\cdot481}$ m.

The pressure force on the step is $(w/2)(d_T + z)^2 - wd_T^2/2 = \mathbf{9\cdot65}$ kN/m.

With a long spillway it would be necessary to take into account frictional effects.

9.7.3 *Shaft Spillways.* Site conditions are occasionally unfavourable to the construction of an open spillway. This may be because there is limited space available in a narrow gorge or because a dam is of the earth embankment type. A shaft spillway, or glory-hole spillway as it is termed in America, may then be the answer. As depicted in Fig. 9.25, it consists of a vertical (or inclined) shaft connected by an easy bend to a horizontal tunnel or culvert discharging just above downstream water level. The tunnel serves a useful purpose during dam construction since it may be temporarily utilised as a flow diversion conduit.

Water enters the shaft over a round-crested circular weir and, except in the simplest designs, the inlet is bell-mouthed or trumpet-shaped. When the weir is operating freely the discharge is in accordance with the general weir equation (Eq. (9.8)), the crest length L being the effective circumference, since account must be taken of the radial piers, or ribs, which are normally spaced around the crest in order to counteract any tendency towards vortex motion. Such behaviour would have the effect of seriously reducing the discharge coefficient.

238

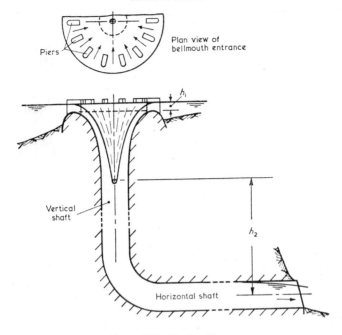

Figure 9.25   Shaft spillway

Three different types of operating condition may be distinguished:

(a) When overspill first commences the quantity of water is insufficient to fill the tunnel portion and it operates as a partially-full conduit with free surface.

(b) As the discharge increases, there is a transition stage between free and pressure flow. The throttling action of the lower bend and tunnel causes the water to back up in the shaft until the head developed h_2 is sufficient to overcome frictional losses. In accordance with the Darcy-Weisbach (λ constant) or Manning formula, the discharge of the pressure conduit is proportional to $h_2^{1/2}$. During both flow conditions (a) and (b) an appreciable volume of air is entrained with the flowing water.

(c) When the discharge has increased to the extent that the water level in the shaft rises above the crest, the normal head-discharge relationship for the weir ceases (Fig. 9.26), and the head h_2 governs the discharge throughout. It follows that any increase of depth over the crest in the drowned condition adds little to the discharge. This is a serious disadvantage and necessitates an ample margin of safety being allowed in design. Fortunately, in the drowned condition, the volume of air entrained is very slight.

Because of the complex flow conditions, model tests are nearly always essential at the design stage.

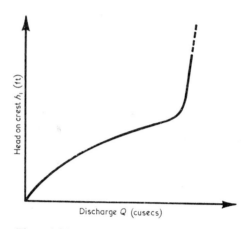

Figure 9.26   Rating curve of a shaft spillway

9.7.4 *Siphon Spillways.* A siphon is essentially a short discharge conduit located above the hydraulic grade line. The existence of sub-atmospheric pressures enables water to be sucked up above the upstream free surface level before it is discharged at a lower level downstream.

The mechanism of operation may be explained by reference to the simple siphon arrangement shown in Fig. 9.27. A gradual rise of water level on the upstream side is supposed. Flow does not commence until water rises above the crest of the siphon, at which stage it spills over in much the same manner as over a weir. A further rise leads to an increase in velocity and a removal of some of the air collected at the summit through entrainment in the flow. Since the outlet is water-sealed there can be no replenishment from the atmosphere. The progressive exhaustion of the air and the associated fall in pressure finally result in the siphon running full bore, at which stage it is said to be *primed*. Any further rise in upstream level now produces only a very slight increase in discharge, since it is the differential head h which is the criterion.

The discharge through the siphon produces a fall in upstream level, but the siphon continues to operate even though this level may have fallen below the crest of the siphon. In fact, it will continue to do so until the inlet is uncovered, whereupon the entry of air breaks the

4 Compound standing-wave flume and recorder house for measuring flow of R. Derwent (Derbyshire)

(*Derwent Valley Water Board*)

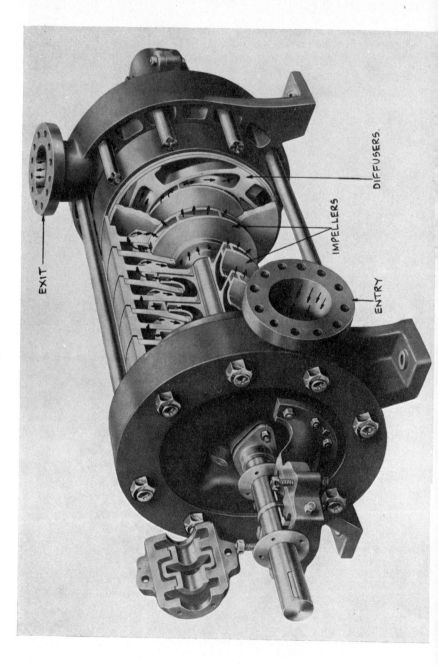

EXIT

DIFFUSERS.

IMPELLERS

ENTRY

siphonic action. Since it is the pressure of the atmosphere acting on the free surface that forces the water up over the siphon it follows that it will only continue to operate provided that the minimum pressure which occurs at the throat does not, at least theoretically, fall below that exerted by the vapour. In practice, there is a tendency for air in solution to be liberated at much higher pressures, promoting flow discontinuity and incipient cavitation. The minimum permissible pressure head is

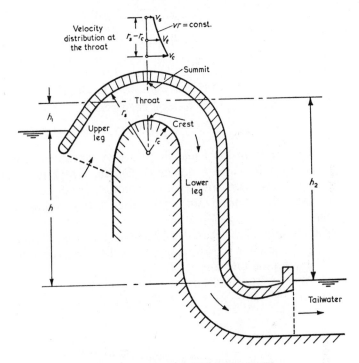

Figure 9.27 Simple type of siphon spillway

usually taken as 3 m absolute (approximately 7.3 m vacuum for installations at or near sea level, there being a slight reduction of atmospheric pressure with altitude).

An expression for siphon discharge may be obtained by applying Bernoulli's equation to the upstream and downstream water levels. Thus

$$h = (K_1 + K_2 + K_3 + K_4)\frac{V_t^2}{2g} \qquad (9.21)$$

where V_t is the mean velocity at the throat and K_1, K_2, K_3, and K_4 are empirical coefficients covering the energy head losses for inlet, upper leg, lower leg, and outlet, respectively. Actually, the apportionment of energy losses in this way is not very satisfactory, since the flow behaviour in the tortuous passages of a siphon is extremely complex so that the normal pipe resistance formulae are not really applicable. The siphon discharge is given by

$$Q = A_t V_t = \frac{A_t \sqrt{2gh}}{(K_1 + K_2 + K_3 + K_4)^{1/2}} \qquad (9.22)$$

where A_t is the sectional area at the throat.

The value of $1/(K_1 + K_2 + K_3 + K_4)^{1/2}$ is usually between 0·5 and 0·9 so that comparison can usefully be made with the coefficient of discharge of an orifice. A simplified expression results if we replace $\sqrt{2g}/(K_1 + K_2 + K_3 + K_4)^{1/2}$ by a single coefficient C so that

$$Q = C A_t \sqrt{h} \qquad (9.23)$$

The form of the discharge equation points to the desirability of reducing the energy losses to a minimum. Conduit proportions, including entry and exit, merit careful consideration. A rectangular cross-section is normally adopted, since this gives better width-height proportions.

Further application of the Bernoulli equation enables expressions to be derived for the absolute pressure head p_t/w at the throat corresponding to V_t. These expressions[1] are

$$\frac{p_t}{w} = \frac{p_a}{w} - h_1 - (\alpha_t + K_1 + K_2)\frac{V_t^2}{2g} \qquad (9.24)$$

and

$$\frac{p_t}{w} = \frac{p_a}{w} - h_2 - (\alpha_t - K_3 - K_4)\frac{V_t^2}{2g} \qquad (9.25)$$

where p_a/w is the atmospheric pressure head and α_t is the velocity head coefficient for the throat.

The flow in the upper bend of a siphon approximates to a free vortex, in which, as was shown in Ch. 4, Sect. 4.9.3, the filament velocity is inversely proportional to the radius. It follows that theoretically the minimum pressure may be expected to occur at the crest of a siphon. Appropriately utilising Eq. (4.24), (p. 48), which is applicable to

---

[1] Actually, a rigorous analysis would make detailed provision for the non-uniform distribution of pressure and velocity at the throat.

a rectangular conduit bend, and remembering that a siphon is in the vertical plane, we obtain for the difference in pressure head at the summit and crest:

$$\frac{p_s - p_c}{w} = \frac{Q^2}{2g[br_c \ln (r_s/r_c)]^2} \left[1 - \left(\frac{r_c}{r_s}\right)^2\right] - (r_s - r_c) \quad (9.26)$$

where b is the throat width and r_s, r_c are the summit and crest radii, respectively.

In order to avoid a serious lowering of pressure at the crest a reasonably large inner radius is clearly necessary. Neglecting losses and assuming that the crest level and upstream water level are the same, the maximum velocity at the crest, based on a minimum absolute pressure head of 3 m, is 12 m/s. This means that the permissible mean velocity at the throat is appreciably less, a value not exceeding 8 m/s generally being acceptable.

The height of the lower leg h_2 largely governs the pressure head at the throat. This follows from a consideration of Eq. (9.25), the last term being neglected. Of course, h_2 only refers to the pressure conduit portion and there is no reason why the siphon should not discharge into the atmosphere well above the tailwater level, the last part of the spillway being of the open type. Alternatively, by constricting the outlet portion it may be possible to maintain the hydraulic grade line at a sufficiently high level in the critical region.

A well-designed siphon spillway is capable of controlling the upstream water level within very close limits. If practicable, a water-sealed outlet should be provided since this assists the priming action. When the outlet is not water-sealed suitable measures must be taken to prevent air replenishment. The simplest procedure is to ensure that a curtain wall of water is deflected across the downstream limb. In Fig. 9.28(a) this is achieved by means of a kink in the conduit, while in Fig. 9.28(b) an S-shaped bend serves the same purpose. The submerged lip at the entrance prevents any gulping of air when there is wave action or surging. In the structural design, allowance must be made for the presence of negative hydraulic pressures and the possibility of vibration during the action of priming and de-priming.

Priming normally occurs when the upstream water level has risen to not more than about one-third of the throat height. An air vent is incorporated in the structure, its function being to break the siphonic action when the upstream water level falls to the level of the crest or a little below it. The fluctuation of water level between priming and venting is thus quite small.

Both design detail and size vary considerably according to the particular site requirements. Several ingenious types of low-head siphon[1]
have been devised. Hydraulic models are valuable as an aid to design

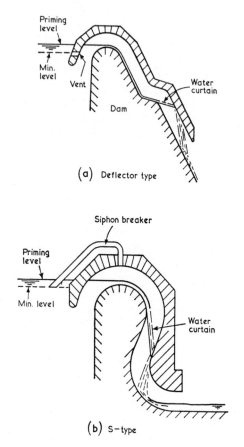

(a) Deflector type

(b) S-type

Figure 9.28   Siphon spillways with a free discharge

(see Ch. 11, Sect. 11.4.3) and the use of transparent perspex enables the
flow pattern to be observed.

The relative merits of the siphon spillway as compared with the more
conventional open spillway are:

[1] For example: (1956) *Experiments on a Self-regulating Spillway Siphon*,
pp. 44–49 of *Hydraulics Research* 1955. H.M.S.O.

ADVANTAGES

(a) Water levels can be controlled automatically within a very small range.

(b) The discharge of a weir is dependent on the small head on the crest whereas a siphon, within certain limits, utilises the much greater head difference between upstream and downstream water levels.

(c) The ability of a siphon to operate at full capacity in immediate response to a rapid rise in upstream water level is particularly useful at an open forebay of a turbine plant, where provision has to be made for a sudden closing of the turbine gates due to rejection of electrical load.

(d) The greater compactness of a siphon is useful when crest length space is limited.

DISADVANTAGES

(a) The rather abrupt priming of a siphon spillway produces a sudden rush of water downstream which in many circumstances would be regarded as objectionable. This disadvantage can be overcome to some extent by the installation of a battery of siphons set to prime at different upstream water levels.

(b) When a siphon is primed, a further rise in water level results in only a small increase in discharge, whereas the discharge rate of a weir increases considerably with the head and there is no upper limit. In the case of important structures, such as dams, the possibility of blockage or excessive surcharge makes it essential to provide auxiliary overflow devices as a safety measure.

(c) The conduits are expensive to construct. In the larger sizes, rectangular sections are much preferred both for this reason and because of the favourable width to height ratio.

Example 9.6

The upper bend of a siphon spillway has a rectangular section 2 m wide and the summit and crest radii are 2·8 m and 1·5 m respectively. Energy losses in the inlet, upper leg, lower leg, and outlet portions are 0·2, 0·2, 0·7, and $0·8V_t^2/2g$, respectively, where V_t is the mean velocity at the throat. The velocity head coefficient α_t for the throat is 1·03. At the particular altitude the atmospheric pressure head is 10·13 m.

Determine the discharge when the reservoir level is 0·3 m above the crest and the tailwater level (water-sealed outlet) is 5 m below it. Also, estimate the minimum pressure head at the throat. What length of open spillway, with coefficient 1·9, would be required to discharge the same quantity under the same upstream head?

From Eq. (9.21), $V_t = [19·61 \times 5·3/(0·2 + 0·2 + 0·7 + 0·8)]^{1/2} = 7·40$ m/s, so that $Q = 7·40 \times 2 \times 1·3 = $ **19·24** cumecs.

Substituting in Eq. (9.25), $p_t/w = 10·13 - 5·65 - (1·03 - 0·7 - 0·8) \times 7·40^2/19·61 = 5·79$ m (abs.), and from Eq. (9.26), $(p_s - p_c)/w = 19·24^2 \times 0·713$ $(19·61 \times 4 \times 0·624^2 \times 2·25) - 1·3 = 2·54$ m.

The minimum pressure head at the crest is thus about **4·5** m (abs.), which is acceptable. In view of the imperfections in the analysis, a more refined procedure taking into account the distribution of pressure and velocity is not justified (for instance, the mean velocity is found at $r_t = Q/[bV_t \ln (r_s/r_c)] = 2·08$ m, instead of 2·15 m).

The open spillway discharge is obtained from $Q = CLh^{3/2}$, whence $L = 19·24/(1·9 \times 0·3)^{3/2} = $ **44·7** m.

9.8 Inflow-Outflow Relationship at a Reservoir

A typical flood hydrograph (discharge-time relationship) for a small river is generally in the form of a curve which rises rapidly to a peak discharge and then declines again more slowly. If such a flood is routed through a reservoir the hydrograph shape downstream may be considerably modified. This is due to temporary storage within the reservoir, the effect of which is always to regulate the flow, prolonging the duration of the flood flow and thereby reducing the peak intensity. The outflow *lag*, as it is usually called, is very beneficial, enabling the spillway and downstream channel to be designed for a lesser maximum discharge than would otherwise be the case. Recommendations as to the design of reservoir spillways in relation to flood discharges are given in an Institution of Civil Engineers' report.[1]

For satisfactory design and operational control it is necessary to determine the relationship with time of the water level in a reservoir under various conditions of inflow. The inflow Q_1 is usually measured at the gauging station just upstream of the reservoir. The outflow Q is based on the spillway rating curve (Eq. (9.18)), due account being taken of any crest gates. The reservoir surface area A is clearly a most important factor in influencing the discharge. In the case of the usual valley type of reservoir, the side slopes are not vertical so that A varies with the head h on the spillway crest.

Let us suppose that during a short time interval Δt the head on the spillway crest varies by a small amount Δh. Then the following relationship must apply:

[1] Interim report of Inst. C.E. Committee (1933) *Floods in Relation to Reservoir Practice*; and later appraisal (1960) *Proc. Inst. C.E.*, **15**, 119.

Inflow volume = increase/decrease in reservoir storage + outflow volume, or

$$Q_1 \Delta t = A \Delta h + CLh^{3/2} \Delta t \qquad (9.27)$$

the appropriate sign being allocated to Δh (positive for filling and negative for emptying). This is known as the *storage equation*, the method of solution depending on the data available and the information required.

Three cases are considered:

(a) $Q_1 = 0$, $A = constant$, $C = constant$.
For an infinitely small element of time, Eq. (9.27) becomes

$$dt = - \frac{A\,dh}{CLh^{3/2}}$$

If $h = h_1$ at time t_1 and $h = h_2$ at time t_2,

$$\int_{t_1}^{t_2} dt = - \frac{A}{CL} \int_{h_1}^{h_2} \frac{dh}{h^{3/2}}$$

or

$$t_2 - t_1 = \frac{2A}{CL} \left(\frac{1}{h_2^{1/2}} - \frac{1}{h_1^{1/2}} \right) \qquad (9.28)$$

From this equation the time of outflow between any two heads on the spillway crest may be determined.

(b) $Q_1 = constant$, $A = \phi'(h)$, $C = \phi''(h)$.
It is assumed that h is given and that it is required to find t.
As before, for an infinitely small element of time:

$$dt = \frac{A\,dh}{Q_1 - CLh^{3/2}} = \phi(h)\,dh$$

so that

$$t_2 - t_1 = \int_{h_1}^{h_2} \phi(h)\,dh \qquad (9.29)$$

Values of $\phi(h)$ are determined for a series of heads. The smaller the head intervals the greater is the degree of accuracy. Computation of the area under the $\phi(h)$ curve is by conventional mensuration method.

Example 9.7

The inset diagram shows the surface area-elevation curve for a reservoir; also included is the spillway coefficient curve. The crest level of the spillway is 186·0 A.O.D. and the crest length is 30 m.

If the inflow to the reservoir is constant at 12 cumecs, estimate how long it will take for the head on the spillway crest to decline from 1·2 to 0·6 m.

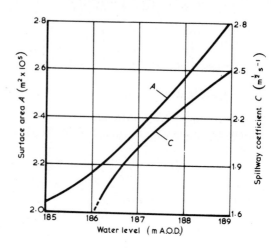

Ex. 9.7

The following table is based on a head interval of 0·1 m:

h (m)	$h^{3/2}$ (m$^{3/2}$)	C (m$^{1/2}$s^{-1})	$CLh^{3/2}$ (cumecs)	$CLh^{3/2} - Q_1$ (cumecs)	A (m$^2 \times 10^5$)	$\phi(h)$
1·2	1·3145	2·06	81·4	69·4	2·39	3450
1·1	1·1537	2·03	70·3	58·3	2·37	4060
1·0	1·0	2·00	60·0	48·0	2·35	4900
0·9	0·8538	1·97	50·4	38·4	2·33	6070
0·8	0·7155	1·94	41·6	29·6	2·31	7800
0·7	0·5857	1·90	33·4	21·4	2·29	10 700
0·6	0·4648	1·86	25·9	13·9	2·27	16 350

Thus

$$t_2 - t_1 = \int_{1·2}^{0·6} \phi(h)\, \mathrm{d}h$$

$$= \frac{0·1}{3}\,[16\,350 + 40(1070 + 607 + 406) + 20(780 + 490) + 3450]$$

$$= 4284 \text{ s or } \mathbf{71·4} \text{ min.}$$

A smaller head interval (particularly at the lesser heads) would yield a more accurate solution.

(c) $Q_1 = \phi'(t)$, $A = \phi''(h)$, $C = \phi'''(h)$.

This is the general case – given the inflow hydrograph and required to determine the outflow hydrograph.

From Eq. (9.27),

$$\Delta h = \frac{(Q_1 - CLh^{3/2})\,\Delta t}{A} \tag{9.30}$$

As Δh is usually very small compared with h, a solution may be obtained by arithmetic integration for successive intervals of time Δt. The procedure is as follows:

(i) Evaluate the head for the steady conditions at commencement, i.e. inflow = outflow.

(ii) Take the mean value of Q_1 for the first time interval direct from the inflow hydrograph.

(iii) Estimate the mean value of Q during the first time interval.

(iv) Determine the rise in water level Δh from Eq. (9.30) – hence obtain h at the end of the first interval.

(v) Calculate Q at the end of the first interval using the spillway equation.

(vi) Check that Q in (iii) above is approximately correct; if not, re-calculate using a more suitable value.

(vii) Repeat the procedure for successive time intervals. A running plot assists in ensuring a good first estimate.

Obviously, a greater accuracy of computation follows from a shorter time interval. However, there is little point in aiming at a higher degree of accuracy than that which the basic data will support and it is generally found that a time interval of $\frac{1}{4}$ or $\frac{1}{2}$ hour suffices.

The procedure is simplified a little if the value of Q in (iii) above, instead of being estimated, is assumed to have the same value as that at the end of the preceding interval. The accuracy suffers slightly and to compensate a shorter interval should be taken.

Example 9.8

The outflow from a reservoir, surface area 182 ha, is over a spillway crest 45 m long with spillway coefficient 1·8. The discharge is steady at 8·5 cumecs when a sudden storm over the drainage basin increases the inflow uniformly over a period of 2 hours to a maximum of 119 cumecs. This peak is maintained for 30 min and the inflow then decreases at a uniform rate of 28·3 cumecs per hour.

Draw the inflow-outflow hydrograph for the first 6 hours from the commencement of increased inflow. State the peak discharge and the time at which it occurs. Use $\frac{1}{2}$ hour time intervals.

The spillway discharge is given by $Q = 1.8 \times 45h^{3/2} = 80.9h^{3/2}$. For the steady condition at $t = 0$, $80.9h^{3/2} = 8.5$, so that $h = 0.223$ m.

With $\frac{1}{2}$ hour time intervals Eq. (9.30) becomes

$$\Delta h = \frac{1800(Q_1 - Q)}{182 \times 10^4} = 0.000\,989(Q_1 - Q)$$

The following is the initial and final portion of the computation table:

Time interval (hr)	Instantaneous time (hr)	Av. inflow during time interval (cumecs)	Instantaneous outflow (cumecs)	Estimated av. outflow during time interval (cumecs)	Net av. inflow during time interval (cumecs)	Rise in W.L. during time interval (m)	Instantaneous head (m)
	0		8·50				0·223
$0-\frac{1}{2}$	$\longrightarrow$	22·3	$\longrightarrow$	9·1 $\longrightarrow$	13·2 $\longrightarrow$	0·013	
	$\frac{1}{2}$		9·3 $\longleftarrow$				0·236
$\frac{1}{2}-1$		49·9		10·5	39·4	0·039	
	1		11·6				0·275
$1-1\frac{1}{2}$		77·5		13·6	63·9	0·063	
	$1\frac{1}{2}$		15·9				0·338
	4		46·2				0·689
$4-4\frac{1}{2}$		69·5		47·3	22·2	0·022	
	$4\frac{1}{2}$		48·5				0·711
$4\frac{1}{2}-5$		55·3		48·8	6·5	0·006	
	5		49·1				0·717
$5-5\frac{1}{2}$		41·1		48·8	−7·7	−0·008	
	$5\frac{1}{2}$		48·3				0·709
$5\frac{1}{2}-6$		26·9		47·5	−20·6	−0·020	
	6		46·4				0·689

The peak discharge over the spillway is found to be **49·1** cumecs occurring **5** hours after the time of commencement of increased inflow. The effect of the reservoir is thus to reduce the peak discharge by 59 per cent and to introduce a time lag of 3 hours.

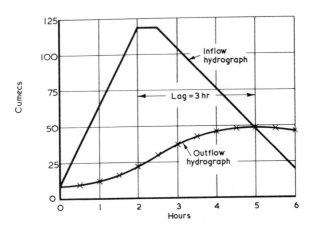

Ex. 9.8

Further Reading

ADDISON, H. (1946) *Hydraulic Measurements*. Chapman and Hall (2nd Edition).

BONNYMAN, G. A. and KEEFE, H. G. (1964) 'Spillway and Flood Control Works', Ch. XII of *Hydro-Electric Engineering Practice* (Ed. BROWN, J. G.), vol. 1. Blackie. (2nd Edition).

B.S. 3680, *Measurement of Liquid Flow in Open Channels*, Part 4A (1965): *Thin Plate Weirs and Venturi Flumes*, Part 4B (1969): *Long Base Weirs*, Part 4C (1971): *Flumes*.

BUREAU OF RECLAMATION (1953) *Water Measurement Manual*. U.S. Govt. Printing Office.

BUREAU OF RECLAMATION (1960) *Design of Small Dams*. U.S. Govt. Printing Office.

CREAGER, W. P., JUSTIN, J. D. and HINDS, J. (1945) *Engineering for Dams*, 3 vols. Wiley.

DAVIS, C. V. and SORENSEN, K. E. (Eds.) (1969) *Handbook of Applied Hydraulics*, Sects. 17, 20 and 21. McGraw-Hill (3rd Edition).

ELEVATORSKI, E. A. (1959) *Hydraulic Energy Dissipators*. McGraw-Hill.

KING, H. W. and BRATER, E. F. (1963) *Handbook of Hydraulics*, Sects. 4 and 5. McGraw-Hill (5th Edition).

LELIAVSKY, S. (1965) *Irrigation and Hydraulic Design*, 4 vols. Chapman and Hall.

LINFORD, A. (1961) *Flow Measurement and Meters*. Spon (2nd Edition).

RAO, N. S. G. (1956) *Design of Siphons*. Central Board of Irrigation and Power, Pub. No. 59, New Delhi.

THORN, R. B. (1966) *The Design of Land Drainage Works*, Paper 5 (*Flumes*). Butterworths. (2nd Edition).

TROSKALANSKI, A. T. (1960) *Hydrometry*. Pergamon Press.

WATER RESOURCES BOARD (1970), *T.N. 8 – Crump Weir Design*.

CHAPTER TEN

Pumps and Turbines

10.1 Introduction

The function of a hydraulic machine is to effect an exchange of energy between a mechanical and a fluid system. In civil engineering the only classes of hydraulic machine with which we are directly concerned are pumps and turbines. Even within this limited sphere there is a great diversity of design, reflecting the wide range of operating conditions and requirements. The two classes are conveniently considered together since in many instances there is little basic distinction apart from the direction of flow.

The use of crude mechanical devices for raising small quantities of water dates back into antiquity. Indeed, it was not until the introduction of the steam engine at the beginning of the last century that there became available a prime mover of sufficient power for large-scale energy conversion. Entirely new and improved methods of pumping resulted, to be followed later by a further definite stage of advancement when the steam engine was superseded by the more efficient high-speed diesel engine and electric motor. Pump design has thus kept pace with the increasing demands of civilised society for the transportation of liquids in both large and small quantities. From small beginnings the manufacture of pumping equipment has grown into an extensive, specialised, and highly competitive industry.

The modern hydraulic turbine is a natural development from the simple water wheel, although the physical resemblance is very remote. When attached to a generator it provides an extremely efficient means of converting hydraulic energy to electrical energy. The capital cost of a hydro-electric scheme (dams, reservoirs, pipelines, turbine installations, etc.) is usually much higher than that of an equivalent thermal station, but there are many advantages. These include high efficiency, operational flexibility, ease of maintenance, low wear and tear, and, apart from droughts, an inexhaustible source of energy. One of the principal difficulties in electric power supply is the fluctuating demand and the fact that thermal stations operate much more efficiently at steady output.

252

In a well integrated power supply network the steam plants would contribute towards the base load whilst the hydro-electric stations would meet the swings of the system.

As hydraulic energy is represented by the product of discharge and pressure head, a complementary availability of river flow and potential head is a necessary prerequisite for hydro-electric power to be an economic proposition. Unfortunately, topographical conditions in England are relatively unfavourable, but in the mountainous terrain of Scotland and in many countries abroad the exploitation of the water resources has proved to be fully justified. Some idea of the importance of water power may be gained from the fact that at the present time it contributes about one quarter of the total world production of electricity.

A type of unit that has attracted interest in recent years is the reversible turbine. This machine can operate either as a pump or as a turbine, according to the direction of rotation; unless special measures are taken its efficiency is necessarily somewhat lower than that of the corresponding single-purpose machine. Reversibility is of particular value in pumped storage projects – the nearest approach to the large-scale storage of electrical energy. When electrical demand is slack the surplus electricity is used to pump a large quantity of water to a high-level storage reservoir which is then returned again for power generation during peak load periods. The Ffestiniog (Wales) and Cruachan (Scotland) hydro-electric schemes are of this type, but it is only in the latter case that a reversible turbine is installed.[1]

While the civil engineer is unlikely to be called upon to design either a pump or a turbine, the associated constructional works are very much his concern and the two are closely related. It is most desirable therefore that he should have some knowledge of the machine types, the basic hydrodynamic theory and the performance characteristics. The aim in the present chapter is to present this information in as simplified a manner as possible. No attempt is made to discuss detailed design, which is a matter of considerable complexity. Greater emphasis is given to pumps since these are the more commonly encountered.

10.2 Head

A knowledge of the static head available to a turbine or against which a pump must deliver is insufficient information for the machine designer, since there may be appreciable pipeline losses, especially in a high head scheme. The generally accepted definitions of head are as follows:

[1] HEADLAND, H. (1961) 'Blaenau Ffestiniog and Other Medium Head Pumped Storage Schemes in Great Britain', Proc. Inst. Mech. E., 175, 319.

Pumps. The *total head* on a pump is the excess of the outlet head over the inlet head. Each of these heads may be regarded as being composed of positional head,[1] pressure head, and velocity head.

Assuming that the suction and delivery pipes have the same respective diameters as the inlet and outlet flanges, then the total head (or more simply the *head*) may be expressed as

$$H = \left(\frac{p_d}{w} - \frac{p_s}{w}\right) + \left(\frac{V_d^2}{2g} - \frac{V_s^2}{2g}\right) \qquad (10.1)$$

where p_s/w, p_d/w are the respective suction and delivery pressure heads referred to the positional datum, and V_s, V_d are the respective velocities. The suction pipe is usually slightly larger than the delivery pipe, but as the velocities are not excessive in either, the difference in velocity heads is small and is usually neglected, in which case

$$H = \frac{p_d}{w} - \frac{p_s}{w} \qquad (10.2)$$

Another expression for the head follows from a consideration of external conditions. Thus for the typical case shown in Fig. 10.1:

$$H = H_s + H_d + h_{fs} + h_{fd} \qquad (10.3)$$

where H_s, H_d are the static suction and delivery lifts respectively, and h_{fs}, h_{fd} are the energy head losses (friction + minor) in the suction and delivery branches, respectively. If the pump is situated below the level of the water surface in the suction well H_s is negative.

Applying Bernoulli's equation to the suction and delivery branches in turn we obtain

$$-\frac{p_s}{w} = H_s + \frac{V_s^2}{2g} + h_{fs} \qquad (10.4)$$

and

$$\frac{p_d}{w} = H_d - \frac{V_d^2}{2g} + h_{fd} \qquad (10.5)$$

It will be noted that summing Eqs. (10.4) and (10.5), and utilising Eq. (10.3), we may obtain Eq. (10.1).

The suction pressure head must not exceed a certain negative value, otherwise cavitation and a falling off of efficiency will result. Accordingly, Eq. (10.4) emphasises the importance of keeping V_s and h_{fs} as

---

[1] The positional datum is taken as the elevation of the highest point of the entrance edges of the first stage impeller blades, whatever the disposition of the shaft.

low as possible. This topic is discussed further in Sect. 10.9. In the design of the suction well and intake, careful attention needs to be given to the avoidance of objectionable air-entraining vortices when the pump is operating.

Turbines. According to B.S. 353:1962, *Methods of Testing Water Turbine Efficiency*, the *net head* on a turbine is the head available for doing work, that is to say, the difference between the total head (potential + pressure + velocity head) at inlet and outlet.[2]

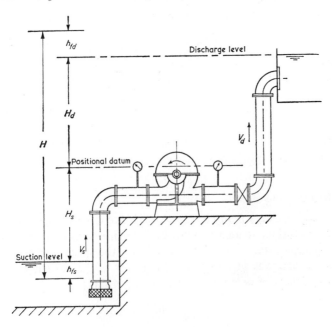

Figure 10.1   Head on a pump

Considering external conditions, the net head (or more simply the *head*) on a reaction turbine situated at some distance from the intake (Fig. 10.2) is given by

$$H = H_G - h_t \tag{10.6}$$

where H_G is the gross head (intake surface level to tailwater level) and h_t is the energy head loss in the supply pipeline. Strictly speaking allowance should be made for the initial and residual velocity heads, but in general these are relatively insignificant.

[2] B.S. 353:1962 includes a number of diagrams explaining the definition of head for the various types of installation.

255

The same expression is applicable to impulse turbines. However, as these machines operate under atmospheric pressure, H_G is measured to an appropriate jet level.

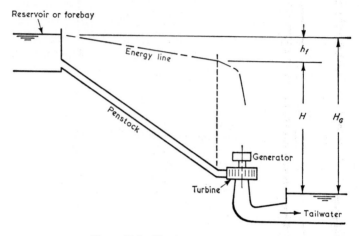

Figure 10.2   Head on a reaction turbine

10.3 Synchronous Speed

Where, as is usual, a pump is direct-coupled to a three-phase induction motor or a turbine to an alternator both machines must necessarily run at the same constant rotational speed. Now the speed n in rev/min of an A.C. machine is related to the frequency of the current f and the number of pairs of pole pieces p by the expression:

$$n = \frac{60f}{p} \qquad (10.7)$$

In the case of motors a small allowance must be made for slip – usually 100 rev/min. As the frequency in Great Britain is 50 Hz, the maximum possible speed of a close-coupled pump unit (with the minimum of one pair of poles) is 2900 rev/min. With four poles it is 1450 rev/min, with six it is 960 rev/min, and so on. The choice of speed towards the upper limit is therefore severely restricted.

As turbines are normally much larger than pumps, speeds are necessarily lower and thus there is much more latitude of choice.

In general it is advantageous to design a unit for the highest permissible speed since this results in a smaller and therefore cheaper installation.

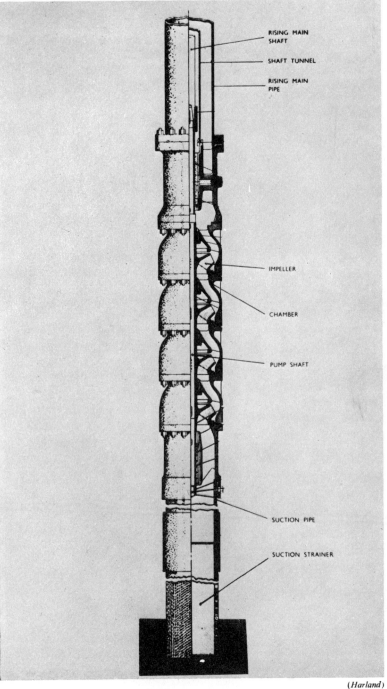

RISING MAIN SHAFT

SHAFT TUNNEL

RISING MAIN PIPE

IMPELLER

CHAMBER

PUMP SHAFT

SUCTION PIPE

SUCTION STRAINER

(*Harland*)

6 Borehole pump

7 Pelton wheel runner

10.4 Types of Pump

10.4.1 *General*

There are three main categories of pump: (a) reciprocating, (b) rotary, and (c) rotodynamic. The first two operate on the principle of positive displacement – that is to say they discharge, apart from any leakage, a definite quantity of fluid irrespective of the head pumped against. Rotodynamic is the title given to the third category because a rotating element, known as an *impeller*, imparts velocity to a liquid and generates pressure. An outer fixed casing, shaft, and driving motor complete the pump unit.

In its usual form the reciprocating pump consists of a ram, plunger, or piston moving to and fro in a cylinder. A suitable valve arrangement enables liquid to be sucked in on one stroke and then on the following stroke forced up the delivery pipe. Consequently, discharge is intermittent unless an air vessel or sufficient number of cylinder units are available to even out the flow. Although the reciprocating pump has now been superseded in most fields of application by the more adaptable rotodynamic pump, it is still employed to advantage in many specialised industrial operations.

In the case of rotary pumps, pressure is generated by means of intermeshing gears or rotors which operate with minimum clearance, the liquid being impelled around within a closed casing. The discharge is uniform and there are no valves. This form of pump is eminently suited to handling small discharges (less than 30 l/s) and viscous liquids. The possible designs are extremely numerous.

The hydraulic theory of reciprocating and rotary pumps is of an elementary nature, being based on simple mechanical concepts, and does not merit further discussion here.

The rotodynamic pump is able to cater for most civil engineering needs and it is very extensively used. Its field of employment ranges from public water supply, drainage, and irrigation to the very special requirements of suction dredging and the transport of concrete or sludge. The various types can be classified under the following headings: (a) centrifugal, (b) multi-stage, (c) borehole, (d) axial flow, and (e) mixed flow.

10.4.2 *Centrifugal Pumps*

These are by far the commonest type of rotodynamic pump and are so called because of the fact that the pressure head created is largely attributable to centrifugal action. They may be designed to handle as

little as 4 l/min or as much as 30 cumecs, whilst the head generated
may be anything from a metre or so to 120 m. In the larger sizes
efficiencies may approach 90 per cent.

The impeller (Fig. 10.3) consists of a number of vanes, curved back-
wards from the direction of motion and supported by metal discs,
called shrouds. Water enters at the centre or eye of the impeller, is
picked up by the vanes, and forced outwards in a radial direction. This
acceleration results in an appreciable gain in both velocity and pressure
energy. At exit the fluid motion has both radial and tangential
components.

In order that energy shall not be wasted and efficiency thereby lowered,
it is essential to convert as much as possible of the considerable velocity

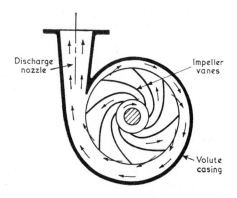

Figure 10.3   Volute type centrifugal pump

head at exit from the impeller into useful pressure head. Normally, this
is achieved by shaping the outer casing in spiral form so that the sectional area of flow around the periphery of the impeller is gradually
expanded. The distribution of velocity in the *volute*, as it is called, is
complex, but is approximately in accordance with the free vortex
relationship (i.e. vr = constant). A further reduction in velocity takes
place in the tapered exit just prior to the delivery flange. The efficiency
of conversion of velocity to pressure head is usually between 40 and 70
per cent.

Fig. 10.4 shows the components of a typical volute pump with single-
suction impeller. For larger flows the double-suction impeller is used;
it is equivalent to two single-suction impellers back to back, enabling the
capacity to be doubled without any increase in the impeller diameter.
It is more expensive to manufacture but has the additional advantage of

overcoming the problem of out-of-balance axial thrust. In both cases the guiding surfaces are carefully smoothed so as to minimise skin friction. Mounting is generally horizontal since this facilitates access for maintenance. However, in order to limit space requirements, some of the larger units are vertically mounted.

There is appreciable latitude in the proportions of impellers, permitting a wide range of operating conditions to be catered for. For instance, liquids with solids in suspension (e.g. sewage) can be handled provided

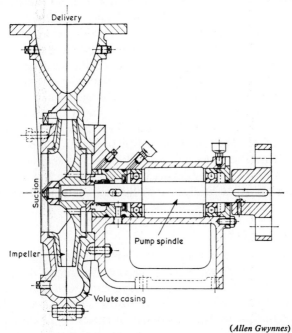

(Allen Gwynnes)

Figure 10.4   Sectional arrangement of a single-suction centrifugal pump

that the flow passages are made sufficiently large; inevitably there is some lowering of efficiency.

Before a centrifugal type pump is in a position to discharge satisfactorily both the suction pipe and the pump itself must be full of water. If the pump is situated below the level of the water surface in the suction well this condition will always exist, but in other cases the air in the suction pipe and pump must first be evacuated and replaced by water – the operation being known as *priming*. The mere rotation of a high-speed impeller is quite insufficient to effect priming and will only result in overheating at the glands.

The two principal methods of priming involve either a foot valve of the flap type near the base of the suction pipe or, for the larger units, the provision of an exhauster pump. In the former case, water from the mains or other source of supply is admitted to the pump casing through a special tapping and air is released at a vent-cock.

10.4.3 *Multi-stage Pumps*

For heads in excess of about 60 m, multi-stage or turbine pumps are usually employed. This type of pump operates on exactly the same principle as the centrifugal and the impeller proportions are very similar. A number of identical impellers are mounted in series, water entering parallel to the shaft and leaving in a radial direction.

Plate 5 shows a typical pumping unit with six stages. The high velocity energy at impeller exit is converted into pressure energy by means of a diffuser ring consisting of diverging guide blades. An S-shaped passage leads the water radially inwards again into the eye of the next impeller. The process is repeated at each stage until the delivery outlet is reached. The total head produced by the pump is the sum of the heads generated at each individual stage. If a sufficient number of stages is employed, up to 1200 m can be attained; in fact, the maximum head is more likely to be dictated by the cost of strengthening the pipe-line rather than by any limitations of the pump.

10.4.4 *Borehole Pumps*

These are of the multi-stage type, vertically mounted and specially designed for raising water from narrow boreholes, deep wells, or drainage pits. They are suitable for operating with bores as small as 150 mm diameter, and in the larger sizes (350 mm and upwards) are capable of raising quantities of water in excess of 1 cumec and from depths as great as 300 m. The impellers are usually designed to discharge in a radial-cum-axial direction, thereby reducing to a minimum the diameter of bore required to house the unit.

The pumping plant (Plate 6) consists of a suction pipe and pump situated below water level and supported from the surface by the rising main pipe and shaft. The shaft passes up through the centre of the pipe and is connected to the power unit at the surface.

When small or moderate quantities of water require to be pumped it is sometimes convenient and economical to site the complete pumping set below water level. The long rising shaft is thus dispensed with, but there is of course the disadvantage that the motor is relatively inaccessible for maintenance purposes.

10.4.5 *Axial Flow Pumps*

This type of pump is well suited to situations where a large discharge is required to be delivered against a low head. Irrigation, land drainage and sewerage are thus its principal fields of employment. The pump efficiency is comparable with that of the centrifugal, and its higher relative speed permits smaller and cheaper pumping and driving units to be provided.

The maximum operating head is between 9 and 12 m. However, two or possibly three impeller stages enable higher heads to be attained, but this is rarely an economical procedure. With large pumps, vertical

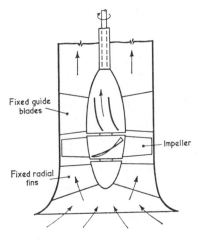

Fixed guide blades

Impeller

Fixed radial fins

Figure 10.5   Axial flow pump

mounting is generally adopted, the spindle passing through the centre of the delivery pipe.

The impeller (Fig. 10.5) is of the open type without shrouds and is shaped like a marine propeller. Water enters axially and the impeller imparts a rotational component, the actual path followed by a particle being that of a helix on a cylinder. Head is developed by the propelling or lifting action of the vanes, centrifugal effects playing no part. It is the function of fixed divergent guide blades to redirect the flow in an axial direction as well as to convert velocity head to pressure head.

In order to avoid creating conditions favourable to the destructive cavitation phenomenon (see Sect. 10.9), the axial flow pump should not be designed with any more than a few feet of suction lift. In fact, it is generally preferable to arrange for the impeller to be immersed,

261

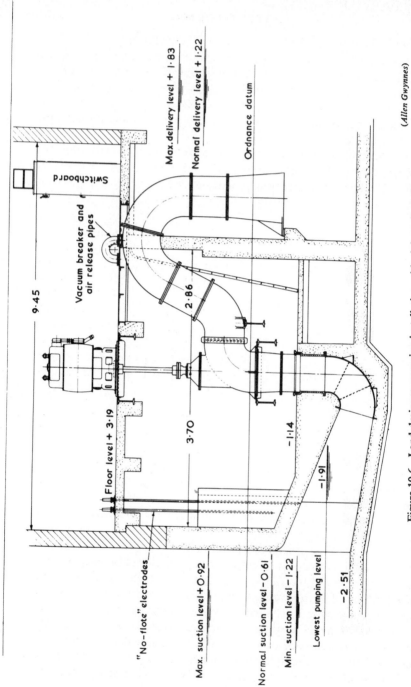

Max. delivery level + 1·83

Normal delivery level + 1·22

Ordnance datum

9·45

Switchboard

Vacuum breaker and
air release pipes

2·86

Floor level + 3·19

3·70

−1·14

−1·91

"No-flote" electrodes

−2·51

Max. suction level + 0·92

Normal suction level − 0·61

Min. suction level − 1·22

Lowest pumping level

Figure 10.6 Land drainage pumping installation with siphon discharge

(Allen Gwynnes)

since in this position the pump is always primed and ready to commence delivery.

Fig. 10.6 shows the layout of a land drainage pumping station equipped with axial flow unit. The purpose of the siphon arrangement is to obviate any risk of the flap valve, which is otherwise necessary, jamming and reverse flow taking place in the pipe, the pump then functioning as a turbine. A butterfly valve is fitted which serves to break the vacuum in the siphon. This valve is slightly off balance towards the open position and directly delivery ceases the valve opens and admits air, thus preventing any reverse flow. The pumping plant may be made fully automatic by means of electrodes which dip into the suction well and control the operation of the pump.

10.4.6 *Mixed Flow Pumps*

The mixed flow pump occupies an intermediate position between the centrifugal and axial flow types. Flow is part radial and part axial, the impeller being shaped accordingly. The path traced by a fluid particle is that of a helix on a cone. The head range is generally up to about 25 m per impeller, and there is a distinct advantage over the axial pump in that the power demand on the motor is nearly constant (see Sect. 10.7.2, p. 282) although there may be a considerable variation of head (e.g. tidal conditions). Pressure head recovery is by diffuser or spiral casing, or a compromise version.

10.5 Types of Turbine

10.5.1 *General*

The possible combination of head and discharge at hydro-electric sites is extremely varied and is reflected in a corresponding diversity of turbine design. For instance, turbines have been constructed for heads as low as 1 m and as much as 2000 m, while the power output ranges up to 200 MW.

There are two main categories of turbine: (a) impulse, and (b) reaction. These descriptions arose because, in the early rudimentary machines, the power was derived either from the forward force of the water striking the rotating blades or from the reaction force of the water leaving them. But the present meaning of the terms has become modified with usage. An *impulse turbine* is now understood to be one in which the pressure energy of the water is converted to velocity energy before it impinges on a rotational element over a limited portion only of the periphery, there being no subsequent change in pressure. Impulse machines today are almost all of the Pelton wheel type and are suitable for high heads.

In a *reaction turbine* the initial pressure-velocity conversion is only partial, so that water enters the rotating element throughout the entire periphery and all the flow passages run full. Modern reaction turbines are either of the Francis or propeller types, catering for medium and low heads respectively.

Unlike pumps, most turbines have to operate for an appreciable portion of the time with outputs differing from the normal, the load variations being met by regulating the quantity of water whilst maintaining constant synchronous speed. The method of control depends on the category of machine, but a rapid response and minimum loss of energy are the primary considerations.

The experience gained over more than a century of development has resulted in present-day peak efficiencies of the order of 90 to 93 per cent. This high standard was attained several decades ago and it is unlikely that it can be improved upon. Future progress is likely to be in the direction of greater operational flexibility and more economical units.

10.5.2 *Pelton Wheel Turbines*

This form of turbine was devised and patented in 1889 by an American named L. A. Pelton. The operating principle is relatively simple, being a logical development from the old water wheel. It shows to best advantage with high heads and, indeed, is the only machine capable of operating under heads in excess of 500 m. Its smooth running and good part-load performance are commendable features.

A typical Pelton wheel arrangement is shown in Fig. 10.7. The nozzle discharges into the atmosphere a high velocity jet which impinges on a series of buckets mounted on the periphery of a wheel, sometimes called a *runner*. The torque exerted by the impact and deviation of the jet causes the wheel to rotate. Its energy usefully expended, water leaves the buckets at a relatively low velocity and is directed towards the discharge channel. Thus the turbine must be set a sufficient height above the maximum flood level if free discharge is to be ensured.

The function of the pear-shaped needle valve is to regulate the flow in the most efficient manner, which it does by varying the diameter of the jet whilst maintaining the velocity. On a sudden rejection of electrical load the wheel tends to accelerate, but this is obviated by the deflector hood moving across the line of the jet, diverting the flow harmlessly to the tailrace. The needle valve then shuts off the flow at a

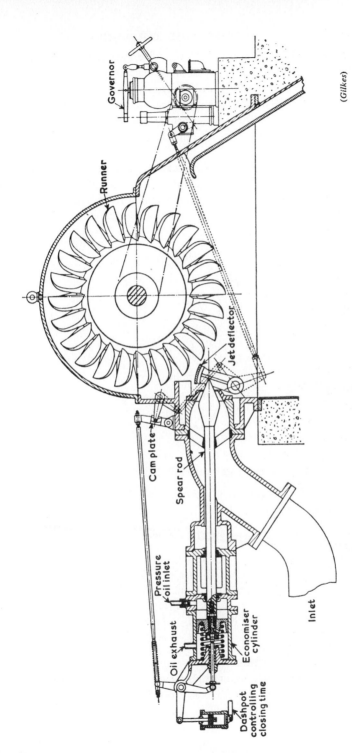

Figure 10.7 Sectional arrangement of a single-jet Pelton wheel with oil pressure governor and automatic spear control

(*Gilkes*)

rate which does not give rise to excessive water hammer pressures (the time of closure usually being of the order of a minute).

The largest practicable size of jet (without deformation) is about 300 mm with 200 mm as the normal maximum. However the power output per wheel may be advantageously increased by the provision of additional nozzles; up to six in total are permissible, spaced equally around the wheel.

Bucket proportions have the greatest influence on the efficiency, and the present design (Fig. 10.8), in the form of a double hemispherical cup, has only been evolved as the result of long experience and painstaking research. The jet strikes tangentially the central fin, called the

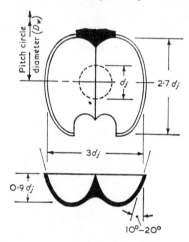

Figure 10.8   Details of Pelton wheel bucket

splitter, and flows equally in either direction, being finally discharged in a direction parallel to that of the shaft (see Fig. 10.18, p. 277). The buckets may be cast integrally with the wheel or fastened separately (Plate 7).

The wheel tends to be fairly large, the diameter being determined by the need to accommodate the requisite size and number of buckets and to avoid splash. It is thus dependent on the size of jet, the ratio diameter of wheel D_w/diameter of jet d_j usually being between 10 and 14. Mounting is generally horizontal and it is sometimes economical to 'double-hang' two Pelton wheels on either side of a central alternator.

10.5.3 *Francis Turbines*

The first successful reaction turbine was built and tested in 1849 by an American engineer named J. B. Francis. His design was superior to that of most earlier forms in that the flow was directed inwards under pressure, so that any tendency towards over-speeding was partly counteracted by the reduction of flow caused by the increase in centrifugal pressure.

The vaned wheel or runner was shaped rather like a centrifugal impeller, flow being predominantly radial with the radii at entry and exit the same for all flow paths. As the need for greater power outputs at higher speeds developed it became necessary to adapt the runner for larger flows without increasing the diameter. This could only be done by arranging for the water to be discharged in a radial-cum-axial direction, the resulting mixed flow type of design (Fig. 10.9) being now the standard practice. Although modern inward-flow turbines bear little resemblance to the original Francis machine, the operating principle is

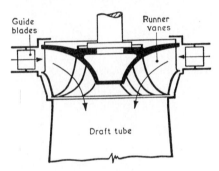

Figure 10.9 Sectional elevation of Francis turbine runner

essentially the same and the name has been retained. The present head range is from 30 m to about 450 m and as this is the most common head availability the machine enjoys a great numerical superiority over the other types. There is certainly a preponderance of Francis machines in Scotland.

Water is directed (with appreciable tangential velocity component) into the runner by means of a spiral casing and a number of aerofoil-shaped blades spaced evenly around the periphery (Fig. 10.10). These guide blades, called the turbine gate, are adjustable, the amount of opening being controlled by the turbine governor through a linkage mechanism. Their role is to guide the flow into the runner with the minimum amount of turbulence, as well as to regulate the discharge and hence the power output.

Because of the converging boundaries, the velocity energy at entry to

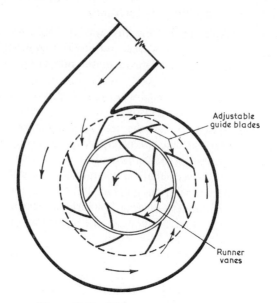

Figure 10.10 Diagrammatic plan of spiral
casing and runner

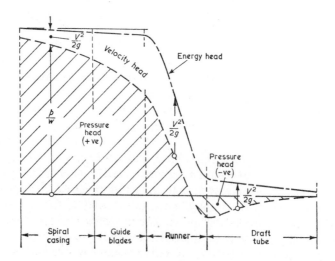

Figure 10.11 Energy head variation in turbine passages

the runner is greater than that in the pipeline and the pressure head correspondingly lower. In the course of its path through the runner there is a further drop in pressure, the water being finally discharged at the centre at a low pressure and with little or no tangential velocity component (Fig. 10.11). The driving torque is derived from the deviation in the direction of flow and the change in pressure and velocity energy. Owing to the problems (e.g. leakage through clearance rings) associated with the high pressures and velocities, there is an upper limit on the head for this type of machine.

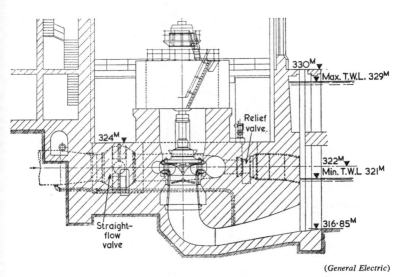

(*General Electric*)

Figure 10.12 Sectional elevation of a Francis turbine installation

The velocity head at discharge from the runner may amount to 20 per cent, or perhaps more, of the available head and as with pumps it is clearly important to convert as much as possible of this otherwise wasted energy to useful pressure head. This is accomplished by means of an expanding passage, called a *draft tube*, which finally discharges the water at a relatively low velocity to the tailwater. A straight conical tube of adequate length would entail expensive excavation and an elbow shape is accordingly preferred; submergence of the exit is essential. Draft tube design cannot be divorced from that of the turbine proper; Fig. 10.12 shows the cross-section of a typical complete installation. The vertical mounting results in considerable economy in space demands.

By recovering pressure head in the draft tube (about 70 per cent conversion is possible), the pressure at exit from the runner is reduced below atmospheric, so that water is in effect sucked through the turbine. Thus the full head above tailwater level is potentially available for useful work – an important consideration with a medium or low head plant, and a distinct advantage over the Pelton wheel. Moreover, there is some welcome flexibility in the level at which the runner may be set, the upper limit being governed by the need to avoid cavitation (see Sect. 10.9).

If the supply pipeline is of appreciable length, a pressure relief valve is fitted in a short conduit leading directly from the pipeline to the tailwater. It performs the same function as the deflector hood on the nozzle

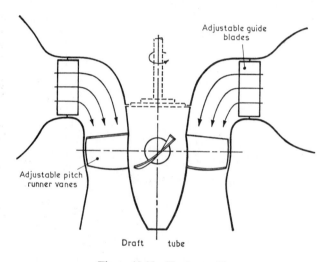

Figure 10.13 Kaplan turbine

of a Pelton wheel. The valve is opened on a sudden reduction of load, thus by-passing the flow, and slowly closed again when the gate movement is completed. The diversion conduit and valve are clearly shown in Fig. 10.12.

10.5.4 *Propeller Turbines*

This type of reaction turbine is the counterpart of the axial flow pump, the rotating element being very similar. The propeller-shaped runner is a logical evolution from the Francis mixed flow design, fulfilling the need for a faster unit and a greater relative flow capacity. The latter consideration is of some importance because in a hydro-electric installation lack of head must be compensated for by enhanced discharge. A

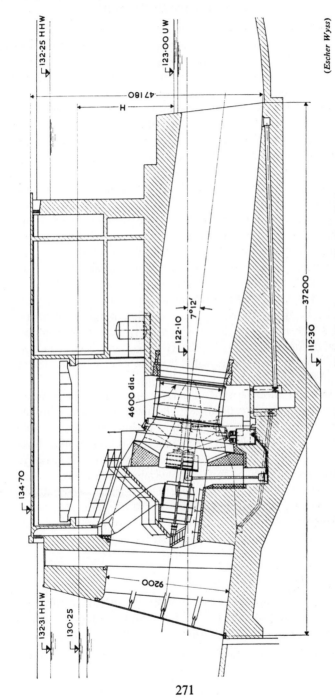

Figure 10.14 Sectional elevation of a bulb type turbine installation

(*Escher Wyss*)

271

head range from about 3 m to 40 m is general, with 60 m as the maximum for small machines. Separation and cavitation are unavoidable in the flow passages of propeller turbines operating with very high velocities, and it is the onset of cavitation that dictates the maximum permissible head.

Unfortunately, the power-efficiency curve for the fixed-blade propeller turbine is very peaked (see Fig. 10.23, p. 287), indicating a poor performance at part load. Victor Kaplan, a professor in Czechoslovakia, overcame the problem in the early 1920's by developing a turbine in which the pitch of the propeller blades was automatically adjusted by the governor to suit the particular load, the actuating mechanism being accommodated in the boss and the hollow bore of the vertical shaft. Although the Kaplan design is much more costly, the improvement to part-load performance is such that the fixed propeller unit is now only installed at sites where the head and load are constant.

The general arrangement does not differ greatly from that of a Francis installation. A whirl component is imparted by the adjustable guides and this is followed by a deviation to the axial direction (Fig. 10.13). In the approach to the runner the fluid motion approximates to that of a free spiral vortex ($vr = $ constant). The whirl component is removed by the runner so that velocity at discharge is entirely axial. The draft tube serves the same essential purpose as in a Francis installation and is of similar form.

The bulb, or tubular, type of turbine installation is an interesting development in the low head field. It comprises a fixed propeller or Kaplan type runner set axially within a short pressure conduit; the generator is close-coupled to the runner and is housed in a 'bulb' surrounded by the flowing water. The entry and exit passages are appropriately designed with a view to minimising energy losses. Fig. 10.14 shows a typical arrangement. The capital cost of a complete installation has been stated[1] to be as much as 40 per cent less than that of the corresponding Kaplan lay-out. This is of particular significance for the very low head projects (e.g. tidal power).

10.6 Elementary Theory

The velocities within a pump impeller or turbine runner may be depicted vectorially. At any point the absolute velocity v is compounded of the velocity relative to the vanes v_r and the vane speed u. Typical

[1] DANEL, P. (1959) 'The Hydraulic Turbine in Evolution', *Proc. Inst. Mech. E.*, **173**, 36.

vectors at entry and exit of a centrifugal impeller are shown in Fig. 10.15. Subscripts 1 and 2 denote entry and exit conditions respectively.

Now in the case of flow between stationary curved boundaries the momentum equation (Ch. 4, Sect. 4.7) was found to afford a ready means of determining the resultant external force, and the same artifice may be usefully extended to a determination of the torque exerted by or on rotating boundaries. Pressures at the periphery are radially applied and therefore have no moment about the centre; also it is reasonable to neglect the moment due to skin friction on the impeller surfaces, since this is relatively insignificant. Thus the torque is equal to the change of

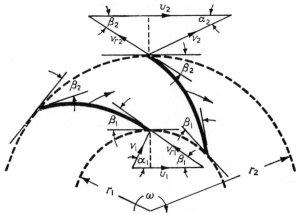

Figure 10.15   Velocity vectors for vanes of a pump impeller

moment of momentum of the flow through the rotating passages, which for a centrifugal impeller is given by

$$T = \frac{1}{g}\left[\int r_2 v_2 \cos \alpha_2 \, dW - \int r_1 v_1 \cos \alpha_1 \, dW\right] \qquad (10.8)$$

where dW is the elemental discharge weight per second.

Since the work done per second is the product of the torque and the angular velocity, we have

$$\text{Work done} = T\omega = \frac{1}{g}\left[\int u_2 v_2 \cos \alpha_2 \, dW - \int u_1 v_1 \cos \alpha_1 \, dW\right]$$

and the work done per unit weight of liquid – in other words the head imparted H_0 of liquid – is

$$H_0 = \frac{1}{Wg}\left[\int u_2 v_2 \cos \alpha_2 \, dW - \int u_1 v_1 \cos \alpha_1 \, dW\right] \qquad (10.9)$$

273

This expression is exact in the sense that it takes account of the variation in magnitude and direction of the velocity at the impeller periphery. But the head actually produced by the pump as a unit is rather less due to the hydraulic losses enumerated later. If opposite signs are applied to the integrals in the brackets – corresponding to the reversed direction of flow – H_0 will represent the head utilised by a reaction turbine.

Evaluation of Eq. (10.9) is not possible without a knowledge of the velocity distribution, which of course is difficult to ascertain. A simplified approach may however, be made if we assume that there are an infinite number of vanes so that the velocity distribution is uniform at any given radius. As the angles α_1 and α_2 are then constant, with $v_1 = V_1$ and $v_2 = V_2$, Eq. (10.9) becomes

$$H_0 = \frac{u_2 V_2 \cos \alpha_2 - u_1 V_1 \cos \alpha_1}{g} \qquad (10.10)$$

This is known as *Euler's equation* and is appropriate to all turbo-machines.

From a consideration of vector geometry, Eq. (10.10) may be expanded to

$$H_0 = \frac{u_2{}^2 - u_1{}^2}{2g} + \frac{V_2{}^2 - V_1{}^2}{2g} + \frac{V_{r1}^2 - V_{r2}^2}{2g} \qquad (10.11)$$

The first term $(u_2{}^2 - u_1{}^2)/2g$ may be identified as the pressure head gain due to the centrifugal forces in the forced vortex. The second term $(V_2{}^2 - V_1{}^2)/2g$ is the gain in kinetic energy, whilst the third term $(V_{r1}^2 - V_{r2}^2)/2g$ is the change in pressure head due to the change in relative velocity. However, as the direction of particle travel is not accounted for, there is little physical meaning in this interpretation.

It might be inferred from the foregoing that the head developed when pumping against a closed delivery valve is $(u_2{}^2 - u_1{}^2)/2g$. Actually, owing to circulatory flow, the head developed is somewhat greater, being more nearly equal to $u_2{}^2/2g$.

Eq. (10.10) may be further simplified, since it is normally arranged that water enters a pump impeller in a radial (or axial) direction, so that

$$H_0 = \frac{u_2 V_{w2}}{g} \qquad (10.12)$$

where $V_{w2} (= V_2 \cos \alpha_2)$ is the tangential or whirl component of the absolute velocity.

Likewise, discharge from the runner of a reaction turbine is radial (or axial) since this minimises eddy losses in the draft tube. Thus

$$H_0 = \frac{u_1 V_{w1}}{g} \qquad (10.13)$$

It is instructive to examine the significance of the vane angle β_2 at impeller exit in relation to pump performance. Firstly, we must re-arrange Eq. (10.12) in the form

$$H_0 = \frac{u_2}{g}(u_2 - V_{f2} \cot \beta_2)$$

Now, for a given impeller and rotational speed, β_2 and u_2 are constants. Also, we know that the discharge Q is equal to the product of the peripheral area and the velocity of flow V_{f2}, which is the component of the absolute velocity at right angles to the direction of vane motion; thus $V_{f2} \propto Q$. Hence we can write

$$H_0 = A - BQ \qquad (10.14)$$

where A and B are constants, B being directly dependent on β_2. Proceeding further, the power developed is given by $P_0 = wQH_0$, so that $P_0 \propto QH_0$ or

$$P_0 = A'Q - B'Q^2 \qquad (10.15)$$

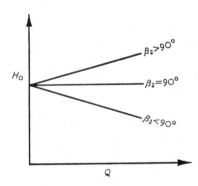

Figure 10.16   Influence of impeller vane exit
angle on H_0-Q relationship (n constant)

These linear and quadratic relationships (Eqs. (10.14) and (10.15)) are illustrated graphically in Figs. 10.16 and 10.17. The influence of β_2 is indicated and it will be noted that when the vanes are forward inclined ($\beta_2 > 90°$) the power input rises steeply with the discharge. This is an undesirable situation and a further objection is that the resulting high velocity at impeller exit is difficult to transform to pressure head. Whilst these theoretical deductions are generally valid in practice, it is important to note that owing to hydraulic losses the actual performance may differ quite appreciably (see Fig. 10.19, p. 283). The optimum value of β_2 for a centrifugal impeller is normally between 20 and 25 degrees.

In practice, the head imparted by a pump impeller is found to be less than that indicated by Eq. (10.12). The reason lies in the fact that the velocity distribution at an impeller periphery is not uniform; nor indeed is the relative velocity parallel to the vane angle. A moment's reflection will reveal that if the assumption of uniform velocity at each radius were

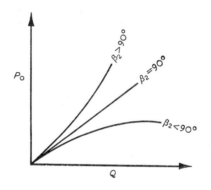

Figure 10.17   Influence of impeller vane exit
angle on P_0-Q relationship (n constant)

correct, then there would be no difference between the pressures on the impelling and trailing faces of the blades, and this is obviously essential if any force is to be exerted. Thus for design purposes it is necessary to multiply the theoretical head H_0 by an empirical correction factor, usually between 0·7 and 0·85. In the case of turbines, the discrepancy is much less since the flow is converging and therefore the velocity at entry more nearly uniform.

Axial flow propellers may be designed on the assumption that the head is constant at all radii (free vortex relationship). As the blade speed varies between the boss and the outer diameter, the head can only be maintained by modifications to the whirl component. This means that the inclination of the blade must be steeper towards the boss. The matter is considered in Ex. 10.1.

More elaborate theoretical treatments have been devised and are described in specialist literature.[1] Generally, the vanes (particularly those of an axial flow machine) are regarded as aerofoils. The design of mixed-flow turbomachines is complicated by virtue of the variation of radius along the vane tips and the three-dimensional nature of the problem. A meridional flow path is normally investigated. Design procedure is necessarily dependent upon empirical coefficients and dimensionless constants, the values of which have been assessed on the basis of considerable research and practical experience. The profiling of vanes is primarily a drafting board problem.

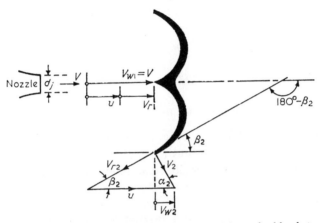

Figure 10.18   Entry and exit vectors for a Pelton wheel bucket
(not to scale)

Pelton wheel turbines present a rather special case because the runner operates at atmospheric pressure. The velocity vectors for a single bucket are depicted in Fig. 10.18. Peripheral speed at entry and exit is the same and the absolute velocity at exit has little or no whirl component. Due to frictional effects there is a diminution in the relative velocity of the water in passing over the buckets, so that at exit $V_{r2} = k_b V_{r1}$, where the coefficient k_b has a value about 0·9.

Thus the expression for the head utilised (Eq. (10.10) with reversed signs) becomes

$$H_0 = \frac{u(V_{w1} - V_{w2})}{g}$$

[1] For example: WISLICENUS, G. F. (1947) *Fluid Mechanics of Turbomachinery.* McGraw-Hill.

Now $V_{w1} = V$ and $V_{w2} = u - k_b(V - u) \cos \beta_2$, so that

$$H_0 = \frac{u}{g}(V - u)(1 + k_b \cos \beta_2) \tag{10.16}$$

The second bracketed term has its maximum value when β_2 is zero, but this is impractical since the water would not clear the succeeding bucket. A value of β_2 equal to about 15 degrees is found to give the best results.

Differentiating in order to determine the peripheral speed for maximum head utilised (or maximum power output or efficiency), we obtain

$$\frac{\mathrm{d}H_0}{\mathrm{d}u} = V - 2u = 0$$

or

$$u = \frac{V}{2} \tag{10.17}$$

In practice, however, it is found that the optimum peripheral speed is when

$$u = 0{\cdot}45 \text{ to } 0{\cdot}47\sqrt{2gH} \tag{10.18}$$

where H is the available head at the entry to the nozzle.

The term $u/\sqrt{2gH}$ is known as the *speed factor* or *peripheral coefficient*, and is a useful dimensionless parameter for all classes of turbomachinery. Owing to the somewhat rigid values prescribed both for this term and the wheel ratio (D_w/d_j) there is little latitude possible in deciding Pelton wheel proportions. In contrast, the range of speed factor values applicable to rotodynamic pumps and reaction turbines affords scope for a welcome flexibility in design.

Example 10.1

An axial flow pump is to discharge 0·15 cumecs at an operating speed of 1450 rev/min. The inner and outer diameters of the impeller are to be 0·14 m and 0·28 m respectively. If the head developed according to the vector triangles is 6 m, determine the angle of the vane tips at the inner and outer diameters and the corresponding diffuser blade angles.

The velocity of flow, constant throughout, is obtained from $Q = (\pi/4) \times (0{\cdot}28^2 - 0{\cdot}14^2)V_f$, whence $V_f = 3{\cdot}25$ m/s, while the inner and outer peripheral speeds are given by $u' = \pi \times 0{\cdot}14 \times 1450/60 = 10{\cdot}62$ m/s and $u'' = 21{\cdot}24$ m/s.

Assuming a constant head H_0 at all radii, we have $H_0 = u'V_{w2}'/g = u''V_{w2}''/g$, whence $V_{w2}' = 9{\cdot}81 \times 6/10{\cdot}62 = 5{\cdot}54$ m/s and $V_{w2}'' = 2{\cdot}77$ m/s.

From the vector triangles:

$$\tan \beta_1' = V_f/u' = 3{\cdot}25/10{\cdot}62 = 0{\cdot}306, \text{ so that } \beta_1' = \mathbf{17° \ 1'}$$
$$\tan \beta_2' = V_f/(u' - V_{w2}') = 3{\cdot}25/(10{\cdot}62 - 5{\cdot}54) = 0{\cdot}640$$
$$\text{so that } \beta_2' = \mathbf{32° \ 37'}$$

Similarly, $\beta_1'' = \mathbf{8° \ 42'}$ and $\beta_2'' = \mathbf{9° \ 59'}$

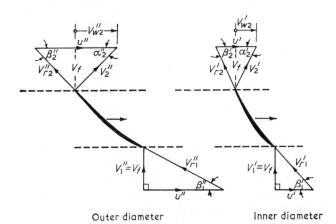

Outer diameter   Inner diameter

Ex. 10.1

For the diffuser blade angles, $\tan \alpha_2' = V_f/V_{w2}' = 3{\cdot}25/5{\cdot}54 = 0{\cdot}587$, or $\alpha_2' = \mathbf{30° \ 24'}$. Similarly $\alpha_2'' = \mathbf{49° \ 34'}$.

10.7 Performance

10.7.1 Losses and Efficiencies

The *overall efficiency*, or simply the *efficiency*, η, of a pump or turbine is the ratio of the useful power output to the power input or available.

Thus for pumps and turbines, respectively:

$$\eta = \frac{wQH}{P_i} \quad \text{and} \quad \eta = \frac{P}{wQH} \tag{10.19}$$

where P_i is the power input at the shaft to a pump and P the corresponding output from a turbine. Pump efficiencies are usually of the order of 80 per cent whereas turbine efficiencies are rarely less than 90 per cent, the difference being largely accounted for by the generally greater size of turbines and the more efficient flow passages.

The energy losses that occur within a pump or turbine are attributable to volumetric, mechanical, and hydraulic losses.

Volumetric loss arises because of the small clearances that must be provided between the rotating element and the casing. A slight leakage Q_L from the high pressure side to the low pressure side is unavoidable. Thus the impeller passages of a pump are handling more water than is actually delivered, whilst the runner passages of a turbine are handling less than is available. The amount of leakage depends on the pressure difference and the clearance sectional area. The *volumetric efficiency* η_v of a pump and reaction turbine respectively is given by:

$$\eta_v = \frac{Q}{Q + Q_L} \quad \text{and} \quad \eta_v = \frac{Q - Q_L}{Q} \tag{10.20}$$

It is not possible to scale down clearances to any extent so that this is a major reason for the lower overall efficiency of small units. With large and medium size units ($Q > 0.3$ cumecs) η_v is rarely less than 98 per cent.

The power input at the shaft to a pump is greater than that indicated by the vector triangles because allowance must be made for the power consumed in overcoming mechanical friction at bearings and stuffing boxes, as well as disc friction on the outer side of the shrouds and fluid shear in the clearances. Similar considerations apply to the output of turbines, excepting that in the case of Pelton wheels disc friction is replaced by windage. Thus the *mechanical efficiency* η_m of a pump and turbine respectively is given by:

$$\eta_m = \frac{w(Q + Q_L)H_0}{P_1} \quad \text{and} \quad \eta_m = \frac{P}{w(Q - Q_L)H_0} \tag{10.21}$$

Values of η_m are relatively high, usually between 95 and 98 per cent.

The head actually produced or utilised is less than that available because of frictional and eddy losses in the flow passages, both rotating and stationary. Thus for pumps and turbines respectively:

$$H = H_0 - \text{losses} \quad \text{and} \quad H = H_0 + \text{losses} \tag{10.22}$$

These hydraulic losses are approximately proportional to Q^2. Thus in the case of a pump the theoretical head-discharge relationship given by Eq. (10.14) is more nearly represented in practice by

$$H = A - BQ - CQ^2 \tag{10.23}$$

It follows that the *hydraulic efficiency* η_h of a pump and turbine respectively is given by

$$\eta_h = \frac{H}{H_0} \quad \text{and} \quad \eta_h = \frac{H_0}{H} \tag{10.24}$$

For large machines η_h may approach 95 per cent, being rather higher for turbines than for pumps.

Referring to Eqs. (10.19), (10.20), (10.21) and (10.24), we find that

$$\eta = \eta_v \times \eta_m \times \eta_h \tag{10.25}$$

A large part of the internal hydraulic energy loss occurs in the expanding passage of the volute casing or draft tube. Applying Bernoulli's equation to the impeller exit and discharge flange of a pump, we obtain

$$\frac{p_2}{w} + \frac{V_2^2}{2g} = \frac{p_d}{w} + \frac{V_d^2}{2g} + h_{fv}$$

where h_{fv} is the energy head loss in the volute casing. Then the *efficiency of pressure head recovery* η_c is given by

$$\eta_c = \frac{p_d - p_2}{w} \div \frac{V_2^2}{2g}$$

$$= 1 - \frac{V_d^2}{V_2^2} - \frac{2gh_{fv}}{V_2^2} \tag{10.26}$$

With a well-designed volute, η_c may be as high as 70 per cent.

Example 10.2

A centrifugal pump runs at 1450 rev/min and delivers 95 l/s against a head of 25 m. The impeller diameter at exit is 0·3 m and the area of flow, constant throughout, is 0·022 m². Vane angle at exit is 23°. Energy head loss in the impeller is estimated to be 1·1 m. There is no whirl at entry. Outlet flange and delivery pipe are 203 mm (8 in.) diameter.

Estimate the pressure head rise across the impeller, and the pressure head rise and loss of energy head in the casing. Assume uniform velocity distributions.

$Q = 95$ l/s; $V_d = 0.095 \times 4/(\pi \times 0.203^2) = 2.93$ m/s; $u_2 = \pi \times 0.3 \times 1450/60 = 22.78$ m/s; $V_1 = V_{f1} = V_{f2} = 0.095/0.022 = 4.32$ m/s.

From the velocity vectors at exit, $V_{w2} = u_2 - V_{f2}/\tan 23 = 22.78 - 4.32/0.4245 = 12.60$ m/s. Thus $V_2^2 = V_{f2}^2 + V_{w2}^2 = 177.4$ and $H_0 = u_2 V_{w2}/g = 29.3$ m.

Applying Bernoulli's equation to the impeller entry and exit,

$$\frac{p_1}{w} + \frac{V_1^2}{2g} = \frac{p_2}{w} + \frac{V_2^2}{2g} - H_0 \; 1.1$$

from which $(p_2 - p_1)/w = 29.3 - 1.1 + (18.7 - 177.4)/19.61 = \mathbf{20.1}$ m.

Substituting in Eq. (10.22) and neglecting losses between the suction flange and impeller, we obtain $h_{fv} = 29\cdot3 - 1\cdot1 - 25 = 3\cdot2$ m. Thus $(p_d - p_2)/w = (V_2^2 - V_d^2)/2g - h_{fv} = \mathbf{5\cdot4}$ m.

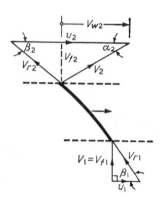

Ex. 10.2

10.7.2 Characteristics

The performance of pumps and turbines may be ascertained by tests on site or, more conveniently, in a laboratory where accurate testing facilities are available. In the case of large units, model tests (see Ch. 11, Sect. 11.4.7) may be accepted in lieu of those on the full scale. Testing procedure is laid down in the relevant British Standard specification.

Pump and turbine characteristics are best described separately:

Pumps. As the discharge is nearly always the primary factor it is customary for the *performance curves* to consist of the three curves of head, power input, and efficiency, drawn to a common baseline of discharge. In the usual case of a fixed speed A.C. motor being employed, only the one set of curves for the given speed is required.

Each design of pump has its own characteristic behaviour. Fig. 10.19 shows the performance curves for two contrasting types – a centrifugal and an axial flow pump. These are representative only and by adjustments to the impeller almost any head-discharge relationship in the intermediate range may be obtained. For example, the effect of varying the blade angle at exit has been discussed earlier.

A flat head-discharge curve is known as an 'unstable' characteristic, since for any given head a number of discharges are possible and the pump may choose to deliver any one of these; there is consequently some uncertainty in predicting the actual discharge. On the other hand a sloping head discharge curve is 'stable' since only one discharge is possible for a given head; this condition should always pertain over the likely operating range.

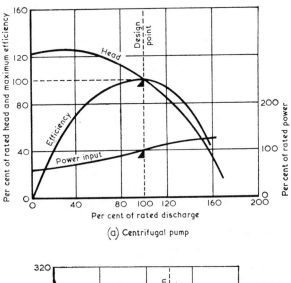

(a) Centrifugal pump

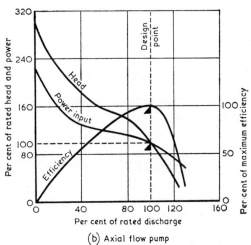

(b) Axial flow pump

Figure 10.19 Pump characteristics

Centrifugal pumps are said to be self-regulating since at low discharges the power input is much less than in the region of the design point. There is thus no risk of overloading the motor. In contrast, the power input to the axial flow pump is much greater under closed delivery valve conditions than at the design point. To avoid overloading the motor, the delivery valve is often omitted or cut-out equipment fitted.

If a pump is tested at a representative number of fixed speeds the information so obtained will enable the *iso-efficiency curves* to be plotted as shown in Fig. 10.20. This single diagram depicts the performance of the pump over the entire range of operation and enables the speed and efficiency to be determined for any given head and discharge.

It is found that over a reasonable head range the optimum efficiency remains approximately the same, provided that the speed variation

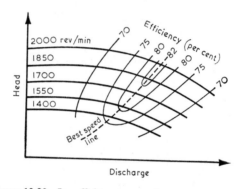

Figure 10.20   Iso-efficiency curves for a centrifugal pump

thereby entailed is not excessive. Thus limiting our consideration to the hydraulic efficiency, we have

$$\eta_h = \left(\frac{gH}{V_{w2}u_2}\right)_{n_1} = \left(\frac{gH}{V_{w2}u_2}\right)_{n_2} = \left(\frac{gH}{V_{w2}u_2}\right)_{n_3} \quad \text{etc.}$$

Now as the impeller geometry is fixed (β_2 constant) and the water is to enter and leave without shock, it follows that for the same efficiency under various operating conditions the velocity triangles must be similar, or $(u_2/V_{w2})_{n_1} = (u_2/V_{w2})_{n_2}$, and so on for the other vectors. Hence any velocity in the vector triangles may be expressed as

$$\text{velocity} = \text{constant} \times H^{1/2} \qquad (10.27)$$

and since $u \propto n$, we have

$$n = \text{constant} \times H^{1/2} \qquad (10.28)$$

Thus for any given pump over a limited range of operating speed:

$$\frac{(H)_{n_1}}{(H)_{n_2}} = \left(\frac{n_1}{n_2}\right)^2 \tag{10.29}$$

$$\frac{(Q)_{n_1}}{(Q)_{n_2}} = \frac{(V_f)_{n_1}}{(V_f)_{n_2}} = \frac{(H)_{n_1}^{1/2}}{(H)_{n_2}^{1/2}} = \frac{n_1}{n_2} \tag{10.30}$$

$$\frac{(P_i)_{n_1}}{(P_i)_{n_2}} = \frac{(QH)_{n_1}}{(QH)_{n_2}} = \left(\frac{n_1}{n_2}\right)^3 \tag{10.31}$$

The information given by Eqs. (10.29), (10.30), and (10.31) enables the approximate performance to be predicted at speeds different from the original.

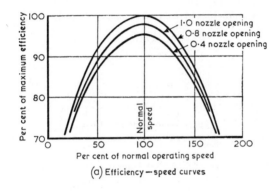

(a) Efficiency – speed curves

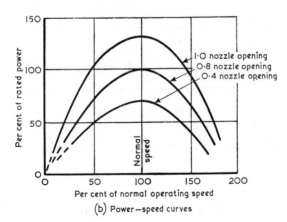

(b) Power – speed curves

Figure 10.21   Performance curves for a Pelton wheel

285

Turbines. Tests are carried out on the actual turbine or its replica at a variety of speeds and gate openings. As turbine output must be varied to suit the electrical demand it is customary to design the machine so

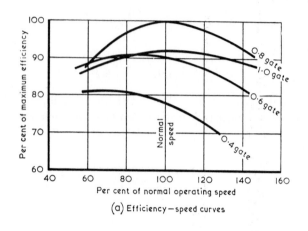

(a) Efficiency—speed curves

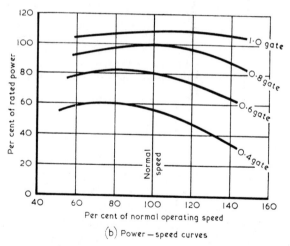

(b) Power—speed curves

Figure 10.22   Performance curves for a Francis turbine

that optimum efficiency occurs at about three-quarters of full load. Efficiency and power output are usually plotted against speed for a constant head.

Fig. 10.21 shows the typical performance curves for a Pelton wheel. It will be noted that the curves are approximately parabolic in shape, thus conforming to theory; also that the optimum efficiency occurs at approximately 50 per cent of no load or runaway condition. The corresponding curves for a representative Francis turbine (Fig. 10.22) exhibit a much greater falling off in efficiency either side of the optimum. This is because there is shock loss at entry. Iso-efficiency curves may be drawn in a similar manner to those for pumps.

A comparison of the part-load performance curves of the various types of turbine is instructive. The typical curves in Fig. 10.23 indicate

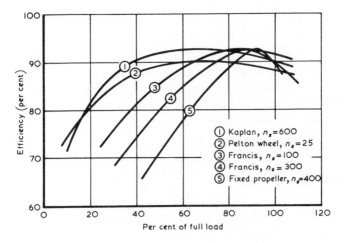

Figure 10.23   Part-load performance curves

that the Kaplan and Pelton wheel are much superior in this respect to the low head Francis and in particular the fixed blade propeller turbine. The additional cost of variable pitch blades could well be recouped in a very short period by more efficient running.

The equations derived for pumps are equally applicable to turbines, and in this case it is easier to appreciate the validity of the velocity-head relationship (vel. $\propto H^{1/2}$), for the turbine passages are but a form of orifice.

10.8 Specific Speed

It is useful to have a common basis on which different types of pump or turbine design can be compared irrespective of size. The parameter

287

known as *specific speed* has been introduced for this purpose, and the respective definitions could be as follows:

(a) The specific speed of a pump is the speed in rev/min of a geometrically similar pump of such a size that it delivers one cumec against 1 metre head.

(b) The specific speed of a turbine is the speed in rev/min of a geometrically similar turbine of such a size that it develops one kilowatt under 1 metre head.

These definitions recognise the significant performance parameters. In the case of pumps it is the discharge that is important, whilst for turbines it is the power output.

Expressions for the specific speed n_s may be derived by dimensional analysis, or in the following manner:

For a rotodynamic pump or reaction turbine[1] and their specific counterparts, the ratio of discharges is given by $Q/Q_s = AV_f/A_sV_{fs}$, where A is the peripheral area and V_f the velocity of flow. For similarity of shape and hydraulic behaviour, we have $A/A_s = (D/D_s)^2$ and $V_f/V_{fs} = (H/H_s)^{1/2}$, whence

$$\frac{Q}{D^2 H^{1/2}} = \frac{Q_s}{D_s^2 H_s^{1/2}} = \text{dimensionless term} \qquad (10.32)$$

Also since $u/u_s = Dn/D_s n_s = (H/H_s)^{1/2}$, we have

$$\frac{Q}{D^3 n} = \frac{Q_s}{D_s^3 n_s} = \text{dimensionless term} \qquad (10.33)$$

Combining Eqs. (10.32) and (10.33) in order to eliminate D, we obtain

$$n_s = n\left(\frac{Q}{Q_s}\right)^{1/2}\left(\frac{H_s}{H}\right)^{3/4} \qquad (10.34)$$

For the specific pump, $H_s = 1$ m and $Q_s = 1$ cumec, so that

$$n_s = \frac{nQ^{1/2}}{H^{3/4}} \qquad (10.35)$$

where Q is in cumecs.

Similarly for turbines:

$$n_s = n\left(\frac{QH}{Q_s H_s}\right)^{1/2}\left(\frac{H_s}{H}\right)^{5/4} = n\left(\frac{P}{P_s}\right)^{1/2}\left(\frac{H_s}{H}\right)^{5/4} \qquad (10.36)$$

[1] It may be shown that the specific speed expression (Eq. (10.37)) is also applicable to a Pelton wheel.

Putting $P_s = 1$ kilowatt and $H_s = 1$ m, we obtain

$$n_s = \frac{nP^{1/2}}{H^{5/4}} \qquad (10.37)$$

It will be noted that the specific speed is independent of the dimensions and therefore relates to shape rather than size. All pumps or turbines of the same shape have the same specific speed.

The values of n, H, Q, and P are those for the normal operating condition (the design point), which would generally coincide with the optimum efficiency. It is the usual (but by no means universal) convention to make no distinction between single-suction and double-suction

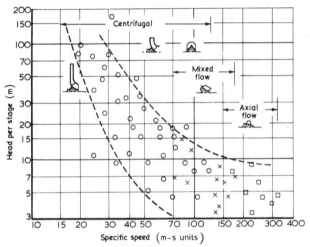

Figure 10.24   Relationship between head and specific speed for pumps

impellers, the total discharge being the only criterion. However, with multi-stage impellers the specific speed is that appropriate to each single stage; thus the value of H in Eq. (10.35) is taken as the total head divided by the number of stages.

The form of Eqs. (10.35) and (10.37) indicates that specific speed is proportional to the speed of rotation. Thus propeller type units might be expected to have a higher specific speed than relatively slower machines such as centrifugal pumps or Pelton wheels, and this is indeed so. The equations also show that specific speed is inversely related to the head. In fact this parameter is a useful first guide to the most suitable type of machine. Furthermore the need to obtain maximum possible efficiency restricts the range of specific speed appropriate for each type. Figs 10.24 and 10.25 show the general H–n_s relationship for pumps

289

and turbines respectively. The distance apart of the enveloping lines and the overlapping of machine types give some indication of the degree of flexibility possible in design. In these diagrams the representative

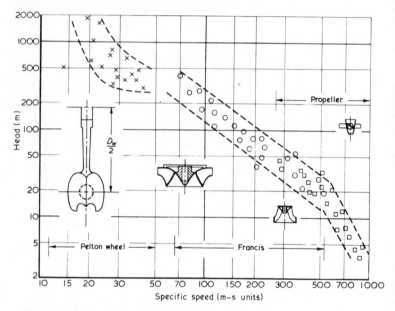

Figure 10.25   Relationship between head and specific speed for turbines

impellers and runners are drawn to the same scale in the sense that they will handle about the same discharge and head if the speed of rotation is changed proportional to the specific speed, thereby giving striking evidence of the greater economy associated with higher specific speed.

Example 10.3

A two-stage centrifugal pump has a rated duty of 22 l/s at 1450 rev/min against 40 m head. Determine the specific speed and the number of additional stages required to obtain a specific speed of about 30.

Assuming that a motor operating at 1200 rev/min is available, estimate for the original pump the revised head and discharge at optimum efficiency.

$$n_s = nQ^{1/2}/H^{3/4} = 1450 \times 0.022^{1/2}/20^{3/4} = \mathbf{22 \cdot 7}$$

For $n_s = 30$, $n = 1450$ rev/min:

$H = [1450 \times 0.022^{1/2}/30]^{4/3} = 13.8$ m so that the number of stages $= 40/13.8 = 2.9$. Hence **one** additional stage is required.

For $n_s = 22 \cdot 7$, $n_1 = 1450$ rev/min, $n_2 = 1200$ rev/min:

From Eq. (10.29), $H_2 = 20 \times (1200/1450)^2 = 13 \cdot 7$ m so that the total head $= 27 \cdot 4$ m. Also, from Eq. (10.30), $Q_2 = 22 \times (1200/1450) = \textbf{18} \boldsymbol{\cdot} \textbf{2}$ l/s.

Example 10.4

A Francis turbine is to be installed at a site where the average available head and discharge are 170 m and 16 cumecs, respectively. The design assumptions are as follows: (a) Specific speed not to exceed 135, (b) rotational speed to be synchronous with alternator operating at 50 cycles/sec, (c) speed factor 0·75, (d) ratio peripheral width/peripheral diameter not to exceed 0·17, (e) blade thickness factor 0·92.

Making appropriate assumptions concerning efficiencies, determine the following:

(i) Power output at the shaft, (ii) rotational speed, (iii) peripheral diameter and width, (iv) runner vane angle at periphery, (v) guide blade angle at entry to runner.

Assuming $\eta = 0 \cdot 93$, we may substitute in Eq. (10.19) and obtain $P = 0 \cdot 93 \times 9 \cdot 81 \times 16 \times 170 = \textbf{24 800}$ kW.

Substituting in Eq. (10.37), $n = 135 \times 170^{5/4}/24\,800^{1/2} = 526$ rev/min. Thus the nearest synchronous speed is $60 \times 50/6 = \textbf{500}$ rev/min.

$$u_1 = 0 \cdot 75\sqrt{2gH} = 0 \cdot 75 \times \sqrt{2g} \times 170^{1/2} = 43 \cdot 3 \text{ m/s}$$

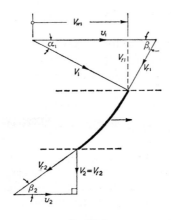

Ex. 10.4

But $u_1 = \pi D_1 n/60$, so that $D_1 = 60 \times 43.3/500\pi = \textbf{1} \boldsymbol{\cdot} \textbf{65}$ m. Hence $b_1 = 0 \cdot 17 D_1 = \textbf{0} \boldsymbol{\cdot} \textbf{28}$ m.

Since $Q = A_1 V_{f1} = \pi D_1 b_1 k_1 V_{f1}$, we find $V_{f1} = 16/(\pi \times 1 \cdot 65 \times 0 \cdot 92$

291

× 0·28) = 12·0 m/s. Assuming η_h = 0·95, then from Eq. (10·24), V_{w1} = 0·95 × 9·81 × 170/43·3 = 36·6 m/s. From the elementary vector triangles at entry to the runner:

$$\tan \beta_1 = V_{f1}/(u_1 - V_{w1}) = 12\cdot0/(43\cdot3 - 36\cdot6) = 1\cdot791, \text{ whence } \beta_1 = \mathbf{60° \, 49'}$$

$$\tan \alpha_1 = V_{f1}/V_{w1} = 12\cdot1/36\cdot6 = 0\cdot328, \text{ whence } \alpha_1 = \mathbf{18° \, 10'}$$

10.9 Cavitation Considerations

10.9.1 *General*

The primary reason why high specific speed pumps and turbines cannot be used in connection with high heads is the need to avoid the harmful phenomenon known as cavitation, the general nature of which has been described earlier (Ch. 4, Sect. 4.11.4). In addition to influencing the design it also imposes severe limitations on the machine setting, that is

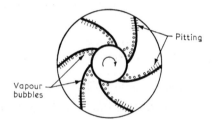

Figure 10.26   Cavitation in a centrifugal impeller

to say the permissible suction lift in the case of a pump and the height above the tailwater in the case of a turbine.

With centrifugal impellers, the most vulnerable points for attack are the vane tips at discharge. The pressure is lowest on the underside of the vanes near entry and this factor, combined with the disturbance caused by the rapid deviation and acceleration in flow, is conducive towards separation. It is here that vapour bubbles tend to form, as in Fig. 10.26, which are then carried forward by the flow to a region of higher pressure near the exit where they collapse violently, causing pitting and severe damage to the metal surface. With axial flow impellers, points a short distance from the outer diameter sometimes suffer the most severe damage; in any event this type of pump is more susceptible to attack by virtue of its higher relative speed.

Cavitation manifests itself in a similar manner with turbines. Each design has its own characteristic susceptibility. Eddy pockets with fluctuating pressures tend to form along the rear face of the runner blades near exit, and this is consequently a vulnerable area for cavitation attack.

Apart from the physical damage caused by cavitation, the reduction of the effective volume of the flow passages due to the presence of water vapour results in a smaller discharge and a sharp drop in efficiency. Additional evidence is the noise and vibration produced by the collapse of the vapour bubbles.

10.9.2 *Pump Setting*

In order to avoid cavitation we must ensure that the pressure at the inlet to the pump does not fall below a certain minimum, dictated by the further reduction in pressure that occurs within the pump passages.

From Eq. (10.4), the absolute pressure at inlet p_{sa} is given by

$$\frac{p_{sa}}{w} = H_a - \left(H_s + \frac{V_s^2}{2g} + h_{fs}\right) \tag{10.38}$$

where H_a is the atmospheric pressure head.

The term *net positive inlet head* (or *net positive suction head*), H_{sv}, has been introduced. It represents the head required to force the water from the inlet into the impeller and by definition is equal to the inlet head minus the vapour pressure head, or

$$H_{sv} = \left(\frac{p_{sa}}{w} + \frac{V_s^2}{2g}\right) - \frac{p_v}{w} \tag{10.39}$$

Utilising Eq. (10.38), we obtain

$$H_{sv} = H_a - \frac{p_v}{w} - H_s - h_{fs}$$

But $H_a - p_v/w$ is equal to the height of the water barometer H_b and $H_s + h_{fs}$ is the effective suction lift H_s', so that

$$H_{sv} = H_b - H_s' \tag{10.40}$$

The required minimum value of H_{sv} is dependent on the design of each individual pump and the conditions under which it operates. For any given total head, speed, and discharge, it may be ascertained by testing the pump with increasing effective suction lifts until a point is reached where there is a marked drop in efficiency, signifying the onset of cavitation.

For design purposes, a cavitation parameter known as *critical sigma*, σ_c, is employed, where

$$\sigma_c = \frac{H_{sv}}{H} \qquad (10.41)$$

Now σ_c is more closely related to the specific speed than any other factor and research has enabled appropriate relationships to be established. As it is prudent to introduce a small factor of safety, the minimum permissible σ is somewhat greater than σ_c. Design charts enable the effective suction lift appropriate to various operating heads and classes of pump to be readily ascertained. For instance in the case of a single-suction impeller with $n_s = 32$ and $H = 40$ m the value of σ is 0·12 so that H_s' is approximately 5 m. Incidentally, the maximum value of H_s' is usually taken as 7·5 m. Atmospheric pressure diminishes with altitude so that permissible suction lifts are less at elevated sites.

Axial flow pumps, of course, are normally sited with their impellers immersed, but as the velocity has to be increased from zero to as much as 6 m/s in a very short distance there is a low upper limit to the maximum total head if cavitation is to be avoided. Indeed, it is possible for H_s' to be negative, in which case the impeller must not be set at a lesser depth than that indicated below the level of the water surface in the suction well.

Similar considerations apply to the setting of reaction turbines. The avoidance of cavitation generally demands that the machine be set within a few feet of (above or below) the tailwater level.

Example 10.5

A Kaplan turbine, with runner 1·5 m diameter, discharges 14 cumecs. In order to avoid cavitation the pressure head at entry to the draft tube must not be more than 1·5 m below atmospheric. The efficiency of the draft tube is 70 per cent.

Estimate the maximum height at which the runner may be set relative to the tailwater level.

The velocity at entry to the draft tube (just below runner boss) is V_e = 14 × 4/2·25π = 7·92 m/s

Neglecting the small residual velocity head at discharge, the draft tube efficiency (cf. Eq. (10.26)) is $\eta_d = 1 - 2gh_{fd}/V_e^2$, so that $h_{fd} = 3.20(1 - 0·7)$ = 0·96 m.

For the limiting case, $p_e/w + V_e^2/2g = h_s + 0·96$, where h_s is the height of the tailwater surface above the entry to the draft tube. Thus $h_s = -1·5 + 3·20 - 0·96 \simeq 0·7$ m.

10.10 Design of a Pumping Main[1]

The piping layout is an important feature of any pumping installation and considerable care requires to be given to the provision of an efficient design.

On the suction side the aim should be to reduce head losses to a minimum so as to keep the total suction lift within the prescribed limits. This usually means that the pump must be located adjacent to the suction well and a suction pipe of fairly liberal diameter provided – generally such as to give pipe velocities of between 1·5 and 2·5 m/s. Sluice valves on the suction side are rarely necessary and in no circumstances should they be used for regulating the flow.

Turning now to the delivery branch, it is customary in the case of centrifugal pumps to control the flow by means of a sluice or gate type valve situated in close proximity to the pump. Normally the pump is started with this valve closed and then when the unit is running at normal speed it is opened the required amount, preferably to the full extent if energy is not to be wastefully expended. When the pumping main is of considerable length, a reflux valve is usually fitted on the pump side of the delivery valve. This is a flap type valve, arranged so as to open in the direction of flow only. On the cessation of pumping, and immediately the water column reverses in direction, the reflux valve closes, thus protecting the pump and suction piping from any harmful pressure rise.

Economic considerations are involved in deciding the best diameter for the delivery pipe. The larger the diameter the higher is the capital cost, but on the other hand the lower is the running cost because of reduced frictional losses. Several alternative pipe sizes may need to be investigated. Usually the pipe sizes selected result in velocities of between 1·5 and 3 m/s.

The piping system cannot be considered in isolation to the pump in an economic analysis, since the pump performance and the external pipe characteristics are inter-related. It is, of course, important that not only should the pump be capable of delivering the stipulated discharge, but that it should do so at or near to its peak efficiency.

Now the pump can only deliver in accordance with its H-Q characteristic curve. The total head pumped against, however, comprises the static lift plus the pipe system losses. The latter may all be expressed in terms of rQ^m (see Ch. 6, Sect. 6.5.3); if $m = 2$ this is a quadratic relationship. Thus a family of parabolic curves based on

---

[1] Pipelines supplying turbine installations have been discussed in Ch. 6, Sect. 6.2. Correlation of pipeline and turbine characteristics is of course necessary.

different pipe diameters may be drawn. By superimposing these on the performance curves of the pump, as in Fig. 10.27, and noting the relevant intersection points the actual discharge and efficiency may be determined for any given diameter. If purpose-made plant is to be employed the pipe-system curves can be used as a basis for pump design.

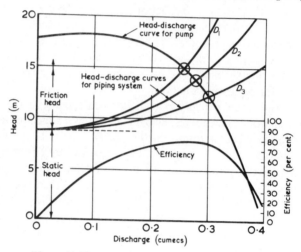

Figure 10.27   Pump and pipe-system characteristics

Further Reading

ADDISON, H. (1966) *Centrifugal and Other Rotodynamic Pumps.* Chapman and Hall (3rd Edition).

ADDISON, H. (1958) *The Pump Users' Handbook.* Pitman.

ANDERSON, H. H. (1962) *Centrifugal Pumps.* Trade and Technical Press.

BROWN, J. G. (Ed.) (1958) *Hydro-Electric Engineering Practice,* vol. 2. Blackie.

CREAGER, W. P. and JUSTIN, J. D. (1950) *Hydro-Electric Handbook.* Wiley.

HICKS, T. G. (1957) *Pump Selection and Application.* McGraw-Hill.

KARASSIK, I. and CARTER, R. (1960) *Centrifugal Pumps.* Dodge Corp.

KOVALEV, N. N. (1965) *Hydroturbines.* Oldbourne Press.

KOVÁTS, A. (1964) *Design and Performance of Centrifugal and Axial Flow Pumps and Compressors.* Pergamon Press.

KRISTAL, F. A. and ANNETT, F. A. (1953) *Pumps.* McGraw-Hill.

LAZARKIEWICZ, S. and TROSKOLANSKI, A. T. (1965) *Impeller Pumps.* Pergamon Press.

MOSONYI, E. (1957) (1961) *Water Power Development,* vol. 1, vol. 2. Hungarian Academy of Sciences.

NECHLEBA, M. (1957) *Hydraulic Turbines, Their Design and Equipment.* Constable.

—— (1961) *Pumping Manual.* Trade and Technical Press.

NORRIE, D. H. (1963) *An Introduction to Incompressible Flow Machines.* Arnold.

STEPANOFF, A. J. (1957) *Centrifugal and Axial Flow Pumps.* Wiley (2nd Edition).

WISLICENUS, G. F. (1947) *Fluid Mechanics of Turbomachinery.* McGraw-Hill ((1964) 2 Vols. Dover revised Edition).

(*Brit. Hydromech. Res. Assn.*)

8 Model (scale 1:200) of spillway barrage on Volta River, Ghana

9 Manometer bank for the investigation of pressures on a spillway profile

(Brit. Hydromech. Res. Assn.

10   Model (scale 1:20) of the spillway at the Lower Drift Reservoir, Cornwall. A prototype
flow of 4000 cusecs is simulated

(Brit. Hydromech. Res. Assn.)

11   Sectional view of 1:20 scale model of Lamaload spillway. The discharge is equivalent
to 2000 cusecs

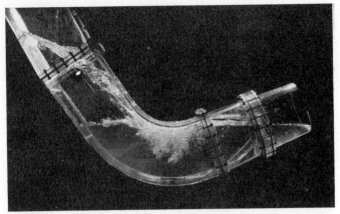

a   Early stage of priming

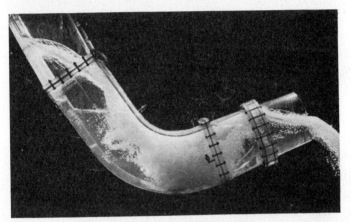

b   Intermediate stage of priming

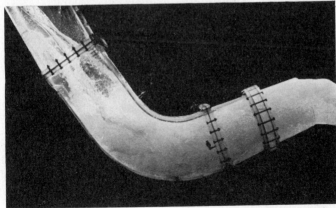

(*Univ. of Aberdeen*)

c   Late stage of priming

12   Priming of a 1:12 scale perspex model siphon

13   Model (scale 1:39·4) of Macagua No. 1 spillweir, showing bed contours after a scour test

14   Sheet-asbestos templates in use for the construction of a model of R. Severn

CHAPTER ELEVEN

Hydraulic Models

11.1 Introduction

In several important branches of engineering the aid of models is invoked in the solution of design problems. The technique has particular relevance to the study of fluid mechanics phenomena where, owing to the complexity of boundary configuration and fluid characteristics, the problems presented are all too often analytically intractable. It is in these circumstances that a model investigation under controlled conditions in a laboratory is likely to prove most valuable. By observing the effects of various design proposals over a wide range of conditions likely to be encountered in practice it is possible to assess their relative merits and in many cases to make useful quantitative predictions. Such an investigation on the full scale would probably be quite impracticable or at least undesirably expensive.

In the sphere of naval architecture the testing of model ships in a towing tank provides valuable guidance on hull design. W. A. Froude pioneered this type of model investigation as early as the middle of the last century. The wind tunnel and water tunnel are more recent adjuncts in model testing. They have an indispensable role in modern aircraft design, and the former is also useful in connection with structural models when it is required to evaluate the pressures and vibrations caused by wind. In the case of hydraulic machinery, notably pumps and turbines, suitable testing facilities are maintained by most large manufacturers. Model tests are of value both for design purposes and as a convenient means of verifying specified performance. The best proportions for suction wells and tailbays are associated problems that are also amenable to model investigation.

Osborne Reynolds, Professor at Manchester University, deserves the credit for being the first[1] to demonstrate the role of models as a design facility in civil engineering hydraulics. Towards the end of the last

---

[1] In 1875 L. J. Fargue constructed a model of a short reach of the R. Garonne, but he was not able to establish scalar relationships.

century he constructed a very small-scale (by present-day standards) model of the Mersey estuary and by simulating the tidal flow was able to obtain a very fair representation of the sandbanks as they existed in nature.[1] Practising engineers of the period, accustomed to designing by instinct, experience, and with the assistance of a limited number of empirical formulae, were sceptical as to the potentialities of this new scientific tool. It was not until several decades later, when more convincing evidence, both abroad and in this country (e.g. Professor A. H. Gibson's[2] classic model studies of the Severn Estuary, 1926–33), had become available, that leading members of the profession were unanimously in favour.

Today it would be true to say that no major hydraulic project, about which there was any uncertainty as to the most efficient design, would be undertaken without the guidance afforded by model studies. There are now many hydraulics laboratories throughout the world – for example the hydraulics research Station, Wallingford, Oxon. – with the equipment and trained personnel to investigate a wide variety of problems concerned with pipelines, canals, rivers, estuaries, harbours, and all kinds of control and transition structures. Hydraulic investigations are not cheap to carry out, but their cost is generally only a small fraction of that of the full-scale proposal and is very often recouped by virtue of economies and improvements which the model studies have indicated as being desirable.

The full-scale system or device to be investigated is called the *prototype*, and the model may be larger, of the same size, or, as is much more common, smaller than the prototype. Generally, the model shape resembles the prototype, although this is not essential so long as the flow patterns are similar over the region to be studied.

A clear understanding of the underlying theory of the phenomenon to be investigated is essential if the model results are to be correctly interpreted. The relationship between model and prototype performance is governed by the *laws of hydraulic similarity*. Owing to the impossibility of achieving simultaneous compliance in respect of all the laws, some discrepancy in extrapolating to the full scale is unavoidable and this is known as *scale effect*. Fortunately, by making the model large enough or taking compensating steps, scale effect may be minimised. However, because of its existence, models should not be looked upon as

[1] REYNOLDS, O. (1887) 'On Certain Laws relating to the Regime of Rivers and Estuaries, and on the Possibility of Experiments on a Small Scale', *Brit. Assoc. Rept.*, 867.

[2] GIBSON, A. H. (1933) *Construction and Operation of a Tidal Model of the Severn Estuary.* H.M.S.O.

calculating machines or computers, capable of providing precise solutions in response to the design data fed to them, but rather as devices dependent upon human skill and ingenuity for their successful operation.

There has been a tendency in recent years for larger model scales to be favoured, thereby reducing scale effect. But of course it must not be lost sight of that both constructional and operational costs increase with size, so that economic considerations demand that the smallest possible model should be constructed consistent with the accuracy of prediction that is required.

Likewise, it follows that model techniques are not a competitor of analytical methods and that the latter should wherever possible be adopted. In fact, recent years have been noteworthy for the development of mathematical models, made possible by high speed systematic calculation performed by computers, generally with large storage capacity. The basis of such a model is a systems program obtained from an application of the appropriate hydrodynamic equations and used to calculate the motion of the fluid in the model.

Mathematical models are being increasingly used to study a wide range of hydraulic problems, and particularly those where simplifying assumptions are admissible. For the more detailed studies of fluid behaviour, physical models are generally required, but a mathematical model may well serve a valuable purpose in defining the boundary conditions.

11.2 Hydraulic Similarity

If the results obtained from a model are to be transferable to the prototype it is necessary for the two flow systems to be hydraulically similar. This entails geometric, kinematic, and dynamic similarity. Consideration will now be given to each of these criteria in turn, the model-prototype relationships for the flow through a control structure (Fig. 11.1) being used in illustration.

(a) *Geometric Similarity*. This is similarity of shape. The ratio of any two dimensions in the model is the same as the corresponding ratio in the prototype, or

$$\frac{(L_1)_m}{(L_2)_m} = \frac{(L_1)_p}{(L_2)_p} \tag{11.1}$$

Thus, if the linear scale of the model is $1:x$, the scalar relationship for area is $1:x^2$ and for volume $1:x^3$.

Complete geometric similarity includes similarity of boundary roughness. On the strictest interpretation this means that the configuration of the surface finish in the model should be an exact reproduction of the prototype, but of course the irregular nature of commercial finishes precludes any such high degree of conformity. The best that can normally be achieved is a scalar roughness, measured in terms of the equivalent excrescence size k as defined by the diameters of sand grain in Nikuradse's experiments on artificially roughened pipes. Thus the ratio $k_m:k_p$ at corresponding points on the boundary surface of model and prototype should be $1:x$.

Even this limited degree of conformity is very often difficult to achieve. For instance, with small-scale models and relatively smooth prototype surfaces (e.g. metal or well-finished concrete) it is not possible to obtain the desired additional degree of smoothness for the model surface. Again, in cases where scour and deposition are to be studied, it is

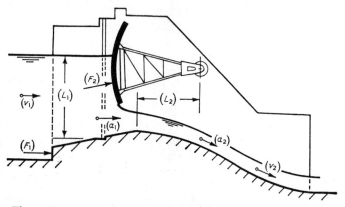

Figure 11.1   Similarity relations for flow at a radial gate installation

customary to provide a mobile bed in the model, but if the sedimentary material found in the prototype were to be reduced according to scale a fine powder would result whose behaviour in the model would be quite different. A more pronounced departure from geometric similarity arises in the case of river and estuary models, where differing vertical and horizontal scales are essential if the model is not to be of excessive size.

When considering the justification for these violations of geometric similarity it must be remembered that it is the hydraulic behaviour arising from the boundary conditions that is important. Other issues are subordinate, and it is in this light that some geometric dissimilarity is very often both necessary and tolerable.

(b) *Kinematic Similarity.* This is similarity of motion, thereby introducing the vector quantity and time factor. The latter is of importance in problems involving unsteady flow, such as tidal phenomena. The velocities (and accelerations) at homologous points and at homologous times in the two systems have the same ratio to each other. Also, the corresponding directions of motion are the same. Thus

$$\frac{(v_1)_m}{(v_2)_m} = \frac{(v_1)_p}{(v_2)_p} \quad \text{and} \quad \frac{(a_1)_m}{(a_2)_m} = \frac{(a_1)_p}{(a_2)_p} \tag{11.2}$$

Clearly, the essential prerequisite of a similar flow pattern is geometric similarity with respect to the boundaries. If any confirmation is needed, it may be recalled that for an ideal fluid the pattern of stream lines is uniquely determined by the boundary geometry.

(c) *Dynamic Similarity.* For dynamic similarity the forces at homologous points in the two systems have the same ratio to each other and act in the same direction. Thus

$$\frac{(F_1)_m}{(F_2)_m} = \frac{(F_1)_p}{(F_2)_p} \tag{11.3}$$

Since the flow pattern is governed by the forces which are acting, it follows that if dynamic similarity exists throughout, kinematic and geometric similarity must also obtain. In other words, the implications of hydraulic similarity and dynamic similarity are identical.

The component forces acting on any element of incompressible fluid may be due to pressure, gravity, viscosity, or surface tension. The conditions for dynamic similarity can thus be expressed algebraically as

$$\frac{(F_P)_p}{(F_P)_m} = \frac{(F_G)_p}{(F_G)_m} = \frac{(F_V)_p}{(F_V)_m} = \frac{(F_{ST})_p}{(F_{ST})_m} \tag{11.4}$$

If the magnitude and direction of the component forces are known, the construction of a vector polygon enables the resultant to be determined. According to Newton's second law ($F = Ma$), the resultant is in fact the inertial force F_I. Vectorially then,

$$F_P + F_G + F_V + F_{ST} = F_I \tag{11.5}$$

Dividing throughout by F_I we obtain

$$\frac{F_P}{F_I} + \frac{F_G}{F_I} + \frac{F_V}{F_I} + \frac{F_{ST}}{F_I} = 1 \tag{11.6}$$

Each of these force ratios will be recognised as the inverse of the relevant dimensionless parameter (see Ch. 4, Sect. 4.12). Thus

$$\frac{K_1}{E} + \frac{K_2}{F} + \frac{K_3}{R} + \frac{K_4}{W} = 1 \tag{11.7}$$

where E, F, R, and W are the Euler, Froude, Reynolds, and Weber numbers respectively, and $K_1, K_2, \ldots$ are dimensionless coefficients.

From a consideration of Eqs. (11.4) and (11.6) it follows that for perfect hydraulic similarity the number must be the same for both prototype and model. No other conclusion could in fact be reached, since it is inconceivable that the resultant forces could be similar but not the components – otherwise there would be some discrepancy evident in the two flow patterns.

11.3 Conformance with Similarity Laws

We will now consider what is involved in achieving similarity of fluid behaviour. In hydraulic model studies the compressibility property of the prototype fluid is rarely of significance and therefore for present purposes will be ignored. The implications of the various similarity criteria or model laws are as follows:

(a) *Euler Law.* As was explained in Ch. 4, Sect. 4.12, p. 71, the Euler number ($E = V/\sqrt{2\,\Delta p/\rho}$) being derived from the fundamental force-momentum concept, describes a simple basic relationship between pressure and velocity. The parameter is of particular relevance to the model of an enclosed fluid system where the turbulence is fully developed; viscous forces are then insignificant in relation to inertial forces and, of course, the forces of gravity and surface tension are entirely absent. Since the applied pressures are now the controlling factor, the pressure force has the role of an independent variable. But this is contrary to most fluid phenomena where the pressure force, being consequent upon the motion, is a dependent variable.

From the Euler law an expression may be derived relating model and prototype velocities, or *corresponding velocities* as they are called. This relationship is

$$\frac{V_p}{V_m} = \frac{\Delta p_p^{1/2} \; \rho_m^{1/2}}{\Delta p_m^{1/2} \; \rho_p^{1/2}} \tag{11.8}$$

It will be observed that the expression does not include a linear dimension (i.e. model scale). In other words, provided that the model is large enough to ensure that the forces other than pressure remain insignificant, the operating speed (or controlling pressure) is within the investigator's discretion. The results of model tests conducted in this way are particularly suited to dimensionless representation. In any event the pressure-velocity relationship defined by Eq. (11.8) is universally applicable wherever inertial forces are of significance.

Pressures are commonly referred to in terms of equivalent head, thus introducing the linear dimension. If it is desired that this linear dimension shall conform to the geometrical scale of the model (1:x), then

$$\frac{V_\mathrm{p}}{V_\mathrm{m}} = \frac{(\rho_\mathrm{p} g h_\mathrm{p})^{1/2}}{(\rho_\mathrm{m} g h_\mathrm{m})^{1/2}} \frac{\rho_\mathrm{m}^{1/2}}{\rho_\mathrm{p}^{1/2}} = x^{1/2} \qquad (11.9)$$

indicating that the model operating speed is now fixed by the scalar relationship.

(b) *Froude Law.* Gravity is the predominant factor influencing fluid motion wherever a free surface gradient is present. The great majority of hydraulic problems amenable to model study are in this category – notably those concerning weirs, spillways, open channels, rivers, and estuaries. Wave motion (except the small, capillary type) is also included, since it is essentially a gravitational phenomenon.

For compliance with the Froude law ($F = V/\sqrt{gL}$) the corresponding velocities must be so related that:[1]

$$\frac{V_\mathrm{p}}{V_\mathrm{m}} = \frac{(g L_\mathrm{p})^{1/2}}{(g L_\mathrm{m})^{1/2}} = x^{1/2} \qquad (11.10)$$

It will be observed that Eqs. (11.9) and (11.10) are identical, as indeed logic would indicate. Model velocities are less than prototype (assuming $x > 1$), which is normally advantageous, since in a laboratory refined measuring instruments are available whereas pumping capacity is limited. The time scale is important in cyclical phenomena such as tidal flow, and this is given by $t_\mathrm{p}/t_\mathrm{m} = (L_\mathrm{p}/L_\mathrm{m})(V_\mathrm{m}/V_\mathrm{p}) = x^{1/2}$. Scalar relationships for other unit quantities are derived in a similar manner by utilising Eq. (11.10) and the unrestricted Euler relationship (Eq. (11.8)). These are listed in Table 11.1.

(c) *Reynolds Law.* All real fluids possess viscosity and the possible influence of viscous shear drag requires consideration in the planning stage of every type of fluid model investigation. A submarine moving through the water at such a depth that no surface waves are developed is a simple instance of where viscosity and inertia are the only effective forces, as gravity forces balance out. Another example is an aircraft travelling below the speed where compressibility of the air needs to be taken into account. Pipelines operating in the 'transition zone' are in the same category, the energy grade line, not the pipeline slope dictating

---

[1] We assume here that the acceleration due to gravity is the same in both prototype and model. This is not invariably so, since structural laboratories are sometimes equipped with a centrifuge (rotating arm) that enables any desired acceleration to be applied to a model under test.

the motion. Because of the relatively low viscosity of water, viscous forces are nearly always a secondary influence in the prototype, but are nevertheless important in view of their role in boundary friction and as the origin of fluid turbulence.

Table 11.1
Scalar Relationships for Models

Quantity	Dimensions	Reynolds law Natural scale 1:x	Froude law Natural scale 1:x	Froude law Distorted scales 1:x horiz.; 1:y vert.
Geometric:				
Length	L	x	x	$\begin{cases} x \text{ (horiz.)} \\ y \text{ (vert.)} \end{cases}$
Area	L^2	x^2	x^2	$\begin{cases} x^2 \text{ (plan)} \\ xy \text{ (sect.)} \end{cases}$
Volume	L^3	x^3	x^3	$x^2 y$
Kinematic:				
Time	T	x^2/ν_r	$x^{1/2}$	$x/y^{1/2}$
Velocity	L/T	ν_r/x	$x^{1/2}$	$\begin{cases} y^{1/2} \text{ (horiz.)} \\ y^{3/2}/x \text{ (vert.)} \end{cases}$
Acceleration	L/T^2	ν_r^2/x^3	1	$\begin{cases} y/x \text{ (horiz.)} \\ y^2/x^2 \text{ (vert.)} \end{cases}$
Discharge	L^3/T	$\nu_r x$	$x^{5/2}$	$xy^{3/2}$
Dynamic:				
Pressure	M/LT^2	$\rho_r \nu_r^2/x^2$	$\rho_r x$	$\rho_r y$ (sect.)
Force	ML/T^2	$\rho_r \nu_r^2$	$\rho_r x^3$	$\rho_r xy^2$ (sect.)
Energy	ML^2/T^2	$\rho_r \nu_r^2 x$	$\rho_r x^4$	$\rho_r xy^3$ (sect.)
Power	ML^2/T^3	$\rho_r \nu_r^3/x$	$\rho_r x^{7/2}$	$\rho_r y^{7/2}$ (sect.)

$$\rho_r = \rho_p/\rho_m \qquad \nu_r = \nu_p/\nu_m$$

For compliance with the Reynolds law ($R = VL/\nu$) the corresponding velocities must be so related that

$$\frac{V_p}{V_m} = \frac{\nu_p}{\nu_m}\frac{L_m}{L_p} = \frac{\nu_p}{\nu_m}\frac{1}{x} \qquad (11.11)$$

The expression shows that if the same fluid (at the same temperature) is employed in both model and prototype the model velocity must be x times that of the prototype. Fulfilment of this requirement proves difficult. For example, a 1 m/s prototype velocity becomes 10 m/s in a 1:10 model, the production of which would necessitate a very high pressure head. With aircraft models fantastic speeds would be required, but fortunately by circulating compressed air (increasing ρ_m and therefore decreasing ν_m) in a wind tunnel, or using a water tunnel ($\nu_{air} \simeq 13 \times \nu_{water}$), more realistic model velocities are prescribed.

The derived scalar relationships in accordance with the Reynolds law are quoted in Table 11.1. The Euler relationship is of course again incorporated.

(d) *Weber Law.* Surface tension always tends to reduce or equalise surface curvature, but it is only of significance where there is an air–water interface and linear dimensions are small. Particular regard to its influence needs to be given in model studies involving very low weir heads, air entrainment, splash, or spray.

For compliance with the Weber Law ($W = V/\sqrt{\sigma/L\rho}$) the corresponding velocities must be so related that

$$\frac{V_p}{V_m} = \frac{\sigma_p^{1/2}}{\sigma_m^{1/2}} \frac{\rho_m^{1/2}}{\rho_p^{1/2}} \frac{L_m^{1/2}}{L_p^{1/2}} = \frac{\sigma_p^{1/2}}{\sigma_m^{1/2}} \frac{\rho_m^{1/2}}{\rho_p^{1/2}} \frac{1}{x^{1/2}} \qquad (11.12)$$

Thus if the fluid in model and prototype is the same, the model velocities must be $x^{1/2}$ times those in the prototype.

In a problem where pressure, gravity, viscosity, and surface tension forces are all involved, simultaneous compliance with all the laws, that is to say perfect hydraulic similarity, necessitates (in addition to geometric similarity) that

$$x^{1/2} = \frac{v_p}{v_m} \frac{1}{x} = \frac{\sigma_p^{1/2}}{\sigma_m^{1/2}} \frac{\rho_m^{1/2}}{\rho_p^{1/2}} \frac{1}{x^{1/2}} \qquad (11.13)$$

Quite understandably, there is no known fluid in existence with the physical properties that will enable this expression to be satisfied, even in the most approximate sense. Thus it is only where the model scale is unity and the same fluid is employed that perfect similarity is attainable.

Now in the majority of cases, surface tension has little or no influence on prototype behaviour. Thus if we can ensure, either by making the model large enough or by limiting the range of operation, that surface tension is still insignificant at the reduced scale, then it is reasonable to forgo compliance with this law.

For the satisfaction of only the Froude and Reynolds laws, the requirement is that

$$v_m = \frac{v_p}{x^{3/2}} \qquad (11.14)$$

indicating that a model fluid is needed with a kinematic viscosity less than that of the prototype, and appreciably less if the model scale is very small. A suitable combination of fluids may be feasible if the prototype viscosity is relatively high – for example, a heavy oil in the prototype might be represented by water in the model. But in the more general case of water as the prototype fluid, the prospects of finding an acceptable model fluid with the required lesser viscosity are indeed bleak. It is true that a small reduction in viscosity might be effected by using in the model (a) water at a higher temperature, or (b) a volatile low viscosity fluid, such as one of the petro-carbons. However, the objections on experimental or safety grounds are such as to preclude both of these courses.

Fortunately, it is characteristic of most hydraulic problems for which model studies are required that viscosity forces are relatively small in comparison with inertial forces (i.e. the value of R is high), so that a reasonable deviation from the Reynolds law is not likely to seriously prejudice hydraulic similarity. The model may then be operated in accordance with the Froude law, or, in the case of a closed conduit model, under any convenient operating conditions. These admit of air being used as the model fluid, for there is no restriction concerning the type of fluid.

Values of R are normally less in the model than in the prototype. In order to ensure comparable boundary frictional effects, either compensating allowances must be made, as for instance by a tilting of the hydraulic gradients, or more generally the aim is to obtain the same friction factor (λ) in the model as in the prototype. This latter procedure requires a relative smoothing of the model surface, as reference to the standard λ-R diagram (Fig. 5.9, p. 98) will confirm. Moreover, it follows that if the prototype is operating in the rough turbulent zone [R_* ($= v_* k/\nu$) is greater than 60] then model conditions should be arranged to be likewise. Of course, owing to the marked difference in the characteristics of laminar and turbulent flow, very serious scale effects would arise if the flow in the model was predominantly laminar whilst that in the prototype was predominantly turbulent.

Geometrically distorted models of rivers and estuaries are operated in accordance with the Froude law based on the vertical scale. The derived scalar relationships are listed in Table 11.1. As will be explained in a more specific context later, if some vertical exaggeration were not adopted, the secondary forces of surface tension and viscosity would be so disproportionately large in the model that the fluid behaviour would be quite unrepresentative.

Another scalar discrepancy concerns sub-atmospheric pressures. As both model and prototype are operating under atmospheric conditions, pressures relative to atmospheric pressure are reproduced to scale, whereas absolute pressures are not. Vaporisation of water commences when the pressure head falls to within a metre or so of absolute zero, but dissolved air is released from solution well before this stage is reached. With a scalar lowering of pressure these phenomena will occur at a much earlier stage in the prototype than in the model. Considerable care in interpreting the model data is therefore needed if false (and unsafe) predictions are not to be made concerning flow discontinuity and the onset of cavitation. In general, indicated pressures up to 4·5 m below atmospheric are acceptable, since there is a reasonable margin for dissimilarity of surface roughness, vorticity, or turbulence, each of which might produce a greater momentary lowering of prototype pressure. One way of overcoming this pressure relationship problem is to operate the model in a vacuum container, but of course the attendant experimental difficulties are formidable.

The choice of model scale is usually in the nature of a judicious compromise, based on a consideration of such factors as laboratory facilities, economics, and the standards of accuracy required. Small-scale pilot models may give useful guidance in planning a larger model. A valuable means of detecting scale effect is to construct models to different scales and to compare the results.

11.4 Types of Model Investigation

11.4.1 *General Scope*

The hydraulic problems that are amenable to model study cover a wide field and are reflected in a corresponding variety of model size, scale, and operating technique. Certain broad categories may be distinguished, namely – enclosed conduits and appurtenances, hydraulic structures, river channels, estuaries, harbours and coastal structures, and turbo-machinery.

Each of these is considered in turn, but in a work of this kind it is obviously not possible to give more than a brief description of the general characteristic features. The reader who wishes to pursue this fascinating topic further is recommended to consult the extensive specialist literature, which includes many reports and papers describing interesting specific investigations.

11.4.2 *Enclosed Conduits and Appurtenances*

In this category are piping systems, pressure tunnels, and the associated control or measurement devices. Generally, the problem concerns local regions, such as bends or junctions, rather than the overall system. Intakes and outfalls, both of which require careful designing, are included providing that they are sufficiently submerged to nullify free surface or gravitational effects (e.g. air-entraining vortices).

The forces involved are those due to pressure and viscosity, but as in most cases the turbulence in the prototype is highly developed, compliance with the Reynolds law is not essential and indeed is rarely practicable. The only stipulation, then, is that the Reynolds number in the model should not be less than about 1×10^6. Accordingly, it is the practice to construct large-scale models, usually in the range 1:5 to 1:30, and to operate them at velocities higher than in the prototype. Pressure-velocity relationships follow the Euler law. In the case of valves or meters on small pipelines, a model that is larger than the prototype may be desirable in order to enable the hydraulic characteristics to be more readily determined.

Transparent perspex is a highly satisfactory but expensive lining material, enabling the flow behaviour to be clearly observed as well as satisfying the usual requirements for a smoother surface in the model. Disproportionately large boundary friction in the model may be compensated for by increasing pipeline slope or reducing length, the calculations involved being based on empirical formula.

In recent years there has been a tendency in many laboratories to use air in preference to water as the model fluid. The principal advantages are as follows:

(a) The fluid has a much lower density. This permits a lighter form of construction. Also, power requirements are less, air being blown or circulated by a fan at velocities much higher than would be possible with water. The atmosphere serves as a reservoir and storage tanks are not required. Thus for the same cost a larger model scale may be adopted.

(b) The conduct of experiments is simplified. Scientific instruments of a high sensitivity, together with appropriate measuring techniques have been developed for the wider field of aeronautics, so that velocities and pressures are more accurately and more expeditiously determined. Very often it is possible to examine the flow at any particular section by splitting the duct and discharging to atmosphere, making observations at the open end.

The fact that air is compressible need not give rise to scale effect, for it may be shown[1] that errors from this source are insignificant so long as the maximum velocity in the model does not exceed one quarter of the sonic speed, or about 75 m/s. Air cannot of course be used where a surface gradient is present and gravitational forces influence the

[1] BALL, J. W. (1952) 'Model Tests Using Low-Velocity Air' *Trans. Am. Soc. C.E.*, **117**, 821.

motion. Another drawback is that it does not exhibit any discontinuity characteristics at low pressures.

11.4.3 *Hydraulic Structures*[1]

By far the greater proportion of model investigations concerns hydraulic structures. These include weirs, sluices, control gates, spillways, stilling basins, tailbays, flumes, and bridge waterways. Standardised designs are available,[2] but the diversity of site conditions is such that recourse to model studies is frequently necessary. Their purpose is generally to determine the best boundary profile from the point of view of flow efficiency or the minimising of scour. Discharge calibration is often a concomitant requirement. Also, problems which arise during the construction stage may be usefully studied.

Gravitational forces are predominant, so that the Froude law is always the criterion. A natural scale is essential in view of the appreciable conversion from one form of energy to another over a short distance. Energy losses are mainly attributable to turbulent eddying rather than boundary friction. To achieve some correspondence of boundary roughness it is necessary for the model surface to be as smooth as possible. Plywood, perspex, sheet metal, and glass are lining materials that are commonly employed.

In those instances where a structure presents a long uniform or repetitive section transversely to the flow, two models are often desirable – (a) a large-scale sectional model of a representative portion, such as a gate opening or short crest length, and (b) a much smaller scale model of the complete structure, including a short length of the approach and tailwater channels. The former enables the detailed behaviour of the sectional flow stream to be readily observed, whilst the latter is of value in determining the flow pattern in plan and the effects of non-uniform approach conditions.

The published results of verification tests on full-scale structures almost all confirm, special circumstances apart, the high degree of dependability of properly conducted model investigations.[3]

(a) *Weirs, Sluices, and Control Gates.* Model scales of 1:5 to 1:40 are commonly adopted. Sectional models can often be conveniently inserted

---

[1] The hydraulic characteristics of prototype structures are discussed in Ch. 9.
[2] For example: *Corps of Engineers Hydraulic Design Criteria*, U.S. Army Engineer Waterways Expt. Sta., Vicksburg, Miss.
[3] ——, (1966) '*Model and Prototype Conformity*'. Vol. 1 of *Proc. of Golden Jubilee Symposia*, Central Water and Power Res. Sta., Poona.
PETERKA, A. J. (1954) *Spillway Tests Confirm Model-Prototype Conformance.* U.S. Bureau of Reclamation, Eng. Monograph No. 16, Denver.

in a glass-sided flume channel, various profiles being tested. Besides the ascertaining of nappe behaviour, hydrodynamic loading on gates is other useful information that may be obtained.

It is important that the head on a model weir should not be less than 6 mm otherwise surface tension forces may cause serious scale effect (e.g. a false clinging of the nappe).

The discharge characteristics of most types of small control structure can be predicted within ± 5 per cent, which would often be sufficiently satisfactory for hydrometric purposes.

Example 11.1

The discharge characteristics of a proposed weir, crest length 18 m, are to be investigated. The discharge coefficient is thought to be about 1·7, and the maximum and minimum river flows are assessed at 30 and 1·0 cumecs, respectively. Recommend a suitable model scale, assuming that the maximum pumping capacity of the laboratory is 90 l/s.

The head on the prototype weir corresponding to 1·0 cumec minimum discharge is approximately $1·0/(1·7 \times 18)^{2/3} = 0·102$ m. To avoid pronounced scale effects the corresponding head in the model should not be less than 6 mm, so that the minimum permissible scale is given by $x = 0·102/0·006 = 17·0$. A model scale of 1:15 would therefore appear satisfactory. It now remains to check that the higher discharges can be studied. As the scalar relationship for discharges is $x^{5/2}$, the model discharge corresponding to the maximum flow is $30 \times 1000/15^{5/2} = 34·5$ l/s, which is well within the pumping capacity. Thus a scale of **1:15** is recommended.

The model will include a short length of the river channel upstream and downstream. Calibration will be effected by operating the model at a series of known discharges and measuring the head. If a more detailed examination of flow behaviour is required, a sectional model of scale say **1:5** might be constructed. The maximum prototype discharge is then $0·09 \times 5^{5/2} = 5·04$ cumecs, corresponding to a crest length of $(5·04/30) \times (18/5) = 0·605$ m for the sectional model.

(b) *Spillways.* (i) OPEN SPILLWAYS. Models, both of the sectional and full (three-dimensional) type, are of considerable assistance in designing the open spillway of a dam. Velocity distributions may be obtained by Pitot tube traverse, but flow conditions are not always representative, since the high spillway velocities which cause air entrainment and consequent 'bulking' to occur in the prototype will not do so to the same relative extent, if at all, when reduced to model scale. Pressures on the face of a dam are important, and may be measured by means of piezometer tappings linked to manometers (Plate 9). Pronounced negative pressures indicate the likelihood of undesirable cavitation in the prototype. For the reasons previously stated, only a limited scalar interpretation under these conditions is justified.

Full models of dam structures usually have scales between 1:20 and 1:100. With these scales, similarity in respect of surface roughness cannot normally be attained. The Manning formula, though not strictly speaking applicable to high velocity flow, serves to illustrate the point. Since a spillway is normally wide relative to the flow depth ($R \simeq d$) and the profile slope is the same in prototype and model, the expression for corresponding velocities is

$$\frac{V_p}{V_m} = \frac{n_m}{n_p} \left(\frac{d_p}{d_m}\right)^{2/3}$$

where n is the roughness coefficient and d the depth of flow. Inserting the Froude scalar relationships,

$$x^{1/2} = \frac{n_m}{n_p} x^{2/3}$$

so that

$$n_m = \frac{n_p}{x^{1/6}} \tag{11.15}$$

Now the prototype surface is usually concrete with n_p about 0·014, whilst the smoothest model surface (sheet metal or perspex) is represented by $n_m = 0·009$. Substituting in Eq. (11.15) we obtain $x \simeq 14$; that is to say the model scale should not be less than about 1:14. But of course in most cases this would be economically impracticable, so that some scalar discrepancy must be accepted. Happily, the role of friction is usually very small in comparison with gravity effects.

One of the most comprehensive hydraulic structure investigations ever carried out was that for the Boulder Dam (re-named Hoover Dam), 215 m high, on the Colorado River.[1] Several models were constructed for the purpose of studying various aspects of the scheme. The largest model had a scale of 1:20, with a maximum discharge of 3·2 cumecs – equivalent to a small river. The performance of smaller models, scale 1:60 and 1:100, was found to be in close agreement.

(ii) SHAFT SPILLWAYS. Model tests are of value in designing the bend at the foot of the vertical shaft and checking the hydraulic behaviour generally. Pronounced vorticity in the entry bellmouth is undesirable, and in this respect the best shape and spacing for the piers at the weir

---

[1] Boulder Canyon Project Final Reports (1938–49), Part VI – *Hydraulic Investigations*, Bull. Nos. 1–4, U.S. Bureau of Reclamation, Denver.

crest can be reliably ascertained. Also, the bellmouth may be calibrated and the maximum capacity of the system determined, both with reasonable accuracy.

At the lower discharges air entrainment is often appreciable and cannot be correctly reproduced in the model. Under these conditions only a limited qualitative interpretation of the flow behaviour in the discharge conduit is permissible. In general, scale effect tends to be on the safe side, since the turbulent flow stream in the prototype has a greater relative capacity for evacuating air than that in the model.

(iii) SIPHON SPILLWAYS. Model investigations are useful as a means of observing general flow characteristics and, more particularly, in making comparison between one design proposal and another. Model-prototype conformity is good for the full discharge condition, but it is not possible to reproduce correctly the priming and de-priming processes, which are complicated by virtue of air entrainment. The priming action demands an absolute velocity sufficient to entrain and transport air bubbles, and therefore does not admit of scale reduction. Consequently, priming in the model will occur at a relatively later stage and higher upstream level than in the prototype. Plate 12 shows the successive stages of priming of a 1:12 perspex model of a siphon spillway with jet deflector on the downstream leg.[1]

The functioning of a siphon necessarily depends on the existence of sub-atmospheric pressures, and under these conditions, as has been explained earlier, great care in the measurement (by piezometer tapping and probe) and interpretation of model pressures is required. For instance it would be quite conceivable for a model siphon to continue to operate whereas the prototype would have ceased to do so.

Example 11.2

In a test on the 1:10 model of a siphon spillway it is found that the velocity at the centre line of the throat is 0·95 m/s when the reservoir level is 0·26 m below this point. Comment as to whether the prototype might be expected to operate satisfactorily under corresponding conditions.

In the absence of detailed information concerning pressure distribution and boundary geometry we can only make a rough appraisal of the situation.

Applying Bernoulli's equation to the throat centre line and reservoir level and neglecting energy losses, we have for the model siphon: $0\cdot26 + p_m/w$

[1] TAYLOR, G. A., LITTLEJOHN, A. G. and ALLEN, J. (1961) 'A Scale-model Investigation of the Characteristics of a Siphon Spillway Designed to Prime and De-prime Rapidly', *J. Inst. Wat. E.*, **15**, 299.

15   Model of R. Trent in vicinity of Burton-on-Trent, scales 1:200 horiz., 1:50 vert.

16 View towards Southampton Water in the 1:1250 horiz., 1:100 vert. scale model of the Solent. The apparatus in the fore-
ground is for the purpose of an investigation into seal's water movement

17 Wave model of Marsamxett Harbour, Malta, Scale 1:120

(*Wimpey Central Laboratory*)

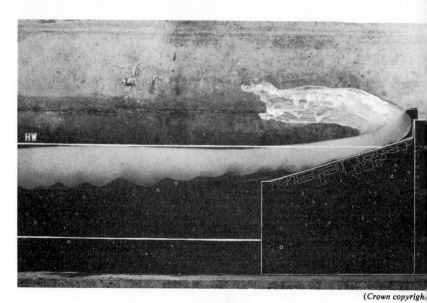

18   Wave breaking against a 1:20 scale model of a sea wall

19   Miniature current meter (set against ½mm gradations)

$+ 0.95^2/2g = 0$, so that $p_m/w = -0.306$ m. The corresponding pressure head p_p/w in the prototype is thus -3.06 m.

Hence, it would appear, even after allowing for frictional loss and non-uniform velocity distribution, that the prototype siphon is likely to operate satisfactorily (see Ch. 9, Sect. 9.7.4).

(c) *Stilling Basins and Tailbays.* These are important downstream features of sluices, weirs, or spillways, and their design is integral to that of the parent structure. Investigations are therefore normally carried out utilising the same model. Plate 13 illustrates such a composite model.

The problem generally is to design a form of exit which will ensure that the high velocity energy is dissipated in the most efficient manner. This means a confining of the turbulence to as short a distance from the structure as possible, with tranquil conditions in the channel downstream.

Owing to scale effect associated with surface tension, the flow behaviour in the model will tend to be relatively less agitated (less splash, spray, etc.) than that in the prototype. Some caution in interpreting the model results is therefore necessary, the normal procedure being to compare the relative performance of various alternative designs.

In the case of the scour of an erodible bed, our present knowledge does not permit a quantitative assessment to be made. Experimental studies of the relative effects of various forms of spillway toe and weir profile do, however, offer valuable guidance. The grain size of bed material in the model is not critical, since the ultimate shape of a scour hole is found to be largely independent of this factor. By using a small gravel the final profile for a given discharge is usually obtained within a reasonably short time.

11.4.4 *River Channels*

River improvement schemes for the purpose of flood control often entail extensive re-alignment and widening works, which tend to be fairly costly. It is therefore important at the design stage to make some quantitative assessment of the benefits that will accrue from the various alternative proposals. This usually involves a prediction of water levels corresponding to sustained high discharges or the passage of transient flood waves. Conditions may well be affected for a considerable distance upstream and downstream of the actual works. Owing to irregular channel configuration (non-uniform section and alignment), analytical methods are usually only capable of giving very approximate results, and in these circumstances a model investigation is advantageous.

In a typical problem, several km of river may require to be reproduced. The adoption of a natural scale will mean either a very large model with reasonable river depths, but one that is extremely costly to construct and operate, or else a model that occupies an economical plan area but with depths that are exceedingly small.

The latter course is unacceptable because:

(a) Depths of flow and differences in water level cannot be accurately measured.

(b) The model surface cannot be made sufficiently smooth (cf. Eq. (11.15)), and in any event there is every likelihood that the flow in the model will be laminar $[R_m (= V_m R_m/\nu) < 500]$ instead of turbulent $(R_m > 2000)$ as in the prototype. Also, surface tension forces will be much too pronounced in the model.

(c) In a case where the model has a mobile bed, the very low channel velocities will provide insufficient tractive force to transport even the most mobile material.

All these difficulties may be overcome by means of geometrical distortion, whereby the vertical scale $1:y$ is made larger than the horizontal scale $1:x$. Actually, some justification for this artifice is to be found in nature, since the width-depth ratio for large rivers is much greater than for small ones, whereas the grain size of the bed material and the general configuration of sandbanks and shoals may not be very different.

There are certain undesirable consequences resulting from distortion.[1] For instance, the sectional distribution of velocity is adversely affected; since turbulence is broadly isotropic, the transverse currents and eddies induced by boundary irregularities and bends are not correctly reproduced. Again, because side slopes are too steep, generally in excess of the natural angle of repose of bed material, scour holes will appear disproportionately large in plan; a partial remedy is to artificially stiffen the slopes.

Whilst scale effects of this nature are not likely to prejudice the prediction of stage-discharge relations over a long length of river, nevertheless they are important locally, and this means that little or no distortion is permissible in those models where detailed flow behaviour is to be studied. Fortunately, modelling of only a short river reach is ordinarily required. A typical example would be an investigation of the

---

[1] FOSTER, J. E. (1975) 'Physical Modelling Techniques Used in River Models', Vol. 1 of Proc. Symposium on Modelling Techniques, *Am. Soc. C.E.*, **540**.

comparative efficacy of various forms of groyne in arresting scour at a bend.

Vertical distortion necessitates a modification of scalar relationships. In accordance with the Froude law, horizontal velocities and vertical dimensions (gravitational influence) are related by $V_p/V_m = y^{1/2}$. The time scale is thus $t_p/t_m = x/y^{1/2}$. Other relationships follow logically and are listed in Table 11.1.

The Manning formula may again be utilised for the purpose of assessing approximate model roughness. Thus:

$$\frac{V_p}{V_m} = \frac{n_m}{n_p}\left(\frac{R_p}{R_m}\right)^{2/3}\left(\frac{S_p}{S_m}\right)^{1/2}$$

Now the hydraulic radius R is dependent upon both horizontal and vertical dimensions, so that R_p/R_m cannot be uniquely represented by a simple scalar ratio. However, as an approximation we may regard the channels in both prototype and model as being wide and shallow, in which case

$$y^{1/2} = \frac{n_m}{n_p} y^{2/3}\left(\frac{y}{x}\right)^{1/2}$$

or

$$n_m = n_p \frac{x^{1/2}}{y^{2/3}} \tag{11.16}$$

In order to obtain satisfactory measurement of water levels a vertical scale not smaller than 1:100 is generally preferred. The horizontal scale is governed by the nature of the problem and by economic considerations; scales between 1:200 and 1:500 are often adopted and are usually associated with vertical exaggerations between 3 and 6.

With regard to the relative roughness, we find that for $x/y = 4$, $n_m/n_p = 2/y^{1/6}$ and for $x/y = 5$, $n_m/n_p = 2.24/y^{1/6}$. Thus the direct influence of y is quite small and, inserting reasonable values, it will be found in general that the model surface is required to have about the same degree of roughness as the prototype. In cases of greater exaggeration the model surface may need to be rougher.

Fixed-bed models are normally moulded in cement mortar, with n_m value about 0·012. Vertical hardboard or sheet metal templates (Plate 14) serve as guides for the screeding and filler material of hardcore or sand. As the value of n_p for natural channels is rarely less than 0·03, some roughening is nearly always necessary. This may be achieved by exposing the aggregate, chipping the surface, or by inserting roughness elements consisting of wire mesh, gravel, or the like, on the bed and flood plain. Even so it is very often difficult to obtain a value much higher than $n_m = 0.035$ without introducing artificial impediments to the flow, such as vertical rods. But attempts to conform to a theoretical roughness are rather unrealistic, firstly because of the inaccuracies inherent in the calculation and secondly because a large proportion of the energy loss in natural channels is accounted for by eddying at bends and abrupt changes of section rather than normal frictional resistance.

It is desirable that the prototype information should include water level data for a number of the larger measured discharges. These discharges are reproduced in the model (circulation by pump and measurement by V-notch or orifice plate) and the water levels noted at corresponding points. In this way the satisfactory performance of the model is verified – the procedure is known as *proving the model*. If, after making adjustments to the surface roughness, conformity is still lacking, it may be necessary to depart somewhat from the Froude law and operate the model in accordance with a discharge scale that is known to produce the correct stage–discharge relations. Alternatively, the longitudinal slope of the model may be adjusted, but with the usual size of model such a trial and error procedure would be impracticable.

Assuming that model operation is in compliance with the Froude law, the scalar head at control structures, such as weirs and sluices, produces scalar discharges, provided that the coefficient of discharge is the same for model and prototype. This may necessitate some modification in the geometrical form of these structures when they are reproduced in a river model.

Models are provided with a mobile bed in those cases where scour and deposition are to be simulated. A variety of bed material is employed. If the tractive force exerted by the flow is sufficient, a sand with a grain size not very different to the prototype may be satisfactory. Otherwise, it is necessary to have recourse to light-weight, easily transported materials, with specific gravity only slightly in excess of unity. Specially processed forms of powdered coal, pumice, crushed plastic, and wood chips are all in this category. It is not possible to reproduce correctly the complicated mechanism of sediment transport, so that the results are necessarily qualitative in character. However, some confidence may be placed in the ability of the model to predict future trends, if it is shown to be capable of reproducing bed changes, in their correct time sequence, that are known to have occurred in the past.

Although it may be argued that there is a marked difference in the behaviour of natural river channels and constant discharge irrigation canals on which regime theory is based (see Sect. 7.7, p. 165), it is nevertheless interesting to note that a simple application of this theory leads to a definite relationship between the horizontal and vertical model scales. This arises from Lacey's empirical equations, $d \propto V^2$ and $b \propto V^3$. Assuming the same coefficients of proportionality for model and prototype (i.e. same bed material), then $b_p/b_m = (d_p/d_m)^{3/2}$ or $y = x^{2/3}$. Comparison may be made with the relationship $y = x^{3/4}$ obtained from Eq. (11.16) with $n_m = n_p$.

Example 11.3

A fixed-bed model is to be constructed of a 10 km length of river (7·1 km direct distance) for the purpose of investigating the passage of flood waves. At the normal winter discharge of 300 cumecs it is known that the average depth and width are 4 m and 50 m, respectively. The roughness coefficient n is estimated to be 0·035. Maximum flood discharge is 850 cumecs.

If the length of laboratory available is 18 m, recommend suitable scales for the model. Comment on the surface roughness of the model and estimate the maximum discharge that the circulating pump will be required to deliver.

The largest horizontal scale is governed by the length of the laboratory. Hence $x = 7100/18 = 394$. Thus a horizontal scale of **1:400** should be satisfactory. A vertical exaggeration of 5 seems reasonable, indicating a vertical scale of **1:80**. This is sufficient to enable surface slopes to be accurately measured.

A check on the Reynolds number ($V_m R_m/\nu$) in the model is desirable. The minimum discharge at which the model will be operated is that corresponding to 300 cumecs in the prototype. For this discharge the approximate prototype mean velocity is $V_p = 300/(4 \times 50) = 1·5$ m/s. Thus $V_m = 1·50/80^{1/2} = 0·168$ m/s. Also, $R_m = 0·0278$ m. The minimum value of R_m is therefore $0·168 \times 0·0278/(1·14 \times 10^{-6}) = 4100$, which confirms that the flow will definitely be turbulent, though not by a great margin.

If we assume the Manning formula to apply, the required model roughness is $n_m = 0·035 \times 400^{1/2}/80^{2/3} = \mathbf{0·038}$. The cement-mortar finish of the model should be left as rough as possible, netting and artificial impediments being introduced until conformity with a known surface slope is obtained.

The maximum model discharge is given by $Q_m = Q_p/xy^{3/2} = 850/(400 \times 80^{3/2}) = \mathbf{2·97}$ l/s.

11.4.5 Estuaries

Estuarial problems usually concern the maintenance of a navigation channel and, in particular, a prediction of the effects on the regime of various proposals, such as dredging, the erection of training walls, or land reclamation. Cooling-water circulation and pollution are other related issues commonly requiring examination. Owing to irregular boundary configuration and the unsteady nature of tidal flow, the problems posed are undoubtedly the most complex of all in the field of hydraulics, and a physical model often offers the only prospect of obtaining a satisfactory solution.

The collection of tidal stream data, sufficient to enable velocity distributions and component discharges to be assessed, is a tedious and expensive procedure and the model designer must ordinarily be content with prototype tidal curves for a limited number of points in the system. This means that in order to obtain the correct discharge-time relationship it is nearly always necessary to reproduce to scale the entire network of waterways as far upstream as the limits of tidal influence. It is permissible, however, for long narrow tidal channels to be represented in labyrinthine form, with consequent saving in floor space. For satisfactory simulation of conditions at the mouth of an estuary the model must include a sufficient expanse of open sea and length of adjoining coastline. The automatic reproduction of any desired tidal curve at the generating point has been made possible by the development of ingenious electronic control equipment.[1]

As with river models, the vertical scale must be large enough to provide adequate water depths and velocities, and also to permit accurate measurement of the variation in tidal level. These requirements indicate a vertical scale of between about 1:50 and 1:150, the final choice depending on tidal range and other factors. Estuaries very often cover an extensive area, so that on economic grounds a considerable vertical exaggeration must generally be accepted – usually between 6 and 17. The corresponding horizontal scale is therefore between about 1:300 and 1:2500.

Surface roughness may be dealt with in a similar manner to that in river models, though generally it is of less consequence. A mobile bed is provided in those cases where scour or deposition is to be studied. It is rarely necessary to make provision for wave simulation, since, in sheltered water, sediment transport is almost entirely attributable to tidal currents.

The periodicity of tidal behaviour arises from the nature of the phenomenon, which is essentially that of the passage of a gravitational wave of very long wave length relative to the depth. As explained in

[1] As, for example, the equipment described in reports of the Hydraulics Research Station, Wallingford, Oxon.

Ch. 8, Sect. 8.10.3, the speed of propagation of this type of wave is proportional to the square root of the water depth, and in the case of a wide rectangular channel it is given by $c = \sqrt{gd}$. Thus estuary models must be operated in accordance with the Froude law, so that scalar relationships are identical with those for distorted river models. It follows that the ratio of vertical velocities (e.g. rate of fall of silt particles) is

$$\frac{(V_\mathrm{v})_\mathrm{p}}{(V_\mathrm{v})_\mathrm{m}} = \frac{y}{t_\mathrm{p}/t_\mathrm{m}} = \frac{y^{3/2}}{x}$$

also that the semi-diurnal tidal period of 12·4 hours in nature is reproduced in a matter of minutes in a model – for instance, with typical values of $x = 480$, $y = 60$, we find that $T_\mathrm{m} = 12$ min. A full cycle of tides (spring – neap – spring) could thus be reproduced in about 6 hours.

Density differences due to the presence of fresh and saline water may have significant effects on flow behaviour or silting phenomena, particularly where the liquids remain stratified. Gravity is the predominant factor influencing the relative motion of interfacial layers, but the gravitational force is now represented by $\Delta w L^3$ instead of $w L^3$ as previously, where Δw is the difference in specific weight of the two liquids. The modified Froude number is thus given by $F = V/\sqrt{\Delta w \, L/\rho}$. Assuming that the model is operated in accordance with the usual Froude relationships, there is now an additional requirement that $\Delta w_\mathrm{m} = \Delta w_\mathrm{p}$. In other words, liquids of the same density must be used in the model as in the prototype. Salt water may be simulated by a saline solution of the same density, or, if corrosion effects are considered undesirable, a stable clay suspension is a useful alternative. In cases where the density difference is produced by a difference of temperature, for example heat transfer in a cooling system, then it must be so arranged that the model operates with the same or equivalent temperature differences.

The vertical exaggeration and smallness of scale of most estuary models inevitably give rise to scalar discrepancies. There is another possible source of error, called *Coriolis effect*, which is the accelerative force created by the earth's rotation. It acts transversely to the direction of flow, and the order of magnitude at the latitude of this country is such as to produce a transverse slope of about 11 mm per km of a mass of water flowing with a velocity of 1 m/s. In a very wide estuary this may well have some influence on the velocity distribution.

For these reasons it is important not to over-estimate the potentialities of estuary models. Their most dependable role lies in the indication

319

of trends, and in assessing the general effects and comparing the relative merits of various alternative design proposals.

11.4.6 *Harbours and Coastal Structures*

A prediction of wave effects is important in the design of harbours and maritime works. The problems that are amenable to model study include the influence of waves on ship mooring and on shoaling at a harbour entrance, also the best location and profile for a spending beach or breakwater. Models are usually constructed in wave basins where oscillatory paddles and ancillary equipment are available.

Waves in the open sea are of the oscillatory gravitational type, and it may be shown that their speed of travel is dependent on both wave length and water depth. It follows that correct reproduction demands a natural scale. In practice, however, a small degree of vertical exaggeration (up to about 3) is sometimes admissible, depending on the nature of the problem.

The best profile for a sea wall or breakwater may be investigated by means of a large-scale model erected in a laboratory wave channel. Since wave form varies with depth, and the reflection from a sloping face must be correctly reproduced, it is essential that models of this type be undistorted. Scales between 1:20 and 1:40 are commonly adopted. For realistic wave simulation, air is blown over the surface.

Example 11.4

A model sea wall in a wave flume has a scale of 1:25. If prototype waves of 3 m amplitude and 8 s period are to be investigated, determine the corresponding values for the model. What would be the maximum pressure on the prototype sea wall corresponding to 7 kN/m^2 for the model?

The wave height in the model is $3/25 = $ **0·12** m and the wave period is $8/25^{1/2} = $ **1·6** s.

The scalar relationship for pressure is $p_p/p_m = (\rho_p/\rho_m) \times (V_p/V_m)^2 = 1·025 \times 25 = 25·6$. Thus the maximum pressure to which the prototype sea wall is likely to be subjected is $25·6 \times 7 = $ **179·2** kN/m^2.

11.4.7 *Turbomachinery*

Laboratory tests on models of pumps and turbines provide valuable information at the design stage. Furthermore, in the case of the larger units, they are commonly relied upon today in lieu of acceptance tests; these performance tests on site are difficult to conduct, particularly with respect to the measurement of large discharges, and the model results are not only obtained more conveniently, but are also likely to be more accurate. The model scale is generally such that the rotor diameter is between 0·3 and 0·4 m.

The close relationship which exists between pressure and velocity in the flow passages of rotodynamic pumps and reaction turbines means that the complete unit must be modelled, and not merely the impeller or runner. For instance the draft tube of a turbine has a significant effect on the pressures within the runner and is of particular relevance in cavitation studies aimed at determining the best setting for the machine.

Velocities in the flow passages are relatively high, so that viscous forces are negligible in comparison with inertial forces. Gravity is also of no consequence since we are dealing with an enclosed system. Models may therefore be operated only with regard to the Euler law.

For hydraulic similarity, velocities at homologous points in the prototype and model systems must correspond both in magnitude and direction. Thus the theoretical vector triangles for the rotor peripheries must be similar, so that in accordance with Eq. (10.33), p. 288,

$$\frac{Q_p}{D_p{}^3 n_p} = \frac{Q_m}{D_m{}^3 n_m}$$

where Q, n, and D refer to the discharge, rotational speed, and peripheral diameter, respectively. Then if $1:x$ is the scale of the model we have

$$\frac{Q_p}{Q_m} = \frac{D_p{}^3 n_p}{D_m{}^3 n_m} = x^3 \left(\frac{n_p}{n_m}\right) \tag{11.17}$$

Similarly, the ratio of pressures or heads is given by

$$\frac{p_p}{p_m} = \frac{H_p}{H_m} = \frac{D_p{}^2 n_p{}^2}{D_m{}^2 n_m{}^2} = x^2 \left(\frac{n_p}{n_m}\right)^2 \tag{11.18}$$

and the ratio of powers (input to pump or output from turbine) is

$$\frac{P_p}{P_m} = \frac{Q_p H_p}{Q_m H_m} = \frac{Q_p D_p{}^2 n_p{}^2}{Q_m D_m{}^2 n_m{}^2} = x^5 \left(\frac{n_p}{n_m}\right)^3 \tag{11.19}$$

It will be noted that these discharge, pressure, and power model-prototype relationships are dependent only on the scale and rotational speed. The model can therefore be operated at any convenient head (within reasonable limits) to suit the test bay facilities. However, for cavitation reasons it is generally considered desirable that the model and prototype submergence should be the same, and that the model speed should be such as to give approximately the same peripheral speed (i.e. $p_p = p_m$ and $H_p = H_m$).

In deriving the power relationship (Eq. (11.19)) it has been assumed that the efficiency η of model and prototype is the same. This is of course theoretically true, but in practice, owing to disproportionate leakage, mechanical friction, and viscosity effects, the model efficiency is lower. Empirical formulae are available which enable the small step-up to be assessed. For reaction turbines one of the best known is that due to Moody[1] – namely

$$\eta_{\mathrm{p}} = 1 - \frac{(1 - \eta_{\mathrm{m}})}{x^{1/5}} \qquad (11.20)$$

11.5 Measuring Instruments and Techniques

The scientific value of hydraulic models is very dependent on the availability of instruments for the accurate measurement of water level, bed level, pressure, current velocity and direction, wave characteristics, temperature, and sediment transport. Only brief mention of some of these is possible here.

Pointer or hook gauges are commonly employed for the determination of water level, the verniers being read to 0·1 mm. Fluctuating levels, as in tidal models, are best measured by automatic water-level follower. Pressures are determined by conventional piezometer connection or, in the case of momentary pressures, by electrical means.

The tracking of miniature floats provides information concerning the magnitude and direction of currents. Photography is a valuable aid, since the paths of a number of floats can be simultaneously recorded; also, an excellent visual picture of the flow pattern results from the spreading of confetti or aluminium dust on the free surface. Immiscible droplets and dye injections are useful means of observing sub-surface currents and flow paths.

Point velocities may be measured by Pitot tube, but this method is rather unsatisfactory owing to the small differential head and the lengthy time of response. The miniature current meter is now used extensively in tidal and river model studies. The instrument developed by the Hydraulics Research Station consists of a small plastic propeller, 10 mm diameter, mounted in jewel bearings with rod suspension (Plate 19). Revolutions are recorded electronically and a calibration curve facilitates conversion to water velocities. The operating range is from about 0·02 m/s to 1·5 m/s.

Ingenious electronic instruments have been devised for measuring

[1] MOODY, L. F. and ZOWSKI, T. (1969) 'Hydraulic Machinery', Sect. 26, pp. 26–46, of *Handbook of Applied Hydraulics* (Eds. DAVIS, C. V. and SORENSEN, K. E.). McGraw-Hill (3rd Edition).

wave heights and periods. Temperatures, too, can be determined by electrical methods, utilising thermistor probes.

Indeed, it would be true to say that some of the greatest progress in hydraulic model research in latter years has been in the field of instrumentation. The hydraulic investigator should be grateful to his mechanical and electronic engineering colleagues for the improved facility and accuracy of experimentation that have resulted.

Further Reading

ALLEN, F. H. (1959) *Hydraulic Model Techniques*. Inst. C.E. Vernon-Harcourt Lecture.

ALLEN, J. (1952) *Scale Models in Hydraulic Engineering*. Longmans (2nd Edition).

BRADSHAW, P. (1964) *Experimental Fluid Mechanics*. Pergamon Press.

GAMESON, A. L. H. (Ed.) (1973) *Proc. of Symposium on Mathematical and Hydraulic Modelling of Estuarine Pollution* (1972). H.M.S.O.

——, (1953) *Hydraulic Laboratory Practice*, U.S. Bureau of Reclamation, Eng. Monograph No. 18, Denver.

——, (1942) *Hydraulic Models*, Am. Soc. C.E. Manual of Eng. Practice No. 25.

——, *Hydraulics Research*, Annual Rept. of the Hydraulics Research Station. H.M.S.O.

IPPEN, A. T. (Ed.) (1966) *Estuary and Coastline Hydrodynamics*. McGraw-Hill.

LANGHAAR, H. L. (1951) *Dimensional Analysis and Theory of Models*. Wiley.

LINFORD, A. (1966) *The Application of Models to Hydraulic Engineering*, Brit. Hydromech. Res. Assn., Rept. No. P. 871.

MURPHY, G. (1950) *Similitude in Engineering*. Ronald Press.

——, (1975) *Proc. of Symposium on Modelling Techniques*, Am. Soc. C.E.

——, (1954) *The Role of Models in the Evolution of Hydraulic Structures: a Symposium* (1952). Central Board of Irrigation and Power, Pub. No. 53, New Delhi.

SORENSEN, R. M. (1978) *Basic Coastal Engineering*. Wiley.

WOOD, A. M. MUIR (1969) *Coastal Hydraulics*. Macmillan.

YALIN, M. S. (1965) 'Similarity in Sediment Transport by Currents', *Hydraulics Research Paper No. 6*. H.M.S.O.

YALIN, M. S. (1971) *Theory of Hydraulic Models*. Macmillan.

Appendix
General Publications Relevant to Hydraulic Engineering

Reference Books

BROWN, J. G. (Ed.) (1 – 1964, 2 – 1958, 3 – 1960) *Hydro-Electric Engineering Practice*, 3 vols. Blackie.

DAVIS, C. V. and SORENSEN, K. E. (Eds.) (1969) *Handbook of Applied Hydraulics*. McGraw-Hill (3rd Edition).

IPPEN, A. T. (Ed.) (1966) *Estuary and Coastline Hydrodynamics*. McGraw-Hill.

KING, H. W. and BRATER, E. F. (1963) *Handbook of Hydraulics*. McGraw-Hill (5th Edition).

LELIAVSKY, S. (1965) *Irrigation and Hydraulic Design*, 4 vols. Chapman and Hall.

———, (1962) *Nomenclature for Hydraulics*. Am. Soc. C.E. Manual of Eng. Practice No. 43.

ROUSE, H. (Ed.) (1950) *Engineering Hydraulics*. Wiley.

SILVESTER, R. (1974) *Coastal Engineering*, 2 vols. Elsevier.

SKEAT, W. O. (Ed.) (1969) (3 vols.) *Manual of British Water Engineering Practice*. Heffer (4th Edition).

SORENSEN, R. M. (1978) *Basic Coastal Engineering*. Wiley.

STREETER, V. L. (Ed.) (1961) *Handbook of Fluid Dynamics*. McGraw-Hill.

WIEGEL, R. L. (1964) *Oceanographical Engineering*. Prentice-Hall.

WOOD, A. M. MUIR (1969) *Coastal Hydraulics*. Macmillan.

Textbooks

ALBERTSON, M. L., BARTON, J. R., and SIMONS, D. B. (1960) *Fluid Mechanics for Engineers*. Prentice-Hall.

BARNA, P. S. (1964) *Fluid Mechanics for Engineers*. Butterworths (2nd Edition).

BATCHELOR, G. K. (1967) *An Introduction to Fluid Dynamics*. Cambridge Univ. Press.

DAILY, J. W. and HARLEMAN, D. R. F. (1966) *Fluid Dynamics*. Addison-Wesley.

324

APPENDIX

DAKE, J. M. K. (1972) *Essentials of Engineering Hydraulics*. Macmillan.

DAUGHERTY, R. L. and FRANZINI, J. B. (1965) *Fluid Mechanics with Engineering Applications*. McGraw-Hill.

DUNCAN, W. J., THOM, A. S., and YOUNG, A. D. (1960) *An Elementary Treatise on the Mechanics of Fluids*. Arnold.

FOX, J. A. (1974) *Introduction to Engineering Fluid Mechanics*. Macmillan.

FRANCIS, J. R. D. (1969) *A Textbook of Fluid Mechanics*. Arnold (S.I. Edition).

LI, W. H. and LAM, S. H. (1964) *Principles of Fluid Mechanics*. Addison-Wesley.

LINSLEY, R. K. and FRANZINI, J. B. (1963) *Water-Resources Engineering*. McGraw-Hill.

MASSEY, B. S. (1970) *Mechanics of Fluids*. Van Nostrand Reinhold (2nd Edition).

OLSON, R. M. (1966) *Essentials of Engineering Fluid Mechanics*. International Textbook (2nd Edition).

PAO, R. H. F. (1961) *Fluid Mechanics*. Wiley.

PAO, R. H. F. (1967) *Fluid Dynamics*. Merrill.

PLAPP, J. E. (1968) *Engineering Fluid Mechanics*. Prentice-Hall.

RAUDKIVI, A. J. and CALLANDER, R. A. (1975) *Advanced Fluid Mechanics*. Arnold.

ROUSE, H. (1961) *Fluid Mechanics for Hydraulic Engineers*. Eng. Soc. Mono. (1938), Dover Pub.

ROUSE, H. and HOWE, J. W. (1953) *Basic Mechanics of Fluids*. Wiley.

SABERSKY, R. H. and ACOSTA, A. J. (1964) *Fluid Flow*. Macmillan.

STREETER, V. L. and WYLIE, E. B. (1975) *Fluid Mechanics*. McGraw-Hill (6th Edition).

SWANSON, W. M. (1970) *Fluid Mechanics*. Holt, Rinehart and Winston.

VENNARD, J. K. (1961) *Elementary Fluid Mechanics*. Wiley (4th Edition).

WALSHAW, A. C. and JOBSON, D. A. (1962) *Mechanics of Fluids*. Longmans.

YIH, C-S. (1969) *Fluid Mechanics*. McGraw-Hill.

YUAN, S. W. (1967) *Foundations of Fluid Mechanics*. Prentice-Hall.

Institution Papers
J. Am. Water Works Assn.
J. Inst. Water Engineers
Proc. Inst. Civil Engineers

Proc. Inst. Mech. Engineers (Fluid Mechanics and Hydraulic Groups)
Proc. Am. Soc. Civil Engineers (Hydraulics, Irrigation and Drainage, Power, Environmental Engineering, Waterway Port Coastal and Ocean, Divisions)
Trans. Am. Soc. Civil Engineers

Conference Papers

Civil Engineering in the Oceans
Coastal Engineering
International Association for Hydraulic Research
International Commission on Irrigation and Drainage
Permanent International Association of Navigation Congresses

Journals

Channel
Civil Engineering and Public Works Review
Coastal Engineering
Dredging and Port Construction
Engineering
Irrigation and Power (India)
Journal of Fluid Mechanics
Journal of Hydraulic Research
La Houille Blanche (France)
Offshore Services
Pumping
The Dock and Harbour Authority
The Engineer
Water and Water Engineering
Water Power

Name Index

Abbot, H. L., 153
Ackers, P., 99, 109, 230
Adams, R. W., 120
Addison, H., 251, 296, 324
Airy, G. B., 3
Albertson, M. L., 324
Allen, F. H., 323
Allen, J., 312, 323
Anderson, H. H., 296
Annett, F. A., 296
Archimedes, 1
Aristotle, 1

Babbitt, H. E., 142
Bakhmeteff, B. A., 4, 54, 109, 167, 169, 188, 210
Ball, J. W., 308
Barna, P. S., 324
Barton, J. R., 324
Batchelor, G. K., 324
Bazin, H. E., 3, 153
Bergeron, L., 141, 142
Bernoulli, D., 2, 33
Blasius, P. R. H., 4, 83
Blench, T., 167
Bonnyman, G. A., 251
Boussinesq, J., 4, 54, 56
Bowman, J. R., 24
Bradshaw, P., 323
Brater, E. F., 167, 175, 210, 251, 324
Bresse, J. A. C., 195
Brown, J. G., 251, 296, 324
Buckingham, E., 74

Camp, T. R., 142
Carter, R., 296
Chézy, A., 3, 151
Chow, V. T., 167, 210
Church, A. H., 296
Cipolletti, C., 220
Cleasby, J. C., 142
Colebrook, C. F., 96, 103–104
Coleman, G. S., 225

Cornish, R. J., 126
Coyne, A., 237
Creager, W. P., 251, 296
Cross, H., 122
Crump, E. S., 229

Dake, J. M. K., 325
Danel, P., 271
Darcy, H. P. G., 3, 78, 81, 153
d'Aubuisson, J. F., 204
Daugherty, R. L., 324
Davis, C. V., 24, 142, 251, 322, 323, 324
Doland, J. J., 142
Drysdale, C. V., 11
Du Boys, P. F. D., 166
Du Buat, P. L. G., 3
Duncan, W. J., 75, 324

Elevatorski, E. A., 183, 210, 251
Engels, H., 4
Euclid, 1
Euler, L., 2, 73, 274

Fair, G. M., 142
Fargue, L. J., 297
Focken, C. M., 75
Forcheimer, P., 4
Fortier, S., 165
Foster, J. E., 314
Francis, J. B., 266
Francis, J. R. D., 324
Franzini, J. B., 325
Freeman, J. R., 4
Frontinus, 1
Froude, W. A., 73, 297

Galileo, 2
Gameson, A. L. H., 323
Ganguillet, E. O., 153
Gauckler, P. G., 154
Gerstner, F. J. von, 3
Geyer, J. C., 142

Gibson, A. H., 4, 106, 298
Gibson, N. R., 133
Goldstein, S., 75
Goncharov, V. N., 167

Hagen, G. H. L., 3, 76
Hagenbach, E., 76
Harrison, A. J. M., 230
Hazen, A., 100
Headland, H., 253
Henderson, F. M., 167, 210
Hickox, G. H., 323
Hicks, T. G., 296
Hinds, J., 251
Hinze, J. O., 75
Howe, J. W., 325
Humphreys, A. A., 153

Ince, S., 5
Ippen, A. T., 323, 324
Ipsen, D. C., 75
Ireland, J. W., 325

Kaplan, V., 270
Karassik, I., 296
Kármán, T. von, 4, 85
Keefe, H. G., 251
Kelvin, Lord, 3, 25
King, H. W., 167, 175, 210, 251, 324
Kovalev, N. N., 296
Kováts, A., 296
Kristal, F. A., 296
Kutter, W. R., 3, 153

Jaeger, C., 325
Jobson, D. A., 325
Joukowsky, N. E., 135
Justin, J. D., 251, 296

Lacey, G., 317
Lagrange, J. L., 3
Lamb, Sir H., 3, 25, 75
Lamont, P. A., 104
Landau, L. D., 75
Langhaar, H. L., 323
Lawler, J. C., 142
Lazarkiewicz, S., 296
Leliavsky, S., 167, 210, 251, 324
Li, W. H., 325
Lifshits, E. M., 75
Linford, A., 251, 323

Linsley, R. K., 325
Littlejohn, A. G., 312

Mach, E., 73
Manning, R., 3, 153
Massey, B. S., 325
Mayer, P. R., 24
Moody, L. F., 98, 322
Mosonyi, E., 296
Murphy, G., 323

Nagler, F. A., 204
Navier, L. M. H., 3
Nechleba, M., 296
Neumann, F., 76
Newton, I., 2, 8
Nikuradse, J., 4, 84–85, 87, 90, 91, 93–95, 300
Norrie, D. H., 296

Olson, R. M., 325

Pannell, J. R., 83
Pao, R. H. F., 325
Parmakian, J., 142
Parshall, R. L., 230
Paynter, H. M., 141
Pelton, L. A., 264
Peterka, A. J., 309
Pickford, J., 142
Pitot, H., 3, 35
Plapp, J. E., 325
Poiseuille, J. L., 3, 76
Posey, C. J., 167, 210
Prandtl, L., 4, 54, 59, 75, 85, 86, 89, 95, 97, 109

Rankine, W. J. M., 25
Rao, N. S. G., 251
Raudkivi, A. J., 167
Rayleigh, Lord, 3, 25, 74
Rehbock, T., 217
Reid, L., 233
Reynolds, O., 4, 55, 73, 79–82, 297, 298
Rich, G. R., 142
Robertson, J. M., 75
Rouse, H., 5, 73, 233, 324, 325
Rozovskii, I. L., 167

Sabersky, R. H., 325
Saint-Venant, J. B., 3
Saph, A. V., 83
Schlichting, H., 75
Schoder, E. W., 83
Scobey, F. C., 165
Sedov, L. I., 75
Sellin, R. H. J., 167, 210
Shames, I. H., 75, 325
Silvester, R., 324
Simin, O., 135
Simons, D. B., 324
Singer, C. J., 5
Skeat, W. O., 105, 324
Smith, D., 225
Smith, H., 217
Sorensen, K. E., 24, 142, 251, 322, 324
Sorensen, R. M., 323, 324
Stanton, T. E., 83
Stepanoff, A. J., 296
Stokes, G. G., 3, 25, 75
Streeter, V. L., 324, 325
Strickler, A., 154

Taylor, G. A., 312
Thom, A. S., 324
Thorn, R. B., 251
Torricelli, E., 2
Troskalanski, A. T., 251, 296
Twort, A. C., 142

Vallentine, H. R., 75
Vennard, J. K., 325
Venturi, G. B., 3, 37
Vermuyden, C., 2
Villemonte, J. R., 218
Vinci, L. da, 2

Walshaw, A. C., 325
Weber, M., 73
Weisbach, J., 4, 78
Whitaker, S., 325
White, C. M., 96, 103–104
Wiegel, R. L., 324
Williams, G. S., 100
Wilson, D. H., 75
Wislicenus, G. F., 277, 296
Wood, A. M. M., 323–324
Woodward, S. M., 167, 210

Yalin, M. S., 167, 323
Yarnell, D. L., 204
Young, A. D., 324
Yuan, S. W., 325

Zowski, T., 322

Subject Index

Absolute pressure, 11
Absolute velocity, 271
Acceleration head, 132
Adiabatic process, 136
Adverse pressure gradient, 61
Aerofoil, 65, 277
Afflux, 203–204, 226, 228–231
Aging of pipes, 103–104
Air,
  properties of, xv
  use in models, 308
Air chambers, 131
Air-entraining vortices, 5, 255, 307
Air entrainment, 143, 178, 215, 239, 240, 305, 312
Air vent, 215, 243
Alternate depth, 171
Analogy, electrical, 52, 115, 120
Angle of heel, 22
Angular momentum, 47, 273
Artificial roughness, 84–86, 93–96, 300
Atmospheric pressure, 10–11
Axial flow pump, 261–263, 276, 278–279, 283, 289, 292, 294

Backwater curve, 188–195, 201–203
Baffle piers, 236
Barometer, 10
Barometric pressure, 11, 293
Bed load, 164
Bed slope
  adverse, 188
  critical, 176
  definition, 145
  horizontal, 188
  mild (subcritical), 176
  steep (supercritical), 176
Bend in channel
  behaviour of ideal fluid, 47–48
  behaviour of real fluid, 68
Bend in pipe
  behaviour of ideal fluid, 35, 47–48, 51–52, 243
  behaviour of real fluid, 67

energy head loss, 67, 104
  force on, 43–44
Bernoulli equation
  applications, 33–40, 43, 47, 51, 52, 70, 106, 180, 203, 212, 213, 223, 234, 241, 242, 254, 281, 312
  derivation, 31–33
  modified form of, 56–57
Best hydraulic section, 158–161, 167
Blasius formula, 83–85, 90, 100, 109, 121
Bore (tidal river), 209–210
Borehole pump, 260, Plate 6
Boulder (Hoover) dam, 311
Boundary
  concave, 45, 68, 233, 237
  convex, 45, 60–61, 232
Boundary layer, 58–61
Bourdon gauge, 14–16
Break-away, 61
Bresse function, 195
Bridge piers
  effect on channel flow, 202–204
  flow around, 63–65
Broad-crested weir, 175, 201, 222–223, 228
Bucket
  Pelton wheel, 264–266, 277, Plate 7
  spillway, 236
Buckingham Pi theorem, 74
Bulb (tubular) turbine, 271–272
Bulk modulus of elasticity, 7, 73, 135
Bulking, 143, 178, 310
Buoyancy, centre of, 21

Canals, 143, 159, 165, 204, 227
Capillarity, 10
Capillary rise, 10
Capillary waves, 72
Cavitation
  phenomenon, 57–58
  structures, 232, 236, 307, 311
  turbomachinery, 254, 261, 270, 292–294, 307, 321
Celerity (wave), 134, 206

Centre of buoyancy, 21
Centre of pressure, 16
Centrifugal pumps, 50, 257–260, 281–284, 289, 290, 292
Channel
  best hydraulic section, 158–161, 167
  empirical formulae, 151–156
  entry, 192–193
  exit, 192–193
  flow at bend, 47–48, 68
  laminar flow, 144–145, 149–150, 314
  stable, 164–166
Chézy formula, 3, 69, 151–152, 186, 195
Cipolletti (trapezoidal) weir, 220
Circumferential stress, 136
Classical hydrodynamics, 3, 25
Clinging nappe, 72, 215, 310
Closure, instantaneous valve, 133–141
Coefficient
  Chézy, 69, 152–153
  contraction, 39, 212, 213
  discharge, 37, 39, 212, 217, 219, 220, 227
  drag, 65–67, 75, 165
  Hazen-Williams, 100
  minor loss, 105–109
  orifice plate, 38
  peripheral, 278
  pipe resistance, 120–121
  Pitot-static tube, 36
  pressure, 71
  roughness (Manning), 100–102, 109, 153–156, 166, 311
  siphon, 242
  sluice gate, 212
  spillway, 233, 248
  throated flume, 227, 228
  velocity, 39
  velocity head, 56, 168, 242
  Venturi tube, 37
  weir, 218, 221–224, 310
Colebrook-White transition law, 96–98, 102–103, 109, 149–150
Coleman and Smith weir formula, 225
Compressibility, 7, 134
Compression wave, 134
Computers, 5, 52, 120
Concave boundary, 45, 68, 233, 237
Conduits
  low pressure, 130
  models of, 307–308
  non-circular, 104, 143
Conical connecting piece, 29–30
Conservation of energy, 33, 136

Constrictions (measuring devices), 36–38, 226
Continuity equation, 26, 30–31
Contraction
  abrupt pipe, 107
  coefficient of, 39, 212, 213
  end (weir crest), 215, 217
  jet, 38, 211
Control points, 177, 191–195, 197, 202
Convex boundary, 45, 60–61, 232
Cooling water circulation, 56, 318, 319
Coriolis effect, 319
Corresponding velocities, 302
Crest profile (spillway), 231–233, 310
Criterion for critical depth, 171–173
Critical depth, 170
Critical sigma, 294
Critical slope, 176, 188
Critical velocity
  channel flow, 170, 207
  pipe flow, 80–81
Crump weir, 229
Culverts, 104, 143
Current meter
  field, 146, 221, 230, Plate 1
  miniature, 322, Plate 19
Curves
  iso-efficiency, 284
  performance, 282–283, 296, 320
Curvilinear flow, 44–50
Cylinder, flow around, 34, 62, 66

Dam
  control point, 191
  spillway, 231–246
Darcy-Weisbach formula, 78, 82–83, 99, 103, 109, 112, 113, 121, 148, 239
d'Aubuisson formula, 204
Deep-water wave, 29
Deformation, 6, 8
Density, 6
Density phenomena, 5, 319
Deposition (silting), 68, 74, 164–165, 316, 319
Depth
  alternate, 171
  critical, 170
  hydraulic mean (hydraulic radius), 100, 148, 315
  initial, 180
  mean (hydraulic depth), 172
  normal, 143
  sequent, 180, 235

Design charts, 'Universal', 99
Design head, 232
Design point, 283, 284, 289
Deterioration of pipes, 103–104
Differential head, 36–37
Differential manometer, 15, 37
Dimensional analysis, 68–70, 73–75, 151, 217, 288
Dimensional homogeneity, 68–69
Dimensionless parameters, 70–73, 301
Dimensions
  derived, xi–xiii, 304
  fundamental, 69
Direct jump, 179
Discharge, maximum, 158, 173
Discharge calibration, 68, 213, 221, 230, 310
Discharge coefficient
  orifice, 39, 242
  orifice plate, 38
  sluice, 212
  throated flume, 227, 228
  Venturi tube, 37
  weir, 217–224
Discharge measurement
  constriction, 36–38, 316
  current meter, 146, 221, 230
  Gibson method, 133
  throated flume, 226–231
  pipe bend, 68
  river stage, 156
  weir, 214–221, 316
Discharge under varying head, 113–114
Discontinuity, 61, 309
Distribution mains, 119–120
Diverging flow, separation, 61, 223, 232, 292
Draft tube, 67, 268–271, 281, 294, 321
Drag
  coefficient, 65–67, 75, 165
  deformation, 63, 74–75
  form (pressure), 64–67
  skin friction (surface), 64–65
  total, 64–65
Drawdown curve, 188–195, 200, 201, 234
Drowned condition, 212–213, 218, 226–227
Drum gate, 19–21, 232
Du Boys equation, 166
Dynamic pressure head, 36, 132
Dynamic similarity, 301
Dynamic viscosity, 8

Eddies, wake, 61–62
Eddy viscosity, 54
Efficiency
  hydraulic, 280
  mechanical, 280
  overall, 279
  pipeline, 112
  pressure head recovery, 281
  volumetric, 280
Elastic pipeline, 136–142
Elastic theory, 134–142
Elasticity, bulk modulus of, 7, 73, 135
Electrical analogy, 52, 115, 120
Elevation head, 33
Empirical hydraulics, 3
End contractions (weir), 215, 217, 222
Energy
  conservation of, 33, 136
  potential, 33
  pressure, 33
  specific, 168
  strain, 136
  velocity (kinetic), 33
Energy dissipation
  channel exit, 193
  hydraulic jump, 179–181, 183, 233
  pipe exit, 107
  spillway (tailwater), 223–238
  wake, 61
  weir (tailwater), 222
Energy grade line, 110–111
Energy gradient, 110, 168, 185
Energy head
  loss of, 57
  total, 33, 169
Enlargement, abrupt pipe, 105–106
Entry, channel, 192–193
Entry loss, pipe, 108
Equation
  Bernoulli, 26, 31–40, 43, 47, 51, 52, 56–57, 70, 106, 180, 203, 212, 213, 223, 234, 241, 242, 254, 281, 312
  continuity, 26, 30–31
  Du Boys, 166
  Euler, 274
  general weir, 218, 222–224, 233, 238
  gradually varied flow, 185–186, 195
  Laplace, 51
  momentum, 26, 40–44, 106, 181, 206, 214, 215, 237, 238, 273
  Poiseuille, 76–78, 82, 85, 97, 109
  Newton's viscosity, 8, 77, 144
  Stokes, 75
  storage, 247

turbulent shear, 55
velocity-deficiency, 87, 93
Equipotential lines, 50–51
Equivalent length of straight pipe, 105
Equivalent single pipe, 115–116, 126
Erodible channels, 164–167, 316
Establishment of flow, 133–134, 141
Euler equation, 274
Euler law, 302, 308, 321
Euler number, 71–72, 302
Excrescence size, 84, 91–95, 147, 150, 152–154, 300
Exit, channel, 192–193
Exit loss, pipe, 107
External energy conversion, 57

Fall velocity, 74, 319
Fittings, pipe, 104–105
Flap valve, 17, 263
Flash-boards, 232
Float gauge, 156, 216, 228
Floats, 322
Flood routing through a reservoir, 246–251
Flotation chamber, 19
Flow
 axial, 261, 267, 270, 271, 276
 curvilinear, 44–50
 establishment of, 133–134, 141
 flood plain, 156–158
 irrotational, 27
 laminar, 26, 76, 144
 non-uniform, 29, 143, 168
 radial, 48, 267
 rotational, 27
 steady, 28, 143
 subcritical, 171
 supercritical, 171
 turbulent, 26
 two-dimensional, 31, 50, 214
 uniform, 29, 143
 unsteady, 28, 129, 143, 204
Flow around a cylinder, 34, 62, 66
Flow measurement, 36–38, 68, 133, 146, 156, 211, 214–221, 226–231, 316
Flow net
 conduit bend, 51–52
 construction, 52
 sluice, 213
 weir, 52–53, 218
Fluid
 definition, 6
 ideal (perfect), 7
 real, 7

Flume, throated, 180, 211, 226–231, Plate 4
Force
 centrifugal, 44–45, 67
 drag, 63–67, 164
 elastic, 73
 gravitational, 71, 301, 303
 hydrostatic pressure, 12
 inertial, 70–72, 74, 301
 pressure, 12
 shear, 77
 surface tension, 9, 72, 218, 301, 305
 tension, 42
 tractive, 166, 314, 316
 viscous, 72, 301, 303
Forced vortex, 48–50, 274
Form drag, 64–67
Formula
 Blasius, 83–85, 90, 100, 109, 121
 Chézy, 3, 69, 151–152, 186, 195
 Colebrook-White (transition law), 96–98, 102–103, 109, 149–150
 Coleman and Smith, 225
 Darcy-Weisbach, 78, 82–83, 99, 103, 109, 112, 113, 121, 148, 239
 d'Aubuisson, 204
 Ganguillet and Kutter, 153
 Hamilton Smith, 217
 Hazen-Williams, 100, 102–103, 109, 121
 Manning, 100–103, 109, 121, 151, 153–158, 167, 176, 186, 187, 197, 199, 201, 208, 230, 234, 238, 311, 315, 317
 Moody, 322
 Rehbock, 217
 rough (quadratic) law, 93–97, 101, 109, 149, 150
 smooth law, 90, 94–96, 109, 148, 150
 Villemonte, 218
Formulae
 exponential (pipes), 99–100, 109, 121
 summary of pipe resistance, 109
Francis turbine, 266–270, 286–287, 290, 291
Free condition, 212, 214, 228
Free overfall, 186–187, 193, 200, 222
Free vortex, 46–48, 51, 67–68, 242, 258, 271, 276
Friction factor, 78, 148
Friction gradient (slope), 78, 99, 185, 308
Friction head, 78
Froude law, 303–306, 309, 315, 316, 319

Froude number, 71–72, 172–173, 179, 183, 302, 319

Ganguillet and Kutter formula, 153
Gas, definition, 6
Gate
    drum, 19–21, 232
    movable, 18–21, 211, 221, 232
    radial (Tainter), 19, 212, 232, 300
    roller, 19
    sluice, 30, 174–175, 180, 191, 193–195, 199, 211–214
    turbine, 130, 208, 245, 267, 286
    vertical lift, 232, Plate 3
Gauge
    Bourdon, 14–16
    float, 156, 216, 228
    hook, 216, 322
    pointer, 322
Gauge pressure, 11
General discharge equation (throated flume), 227
General equation of gradually varied flow, 185–186, 195
General weir equation, 218, 222–224, 233, 238
Geometric similarity, 299–300
Gibson method of discharge measurement, 133
Gradient
    energy, 110, 168, 185
    friction, 78, 99 185, 308
    hydraulic, 110–111, 143
    pressure, 61
    velocity, 8
Gradually varied flow, general equation of, 185–186, 195
Gravitational forces, 71, 301, 303
Gravity, specific, 7
Guide blades, 260, 261, 267, 270, 271, 278–279, 291

Hamilton Smith formula, 217
Hardy Cross analysis, 120–129
Hazen-Williams formula, 100, 102–103, 109, 121, 124
Head
    acceleration, 132
    differential, 36–37
    design, 232
    dynamic, 36, 132
    elevation, 33
    friction, 78
    gross, 255
    loss of energy, 57
    net positive inlet, 293
    piezometric, 14, 34
    pressure, 12
    pump, 254
    sluice, 180, 211–213
    specific, 168
    static, 36
    total energy, 33, 169
    turbine, 255
    varying, 113–114
    velocity, 33
    weir, 215
Head balance, 122–124
Head loss, minor, 104–109, 111, 113
Heat transfer, 56, 319
High stage, 171
Homogeneity, dimensional, 68–69
Hoover (Boulder) dam, 311
Horsepower
    input, 279
    output, 279
Hydraulic depth (mean depth), 172
Hydraulic grade line, 110–111, 243
Hydraulic gradient, 110–111, 143
Hydraulic jump, 179–184, 186, 192–195, 199, 207, 228–229, 233–236, 238
Hydraulic radius (hydraulic mean depth), 100, 148, 315
Hydraulic similarity, 298–302
Hydraulics, empirical, 3
Hydrodynamics, classical, 3, 25
Hydro-electric power, 252–253, 263
Hydrostatic pressure, 12

Ideal fluid, 7
Impeller, 50, 58, 257–261, 273–276, 289, 292, 321
Implosion, cavitation, 58
Impulse turbine, 256, 263–264
Incompressible theory, 132–134
Inertial force, 70–72, 74, 301
Inflow-outflow relationship at a reservoir, 246–251
Initial depth, 180
Instantaneous valve closure, 133–141
Intensity
    pressure, 12
    turbulence, 56, 164
Interfacial mixing, 5, 56
Irrotational flow, 27
Iso-efficiency curves, 284

Jet
   nozzle fitting, 42
   orifice, 38–40
   Pelton wheel, 256, 264–266, 277–278
   ski-jump spillway, 237
   sluice, 211
Jet contraction, 38, 211–213
Jet deflector, 264–265
Jet trajectory, 39–40, 214, 232
Joukowsky's law, 135
Jump, hydraulic, 179–184, 186, 192–195, 199, 207, 228–229, 233–236, 238
Junctions, head loss at pipe, 104

Kaplan turbine, 270–271, 287, 294
Kármán-Prandtl pipe formulae, 85–95, 109, 147–149
Kinematic similarity, 300–301
Kinematic viscosity, 9
Kinetic energy, 33
Kutter's n, 153, 154

Lag (reservoir), 246, 251
Laminar flow, definition, 26
Laminar flow in a channel, 144–145, 149–150, 314
Laminar flow in a pipe
   analysis, 76–78
   dye experiments, 79
   pumping main, 82
Laminar sub-layer, 59, 88–96
Land drainage pump, 262–263
Laplace equation, 51
Law
   Colebrook-White transition, 96–98, 102–103, 109, 149–150
   Euler, 302, 308, 321
   Froude, 303–306, 309, 315, 316, 319
   Joukowsky's, 135
   Reynolds, 303–306, 308
   rough channel, 149
   rough pipe (quadratic), 93–97, 101, 109
   smooth channel, 148
   smooth pipe, 90, 94–96, 109
   Weber, 305
Laws
   Newton's, 2, 69
   pipe resistance, 80
Lift, 67
Lines
   equipotential, 50–51

stream, 29–30
Liquid, definition, 6
List
   conversion factors and useful constants, xvi
   general publications relevant to hydraulics, 324–325
   symbols, xi–xiii
Load
   bed, 164
   suspended, 164
Load rejection, 130
Loss
   energy head, 57
   friction head, 78
   minor head (pipe), 104–109, 111, 113, 254
   pipe entry, 108
   pipe exit, 107
Low pressure conduit, 130
Low stage, 171

Mach number, 73
Mains, distribution, 119–120
Manning formula, 100–103, 109, 121, 151, 153–158, 167, 176, 186, 187, 197, 199, 201, 208, 230, 234, 238, 311, 315, 317
Manometer, 10, 14–15, 36, 37, 310, Plate 9
Mean (hydraulic) depth, 172
Mean velocity, 27, 31
Measurement
   discharge, 36–38, 68, 133, 146, 156, 211, 214–221, 226–231, 316
   pressure, 13–16, 65, 310, 322
   velocity, 35–36, 146, 310, 318, 322
Meniscus, 10
Mercury, properties of, xv
Metacentre, 22
Metacentric height, 23
Meter
   current (field), 146, 221, 230, Plate 1
   current (miniature), 322, Plate 19
Minor losses in pipes, 104–109, 111, 113, 254
Mixed flow pump, 263, 289
Mixing length, 55, 86
Models
   enclosed conduit, 307–309
   estuary, 298, 306, 317–320, Plate 16
   geometrically distorted, 300, 306, 314–315, 318–320

Models—*cont.*
  hydraulic structure, 213, 231, 309–313
  instrumentation for, 322
  mobile-bed, 300, 313, 316–317, 318
  open spillway, 233, 310–311, Plate 8, Plate 9
  river, 306, 313–317, Plate 14, Plate 15
  sectional, 309, 310
  shaft spillway, 240, 311–312
  siphon spillway, 244, 312, Plate 12
  stilling basin, 313
  tailbay, 313, Plate 13
  turbomachinery, 297, 320–322
  wave, 320, Plate 17
Modular limit, 229
Modulus, bulk, 7, 73, 135
Momentum
  angular, 47, 273
  correction factor, 41, 168
Momentum equation
  applications, 41–44, 106, 181, 206, 214, 215, 237, 238, 273
  derivation, 40–41
Momentum exchange, 54
Moody step-up formula, 322
Movable gates, 18–21, 211, 221, 232
Multiple reservoirs, 117–119
Multi-stage pump, 260, 289, Plate 5

Nappe, 214–217
Negative surge, 137, 205
Networks, pipe, 119–129
Newtonian fluid, 8–9
Newton's equation of viscosity, 8, 77, 144
Newton's laws, 2, 69
Non-circular conduits, 104, 143
Non-uniform flow, 29, 143, 168
Normal depth, 143
Nozzle
  Pelton wheel, 111, 264, 265
  pipe fitting, 42
Number
  Euler, 71–72, 302
  Froude, 71–72, 172–173, 179, 183, 302, 319
  Mach, 73
  Reynolds, 62, 72, 82–85, 90–101, 109, 114, 148–153, 156, 217, 218, 302, 306, 308, 314, 317
  Reynolds roughness, 94–98, 103, 306
  Weber, 72, 217, 302

Oil, properties of, xv
Optimum size of pipeline, 112, 295
Orifice, 38–40
Orifice plate, 38, 316
Overfall, free, 186–187, 193, 200, 222

Parameters, dimensionless, 70–73, 301
Parshall flume, 230
Pelton wheel turbine, 111, 264–266, 277–278, 285, 287, 289, 290
Penstock, turbine, 130, 256
Perfect fluid, 7
Performance curves, 282, 285–287, 295–297, 320
Perimeter, wetted, 100, 148, 158
Peripheral coefficient, 278
Petrol, properties of, xv
Piers
  baffle, 236
  bridge, 63–65, 202–204
  spillway crest, 238
Piezometer, 14, 65, 310, 312, 322, Plate 9
Piezometric head, 14, 34
Pipe
  abrupt contraction, 107
  abrupt enlargement, 105–106
  constriction, 36–38
  entry, 59–60, 108
  equivalent single, 115–116, 126
  equivalent straight, 105
  exit, 107
  flow at bend, 35, 47–48, 51–52, 67
  laminar flow, 26, 76–85
  turbulent flow, 26, 53–56, 78–79
Pipe fittings, 104–105
Pipe networks, 119–129
Pipe resistance, laws of, 80
Pipe resistance formulae, summary, 109
Pipeline
  elastic, 136–142
  optimum size of, 112, 295
  rigid, 132–136
Pipes
  artificially roughened, 84–86, 93–96, 300
  commercial, 86, 95–97, 161
  deterioration of, 103–104
  head loss at junction of, 104
  minor losses in, 104–109, 111, 113, 254
  non-circular, 104, 164
  parallel, 115–117
  rough, 91–104

series, 114–115
smooth, 83–85, 88–91
transition zone, 95–98, 100
Pi theorem, 74
Pitot tube, 35–36, 310, 322
Plate
  flow parallel to a thin, 59
  orifice, 38, 316
Poiseuille equation, 76–78, 82, 85, 97, 109
Pollutants, dispersion of, 56, 180, 318
Positive surge, 137, 205, 313
Potential energy, 33
Power, maximum, 112, 278, 286
Power transmission, 111–112
Pressure
  absolute, 11
  atmospheric, 10–11
  centre of, 16
  dynamic, 36
  gauge, 11
  hydrostatic, 12
  optimum, 119
  piezometric, 14, 34, 51, 310
  stagnation, 34–35
  sub-atmospheric (vacuum), 11
  vapour, 10
Pressure coefficient, 71
Pressure energy, 33
Pressure gradient, 61
Pressure head, 12
Pressure intensity, 12
Pressure measurement, 13–16, 65, 310, 322
Pressure relief valve, 130, 269–270
Pressure-time diagram, 139–141
Priming
  pump, 259–260, 263
  siphon, 240, 243, 245, 312, Plate 12
Profile evaluation, 195–202
Propeller turbine, 270–272, 276, 287, 290
Properties of fluids, table of, xv
Prototype, 298
Proving of models, 316
Pump
  axial flow, 261–263, 276, 278–279, 283, 289, 292, 294
  borehole, 260, Plate 6
  centrifugal, 50, 257–260, 281–284, 289, 290, 292
  land drainage, 262–263
  mixed flow, 263, 289
  model, 297, 320–322
  multi-stage, 260, 289, Plate 5

reciprocating, 257
rotary, 257
rotodynamic, 257
Pumped storage, 253
Pumping main
  design of, 295–296
  surge pressure in, 131, 141, 295
Pump setting, 293–294

Quadratic (rough pipe) law, 93–97, 101, 109
Quantity balance, 126–127

Radial flow, 48, 267
Radial (Tainter) gate, 19, 212, 232, 300
Radius, hydraulic, 100, 148, 315
Rarefaction wave, 137
Rating curve
  river channel, 156
  spillway, 233, 234, 246
Rayleigh method, 74
Reaction turbine, 255, 264
Real fluid, 7
Reciprocating pump, 257
Rectangular weir, 52–53, 214–220
Reflux valve, 131, 295
Regime theory, 165, 316–317
Rehbock formula, 217
Rejection of load, 130, 245, 264, 270
Reservoirs
  inflow–outflow relationship, 246–251
  multiple, 117–119
Resistance laws, pipe, 80
Reversible turbine, 253
Reynolds law, 303–306, 308
Reynolds number, 62, 72, 82–85, 90–101, 109, 114, 148–153, 156, 217, 218, 302, 306, 308, 314, 317
Reynolds roughness number, 94–98, 103, 306
Rigid pipeline, 132–136
River flood plain, 156–158
Roller gate, 19
Rotary pump, 257
Rotational flow, 27
Rotodynamic pump, 257
Rough channel law, 149
Rough pipe (quadratic) law, 93–97, 101, 109
Roughness
  annual rate of growth of, 104
  artificial, 84–86, 93–96, 300

Roughness—*cont.*
commercial pipe, 86, 95–97
effective, 95–97, 101–102, 147, 154, 300
excrescence size, 84, 91–95, 300
relative, 83–85, 91, 93, 98, 101–102
Roughness coefficient (Manning), 100–102, 154–156, 311
Round-crested (ogee) weir, 221–222
Runner, 58, 264–271, 275, 277, 290–292, 294, 321
Run-off, sheet, 144, 150

Scalar relationships for models, 304
Scale effect, 298, 306–307, 314, 319, 322
Scouring, 68, 164–166, 211, 300, 313, 316, 318
Sea water, properties of, xv
Section, best hydraulic, 158–161, 167
Sediment, transport of, 63, 68, 164–166, 300, 316, 318
Separation, 46, 60–62, 67, 105, 223, 232, 270, 292
Sequent depth, 180, 235
Settling velocity, 74, 319
Sewers, 161–164
Shallow-water wave (small), 173, 208
Sharp-crested weirs, 52–53, 211, 214–221
Shear velocity, 86, 147
Shearing strain, 8
Shearing (viscous) stress, 6–8
Sheet run-off, 144, 150
Side weir, 225
Sigma, critical, 294
Silting (deposition), 68, 74, 164–165, 316, 319
Similarity
concept, 70
dynamic, 301
geometric, 299–300
hydraulic, 298–302
kinematic, 300–301
pipe flow, 82
Siphon, 240–246, 312, Plate 12
Siphonage, 111
Ski-jump, 237
Skin friction (surface) drag, 64–65
Sluice, 30, 174–175, 180, 191, 193–195, 199, 211–214
Smooth channel law, 148
Smooth pipe law, 90, 94–96, 109
Solid weirs, 221–223
Sound, velocity of, 73, 135, 308

Specific energy, 168
Specific gravity, 7
Specific head, 168
Specific speed, 287–290, 294
Specific weight, 6
Speed
rotational, 256
specific, 287–290, 294
synchronous, 256, 264
Speed factor, 278
Sphere, steady rate of fall of, 74
Spillway
open, 231–238, 310–311, Plate 8, Plate 9
shaft (glory-hole), 238–240, 311–312
siphon, 240–246, 312, Plate 12
Spillway coefficient, 233, 248
Spiral vortex, 48, 271
Stability of floating bodies, 21–24
Stable channels, 164–166
Stage
high, 171
low, 171
Stage-discharge relationship, 156, 316
Stagnation point, 34–35, 64
Stagnation pressure, 34–35
Standard λ-R diagram, 98–99, 101, 121, 148–150, 306
Standing wave, 179, 207, 228
Standing-wave flume, 180, 228–231, Plate 4
Steady flow, 28
Stilling basin, 235, 237–238, 313
Stilling pool, 222, 231, 237
Stokes equation, 75
Stop-logs, 232
Storage equation, 247
Strain energy, 136
Stream lines, 29–30
Stream tube, 30–31, 56
Streamlining, 65
Stress
circumferential, 136
shearing (viscous), 6–8
Subcritical flow, 171
Subcritical velocity, 171
Sub-layer, laminar, 59, 88–96
Submergence ratio, 218, 223, 229
Supercritical flow, 171
Supercritical velocity, 171
Surface profiles
analysis, 195–199
classification, 187–191
Surface tension, 9

Surge
  channel, 205, 313
  pipeline, 130
Surge celerity, 134, 206
Surge tank, 130
Suspended load, 164
Symbols (list of), xi–xiii

Tables and lists
  conversion factors and useful constants, xvi
  effective roughness (pipes), 97
  general publications relevant to hydraulic engineering, 324–325
  maximum permissible water velocities and unit tractive force values (stable channels), 166
  principal symbols with derived dimensions, xi–xiii
  properties of fluids, xv
  scalar relationships for models, 304
  summary of pipe resistance formulae, 109
  values of m and r (pipe resistance formulae), 121
  values of n (channels), 155
Tailbay, 222, 313, Plate 13
Tainter (radial) gate, 19, 212, 232, 300
Tapping, piezometer, 14, 65, 66, 310, 312, 322, Plate 9
Temporal mean velocity, 26, 55
Tension force, 42
Thickness of boundary layer, 59–61
Thickness of laminar sub-layer, 88–91
Throated flume, 180, 211, 226–231, Plate 4
Tides, 28, 208, 318, 319
Time of flow establishment, 133–134, 141
Time of outflow, 113–114
Time of valve closure, 130, 133, 139–141, 266
Total head, 33, 169
Tractive force, 165–167, 314, 316
Trajectory
  jet, 39–40, 214, 232
  nappe, 214
Transition (Colebrook-White) law, 96–98, 102–103, 109, 149–150
Transmission of power, 111–112
Transporting power, 164–165, 314
Trapezoidal (Cipolletti) weir, 220
Triangular weir (V-notch), 219–220, 316

Tube
  draft, 67, 268–271, 281, 294, 321
  Pitot, 35–36, 310, 322
  stream, 30–31, 56
  Venturi, 37
Turbine
  bulb (tubular), 271–272
  Francis, 266–270, 286–287, 290, 291
  impulse, 256, 263–264
  Kaplan, 270–271, 287, 294
  model, 297, 320–322
  Pelton wheel, 111, 264–266, 277–278, 285, 287, 289, 290
  propeller, 270–272, 276, 287, 290
  reaction, 255, 264
  reversible, 253
Turbine gate, 130, 208, 245, 267, 286
Turbulence
  general equation of, 56, 86
  intensity, 56, 164
  theory, 53–56
Turbulent flow, definition, 26
Turbulent flow in a pipe, 26, 78
Turbulent rough zone, 91–101, 306
Turbulent shear equation, 55

Undular jump, 179
Uniform flow, 29, 143
Unsteady flow, 28, 129, 143, 204
U-tube, 14

Vacuum pressure, 11
Valve
  butterfly, 263
  closed delivery, 274, 284, 295
  flap, 17, 263
  foot, 260
  head loss at, 104, 295
  needle, 264–265
  pressure relief, 130, 269–270
  reflux, 131, 295
  sluice (gate), 295
Valve closure, 133, 134, 139–141, 266
Valve fluttering, 138
Vane angle, 273–276
Vanes
  impeller, 258, 273–277
  runner, 267–268, 277
Vaporisation, 47, 58, 111, 306
Vapour pressure, 10
Varying head, 113–114

Velocity
  absolute, 271
  approach, 212, 216, 228
  coefficient, 39
  corresponding, 302
  critical (channel flow), 170, 207
  critical (pipe flow), 80–81
  flow, 275
  maximum (channel flow), 160
  maximum permissible (stable channel), 165–167
  mean, 27, 31
  relative, 29, 271
  settling, 74, 319
  shear, 86, 147
  sonic, 73, 135, 308
  spouting, 38
  subcritical, 171
  supercritical, 171
  temporal mean, 26, 55
  vane, 271
  whirl, 274
Velocity distribution
  channel, 146–148, 168, 314, 318
  pipe, 26, 77, 86–90
Velocity energy, 33
Velocity gradient, 8
Velocity head, 33
Velocity head coefficient, 56, 168, 242
Velocity measurement
  current meter (field), 146, 221, 230, Plate 1
  current meter (miniature), 322, Plate 19
  floats, 322
  Pitot tube, 35–36, 310, 322
Velocity vectors, 40, 273
Velocity-deficiency equation, 87, 93
Vena contracta, 38, 107, 199
Vent, air, 215, 243
Venturi tube, 37
Villemonte formula, 218
Viscosity
  dynamic, 8
  eddy, 54
  kinematic, 9
Viscous (shearing) stress, 6–8
V-notch (triangular weir), 219–220, 316
Volute, 258

Vortex
  air-entraining, 5, 255, 307
  forced, 48–50, 274
  free cylindrical, 46–48, 67–68, 242, 258, 276
  free spiral, 48, 271
Vortex street, 62

Wake, 61–62
Water, properties of, xv
Water barometer, 11, 293
Water hammer, 130, 266
Wave
  capillary, 72
  compression, 134
  deep-water, 29
  pressure, 130
  rarefaction, 137
  shallow-water (small), 173, 208
  standing, 179, 207, 228
  surge, 130, 205, 313
Wave celerity, 134, 206
Wave length, 208
Weber law, 305
Weber number, 72, 217, 302
Weight, specific, 6
Weir
  broad-crested, 175, 201, 222–223, 228
  compound, 221
  Crump, 229
  extended, 224
  rectangular, 52–53, 214–220
  round-crested (ogee), 221–222
  side, 225
  submerged, 218
  trapezoidal (Cipolletti), 220
  triangular (V-notch), 219–220, 316
Weir as control point, 191, 201–202, 316
Weir coefficient, 218, 221–224, 310
Wetted perimeter, 100, 148, 158
Wheel ratio, 266, 278

Zone
  transition, 95–98, 100
  turbulent rough, 91–101, 306
Zone of discontinuity, 61